Environmental Soil Chemistry

Second Edition

Environmental Soil Chemistry

Second Edition

Donald L. Sparks

University of Delaware

ACADEMIC PRESS

An imprint of Elsevier Science

Amsterdam • Boston • London • New York • Oxford • Paris • San Diego •
San Francisco • Singapore • Sydney • Tokyo

Senior Publishing Editor	Charles R. Crumly, Ph.D
Senior Project Manager	Julio Esperas
Editorial Assistant	Christine Vogelei
Marketing Manager	Anne O'Mara
Cover Design	Gary Ragaglia
Copyeditor	Charles Lauder, Jr.
Production Services	Lorretta Palagi
Composition	RDC Tech
Printer	China Translation & Printing Services, Ltd.

This book is printed on acid-free paper. ⊗

Academic Press
An imprint of Elsevier Science
525 B Street, Suite 1900, San Diego, California 92101-4495, USA
http://www.academicpress.com

Academic Press
84 Theobald's Road, London, WC1X 8RR, UK
http://www.academicpress.com

Library of Congress Control Number: 2002104258

International Standard Book Number: 0-12-656446-9

PRINTED IN CHINA

02 03 04 05 06 07 CTP 8 7 6 5 4 3 2 1

For Joy and my doctoral advisors, David C. Martens and Lucian W. Zelazny

Table of Contents

Preface

Since the first edition of *Environmental Soil Chemistry* was published in 1995, a number of important developments have significantly advanced the soil and environmental sciences. These advancements were the primary motivation for publishing the second edition. The use of synchrotron-based spectroscopic and microscopic techniques, which employ intense light, has revolutionized the field of environmental soil chemistry and allied fields such as environmental chemistry, materials science, and geochemistry. The intense light enables one to study chemical reactions and processes at molecular and smaller scales and *in situ*. A new multidisciplinary field has evolved, molecular environmental science, in which soil chemists are actively involved. It can be defined as the study of the chemical and physical forms and distribution of contaminants in soils, sediments, waste materials, natural waters, and the atmosphere at the molecular level. Chapter 1 contains a major section on molecular environmental science with discussions on synchrotron radiation and important spectroscopic and microscopic techniques. The application of these techniques has greatly advanced our understanding of soil organic matter macromolecular structure (Chapter 3), mechanisms of metal and metalloid sorption on soil components and soils, and speciation of inorganic contaminants (Chapter 5). This second edition also contains new information on soil and water quality (Chapter 1), carbon sequestration (Chapter 3), and surface nucleation/precipitation (Chapter 5) and dissolution (Chapter 7). Other material throughout the book has been updated.

As with the first edition, the book provides extensive discussions on the chemistry of inorganic and organic soil components, soil solution–solid phase equilibria, sorption phenomena, kinetics of soil chemical processes, redox reactions, and soil acidity and salinity. Extensive supplementary readings are contained at the end of each chapter, and numerous boxes in the chapters contain sample problems and explanations of parameters and terms. These should be very useful to students taking their first course in soil chemistry. The second edition is a comprehensive and contemporary textbook for advanced undergraduate and graduate students in soil science and for students and professionals in environmental chemistry and engineering, marine studies, and geochemistry.

Writing the second edition of *Environmental Soil Chemistry* has been extremely enjoyable and was made easier with the support and encouragement of a number of persons. I am most grateful to the administration at the University of Delaware for providing me with a truly wonderful environment in which to teach and conduct research. I particularly want to thank our great president of the University of Delaware, David P. Roselle, for his fabulous support of me and my soil chemistry program during the last decade. I am also extremely fortunate to have had an extraordinarily bright and dedicated group of graduate students and postdoctoral fellows. The highlight of my career has been to advise and mentor these fine young scientists. I am also deeply indebted to support personnel. I especially want to acknowledge Fran Mullen who typed the entire manuscript, Jerry Hendricks who compiled the figures and secured permissions, and Amy Broadhurst who prepared references and permissions. Without their support, this book would not have resulted. I also am grateful to Dr. Charles Crumly at Academic Press for his support and encouragement. Lastly, I shall be forever grateful to my wife, Joy, for her constant understanding, love, and encouragement.

Donald L. Sparks

1 Environmental Soil Chemistry: An Overview

S oil chemistry is the branch of soil science that deals with the chemical composition, chemical properties, and chemical reactions of soils. Soils are heterogeneous mixtures of air, water, inorganic and organic solids, and microorganisms (both plant and animal in nature). Soil chemistry is concerned with the chemical reactions involving these phases. For example, carbon dioxide in the air combined with water acts to weather the inorganic solid phase. Chemical reactions between the soil solids and the soil solution influence both plant growth and water quality.

Soil chemistry has traditionally focused on the chemical reactions in soils that affect plant growth and plant nutrition. However, beginning in the 1970s and certainly in the 1990s, as concerns increased about inorganic and organic contaminants in water and soil and their impact on plant, animal, and human health, the emphasis of soil chemistry is now on environmental soil chemistry. Environmental soil chemistry is the study of chemical reactions between soils and environmentally important plant nutrients, radionuclides, metals, metalloids, and organic chemicals. These water and soil contaminants will be discussed later in this chapter.

A knowledge of environmental soil chemistry is fundamental in predicting the fate of contaminants in the surface and subsurface environments. An understanding of the chemistry and mineralogy of inorganic and organic soil components is necessary to comprehend the array of chemical reactions that contaminants may undergo in the soil environment. These reactions, which may include equilibrium and kinetic processes such as dissolution, precipitation, polymerization, adsorption/desorption, and oxidation–reduction, affect the solubility, mobility, speciation (form), toxicity, and bioavailability of contaminants in soils and in surface waters and groundwaters. A knowledge of environmental soil chemistry is also useful in making sound and cost effective decisions about remediation of contaminated soils.

Evolution of Soil Chemistry

Soil chemistry, as a subdiscipline of soil science, originated in the early 1850s with the research of J. Thomas Way, a consulting chemist to the Royal Agricultural Society in England. Way, who is considered the father of soil chemistry, carried out a remarkable group of experiments on the ability of soils to exchange ions. He found that soils could adsorb both cations and anions, and that these ions could be exchanged with other ions. He noted that ion exchange was rapid, that clay was an important soil component in the adsorption of cations, and that heating soils or treating them with strong acid decreased the ability of the soils to adsorb ions. The vast majority of Way's observations were later proven correct, and his work laid the groundwork for many seminal studies on ion exchange and ion sorption that were later conducted by soil chemists. Way's studies also had immense impact on other disciplines including chemical engineering and chemistry. Research on ion exchange has truly been one of the great hallmarks of soil chemistry (Sparks, 1994).

The forefather of soil chemistry in the United States was Edmund Ruffin, a philosopher, rebel, politician, and farmer from Virginia. Ruffin fired the first Confederate shot at Fort Sumter, South Carolina. He committed suicide after Appomattox because he did not wish to live under the "perfidious Yankee race." Ruffin was attempting to farm near Petersburg, Virginia, on soil that was unproductive. He astutely applied oyster shells to his land for the proper reason—to correct or ameliorate soil acidity. He also accurately described zinc deficiencies in his journals (Thomas, 1977).

Much of the research in soil chemistry between 1850 and 1900 was an extension of Way's work. During the early decades of the 20th century classic ion exchange studies by Gedroiz in Russia, Hissink in Holland, and Kelley and Vanselow in California extended the pioneering investigations and conclusions of Way. Numerous ion exchange equations were developed to explain and predict binary reactions (reactions involving two ions) on clay minerals and soils. These were named after the scientists who developed them and included the Kerr, Vanselow, Gapon, Schofield, Krishnamoorthy and Overstreet, Donnan, and Gaines and Thomas equations.

Linus Pauling (1930) conducted some classic studies on the structure of layer silicates that laid the foundation for extensive studies by soil chemists and mineralogists on clay minerals in soils. A major discovery was made by Hendricks and co-workers (Hendricks and Fry, 1930) and Kelley and co-workers (1931) who found that clay minerals in soils were crystalline. Shortly thereafter, X-ray studies were conducted to identify clay minerals and to determine their structures. Immediately, studies were carried out to investigate the retention of cations and anions on clays, oxides, and soils, and mechanisms of retention were proposed. Particularly noteworthy were early studies conducted by Schofield and Samson (1953) and Mehlich (1952), who validated some of Sante Mattson's earlier theories on sorption phenomena (Mattson, 1928). These studies were the forerunners of another important theme in soil chemistry research: surface chemistry of soils.

One of the most interesting and important bodies of research in soil chemistry has been the chemistry of soil acidity. As Hans Jenny so eloquently wrote, investigations on soil acidity were like a merry-go-round. Fierce arguments ensued about whether acidity was primarily attributed to hydrogen or aluminum and were the basis for many studies during the past century. It was Coleman and Thomas (1967) and Rich and co-workers (Rich and Obenshain, 1955; Hsu and Rich, 1960) who, based on numerous studies, concluded that aluminum, including trivalent, monomeric (one Al ion), and polymeric (more than one Al ion) hydroxy, was the primary culprit in soil acidity.

Studies on soil acidity, ion exchange, and retention of ions by soils and soil components such as clay minerals and hydrous oxides were major research themes of soil chemists for many decades.

Since the 1970s studies on rates and mechanisms of heavy metal, oxyanion, radionuclide, pesticide, and other organic chemical interactions with soils and soil components (see Chapters 5 and 7); the effect of mobile colloids on the transport of pollutants; the environmental chemistry of aluminum in soils, particularly acid rain effects on soil chemical processes (see Chapter 9); oxidation–reduction (see Chapter 8) phenomena involving soils and inorganic and organic contaminants; and chemical interactions of sludges (biosolids), manures, and industrial by-products and coproducts with soils have been prevalent research topics in environmental soil chemistry.

The Modern Environmental Movement

To understand how soil chemistry has evolved from a traditional emphasis on chemical reactions affecting plant growth to a focus on soil contaminant reactions, it would be useful to discuss the environmental movement.

The modern environmental movement began over 30 years ago when the emphasis was on reducing pollution from smokestacks and sewer pipes. In the late 1970s a second movement that focused on toxic compounds was initiated. During the past few decades, several important laws that have had a profound influence on environmental policy in the United States were enacted. These are the Clean Air Act of 1970, the Clean Water Act of 1972, the Endangered Species Act, the Superfund Law of 1980 for remediating contaminated toxic waste sites, and the amended Resource Conservation and Recovery Act (RCRA) of 1984, which deals with the disposal of toxic wastes.

The third environmental wave, beginning in the late 1980s and orchestrated by farmers, businesses, homeowners, and others, is questioning the regulations and the often expensive measures that must be taken to satisfy these regulations. Some of the environmental laws contain regulations that some pollutants cannot be contained in the air, water, and soil at levels greater than a few parts per billion. Such low concentrations can be measured only with very sophisticated analytical equipment that was not available until only recently.

Critics are charging that the laws are too rigid, impose exorbitant cost burdens on the industry or business that must rectify the pollution problem, and were enacted based on emotion and not on sound scientific data. Moreover, the critics charge that because these laws were passed without the benefit of careful and thoughtful scientific studies that considered toxicological and especially epidemiological data, the risks were often greatly exaggerated and unfounded, and cost–benefit analyses were not conducted.

Despite the questions that have ensued concerning the strictness and perhaps the inappropriateness of some of the regulations contained in environmental laws, the fact remains that the public is very concerned about the quality of the environment. They have expressed an overwhelming willingness to spend substantial tax dollars to ensure a clean and safe environment.

Contaminants in Waters and Soils

There are a number of inorganic and organic contaminants that are important in water and soil. These include plant nutrients such as nitrate and phosphate; heavy metals such as cadmium, chromium, and lead; oxyanions such as arsenite, arsenate, and selenite; organic chemicals; inorganic acids; and radionuclides. The sources of these contaminants include fertilizers, pesticides, acidic deposition, agricultural and industrial waste materials, and radioactive fallout. Discussions on these contaminants and their sources are provided below. Later chapters will discuss the soil chemical reactions that these contaminants undergo and how a knowledge of these reactions is critical in predicting their effects on the environment.

Water Quality

Pollution of surface water and groundwater is a major concern throughout the world. There are two basic types of pollution—*point* and *nonpoint*. Point pollution is contamination that can be traced to a particular source such as an industrial site, septic tank, or wastewater treatment plant. Nonpoint pollution results from large areas and not from any single source and includes both natural and human activities. Sources of nonpoint pollution include agricultural, human, forestry, urban, construction, and mining activities and atmospheric deposition. There are also naturally occurring nonpoint source pollutants that are important. These include geologic erosion, saline seeps, and breakdown of minerals and soils that may contain large quantities of nutrients. Natural concentrations of an array of inorganic species in groundwater are shown in Table 1.1.

To assess contamination of ground and surface waters with plant nutrients such as N and P, pesticides, and other pollutants a myriad of interconnections including geology, topography, soils, climate and atmospheric inputs, and human activities related to land use and land management practices must be considered (Fig. 1.1).

Perhaps the two plant nutrients of greatest concern in surface and groundwater are N and P. The impacts of excessive N and P on water quality, which can affect both human and animal health, have received increasing attention. The U.S. EPA has established a maximum contaminant level (MCL) of 10 mg liter^{-1} nitrate as N for groundwater. It also established a goal that total phosphate not exceed 0.05 mg liter^{-1} in a stream where it enters a lake or reservoir and that total P in streams that do not discharge directly to lakes or reservoirs not exceed 0.1 mg liter^{-1} (EPA, 1987).

Excessive N and P can cause eutrophication of water bodies, creating excessive growth of algae and other problematic aquatic plants. These plants can clog water pipes and filters and impact recreational endeavors such as fishing, swimming, and boating. When algae decays, foul odors, obnoxious tastes, and low levels of dissolved oxygen in water (hypoxia) can result. Excessive nutrient concentrations have been linked to hypoxia conditions in the Gulf of Mexico, causing harm to fish and shellfish, and to the growth of the dinoflagellate *Pfisteria*, which has been found in Atlantic Coastal Plain waters. Recent outbreaks of *Pfisteria* have been related to fish kills and toxicities to humans (USGS, 1999). Excessive N, in the form of nitrates, has been linked to methemoglobinemia or blue baby syndrome, abortions in women (Centers for Disease Control and Prevention, 1996), and increased risk of non-Hodgkin's lymphoma (Ward *et al.*, 1996).

Phosphorus, as phosphate, is usually not a concern in groundwater, since it is tenaciously held by soils through both electrostatic and nonelectrostatic mechanisms (see Chapter 5 for definitions and discussions) and usually does not leach in most soils. However, in sandy soils that contain little clay, Al or Fe oxides, or organic matter, phosphate can leach through the soil and impact groundwater quality. Perhaps the greatest concern with phosphorus is con-

TABLE 1.1.   *Natural Concentrations of Various Elements, Ions, and Compounds in Groundwater[a,b]*

Element	Concentration		Element	Concentration	
	Typical value	Extreme value		Typical value	Extreme value
	Major Elements (mg liter[-1])		Bi	< 20	
Ca	1.0–150[c]	95,000[d]	Br	< 100–2,000	
	< 500[e]		Cd	< 1.0	
Cl	1.0–70[c]	200,000[d]	Co	<10	
	< 1,000[e]		Cr	< 1.0–5.0	
F	0.1–5.0	70	Cu	< 1.0–3.0	
		1,600[d]	Ga	< 2.0	
Fe	0.01–10	> 1,000[d,f]	Ge	< 20–50	
K	1.0–10	25,000[d]	Hg	< 1.0	
Mg	1.0–50[c]	52,000[d]	I	< 1.0–1,000	48[d]
	< 400[e]		Li	1.0–150	
Na	0.5–120[c]	120,000[d]	Mn	< 1.0–1,000	10[b]
	< 1,000[e]		Mo	< 1.0–30	10
NO$_3$	0.2–20	70	Ni	< 10–50	
SiO$_2$	5.0–100	4,000[d]	PO$_4$	< 100–1,000	
SO$_4$	3.0–150[c]	200,000[d]	Pb	< 15	
	< 2,000[e]		Ra	< 0.1–4.0[g]	0.7[d,g]
Sr	0.1–4.0	50	Rb	< 1.0	
	Trace Elements (mg liter[-1])		Se	< 1.0–10	
Ag	< 5.0		Sn	< 200	
Al	< 5.0–1,000		Ti	< 1.0–150	
As	< 1.0–30	4	U	0.1–40	
B	20–1,000	5	V	< 1.0–10	0.07
Ba	10–500		Zn	< 10–2,000	
Be	< 10		Zr	< 25	

[a] From Dragun (1988).
[b] Based on an analysis of data presented in Durfer and Becker (1964), Hem (1970), and Ebens and Schaklette (1982).
[c] In relatively humid regions.
[d] In brine.
[e] In relatively dry regions.
[f] In thermal springs and mine areas.
[g] Picocuries liter[-1] (i.e., 0.037 disintegrations sec[-1]).

tamination of streams and lakes via surface runoff and erosion. Nitrate-N is weakly held by soils and readily leaches in soils. Contamination of groundwater with nitrates is a major problem in areas that have sandy soils.

Major sources of N and P in the environment are inorganic fertilizers, animal manure, biosolid applications, septic systems, and municipal sewage systems. Inorganic N and P fertilizers increased 20- and 4-fold, respectively, between 1945 and the early 1980s and leveled off thereafter (Fig. 1.2). In 1993, ~12 million metric tons of N and 2 million metric tons of P were used nation-wide. At the same time, animal manure accounted for ~7 million metric tons of N and about 2 million metric tons of P. Additionally, about 3 million metric tons of N per year are derived from atmospheric sources (Puckett, 1995).

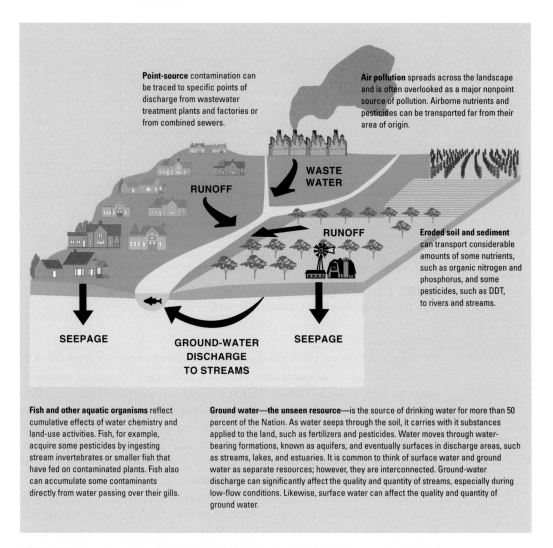

FIGURE 1.1. *Interactions between surface and groundwater, atmospheric contributions, natural landscape features, human activities, and aquatic health and impacts on nutrients and pesticides in water resources. From U.S. Geological Survey, Circular 1225, 1999.*

Pesticides

Pesticides can be classified as herbicides, those used to control weeds, insecticides, to control insects, fungicides, to control fungi, and others such as nematicides and rodenticides.

Pesticides were first used in agricultural production in the second half of the 19th century. Examples included lead, arsenic, copper, and zinc salts, and naturally produced plant compounds such as nicotine. These were used for insect and disease control on crops. In the 1930s and 1940s 2,4-D, an herbicide, and DDT, an insecticide, were introduced; subsequently, increasing amounts of pesticides were used in agricultural production worldwide.

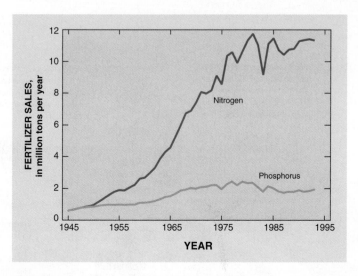

FIGURE 1.2. *Changes in nitrogen and fertilizer use over the decades.*
From U.S. Geological Survey, Circular 1225, 1999.

The benefits that pesticides have played in increasing crop production at a reasonable cost are unquestioned. However, as the use of pesticides increased, concerns were expressed about their appearance in water and soils, and their effects on humans and animals. Total pesticide use in the United States has stayed constant at about 409 million kg per year after increasing significantly through the mid-1970s due to greater herbicide use (Fig. 1.3). Agriculture accounts for 70–80% of total pesticide use. About 60% of the agricultural use of pesticides involves herbicide applications.

One of the most recent and comprehensive assessments of water quality in the United States has been conducted by the USGS through its National Water Quality Assessment (NAWQA) Program (USGS, 1999). This program is assessing water quality in more than 50 major river basins and aquifer systems. These include water resources provided to more than 60% of the U.S. population in watersheds that comprise about 50% of the land area of the conterminous United States. Figure 1.4 shows 20 of the systems that were evaluated beginning in 1991, and for which data were recently released (USGS, 1999). Water quality patterns were related to chemical use, land use, climate, geology, topography, and soils.

The relative level of contamination of streams and shallow groundwater with N, P, herbicides, and insecticides in different areas is shown in Fig. 1.5. There is a clear correlation between contamination level and land use and the amounts of nutrients and chemical used.

Nitrate levels were not a problem for humans drinking water from streams or deep aquifers. However, about 15% of all shallow groundwater sampled below agricultural and urban areas exceeded the MCL for NO_3^-. Areas that ranked among the highest 25% of median NO_3^- concentration in shallow groundwaters were clustered in the mid-Atlantic and Western parts of the

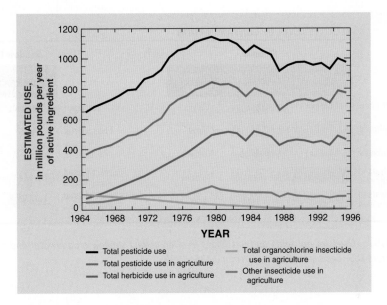

FIGURE 1.3. *Changes in agricultural pesticide use over the decades. From U.S. Geological Survey, Circular 1225, 1999.*

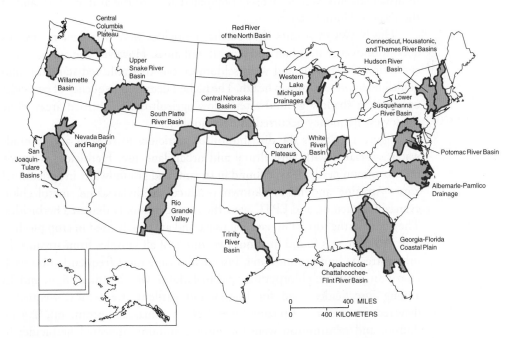

FIGURE 1.4. *Locations of wells sampled as part of NAWQA land-use studies and major aquifer survey conducted during 1992–1995. From U.S. Geological Survey, Circular 1225, 1999.*

RELATIVE LEVEL OF CONTAMINATION

	Streams				Shallow Ground Water	
	Urban areas	Agricultural areas	Undeveloped areas		Urban areas	Agricultural areas
Nitrogen	Medium	Medium–High	Low	Nitrogen	Medium	High
Phosphorus	Medium–High	Medium–High	Low	Phosphorus	Low	Low
Herbicides	Medium	Low–High	No data	Herbicides	Medium	Medium–High
Currently used insecticides	Medium-High	Low–Medium	No data	Currently used insecticides	Low–Medium	Low–Medium
Historically used insecticides	Medium-High	Low–High	Low	Historically used insecticides	Low-High	Low-High

FIGURE 1.5.   *Levels of nutrients and pesticides in streams and shallow groundwater and relationship to land use. From U.S. Geological Survey, Circular 1225, 1999.*

United States (Fig. 1.6). These findings are representative of differences in N loading, land use, soil and aquifer permeability, irrigation practices, and other factors (USGS, 1999).

Total P concentrations in agricultural streams were among the highest measured and correlated with nonpoint P inputs. The highest total P levels in urban streams were in densely populated areas of the arid Western and of the Eastern United States.

The NAWQA studies showed that pesticides were prevalent in streams and groundwater in urban and agricultural areas. However, the average concentrations in streams and wells seldom exceeded established standards and guidelines to protect human health. The highest detection frequency of pesticides occurred in shallow groundwater below agricultural and urban areas while the lowest frequency occurred in deep aquifers.

Figure 1.7 shows the distribution of pesticides in streams and groundwater associated with agricultural and urban land use. Herbicides were the most common pesticide type found in streams and groundwater in agricultural areas. Atrazine and its breakdown product, deethylatrazine, metolachlor, cyanazine, alachlor, and EPTC were the most commonly detected herbicides. They rank in the top 10 in national usage and are widely used in crop production. Atrazine was found in about two-thirds of all samples from streams. In urban streams and groundwater, insecticides were most frequently observed. Diazinon, carbaryl, chlorpyrifos, and malathion, which rank 1, 8, 4, and 13 among insecticides used for homes and gardens, were most frequently detected in streams. Atrazine, metolachlor, simazine, prometon, 2,4-D, diuron, and tebuthiuron were the most commonly detected herbicides in urban streams. These are used on lawns, gardens, and commercial areas, and in roadside maintenance.

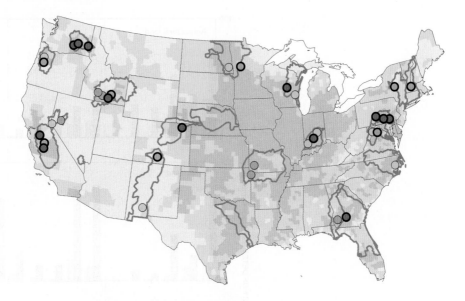

Median concentration of nitrate—in milligrams per liter.
Each circle represents a ground-water study

- ● Highest (greater than 5.0)
- ◐ Medium (0.5 to 5.0)
- ○ Lowest (less than 0.5)

Background concentration

- O Bold outline indicates median values
  greater than background concentration
  (2 milligrams per liter)

FIGURE 1.6.   *Levels of nitrate in shallow groundwater. From U.S. Geological Survey, Circular 1225, 1999.*

Acid Deposition

Much concern about the effects of acid rain on plants, bodies of water, and soils has also been expressed. Acid rain also can cause damage to buildings and monuments, particularly those constructed of limestone and marble, and it can cause corrosion of certain metals.

Acid rain results from the burning of fossil fuels such as coal, which generates sulfur dioxide and nitrogen oxides, and from exhausts of motor vehicles, a main source of nitrogen oxides. These combine in the atmosphere with water and other materials to produce nitric and sulfuric acids that are often carried for long distances by wind and then fall to the earth via precipitation such as rain, snow, sleet, mist, or fog. Acidic deposition has been linked to a number of environmental issues (Table 1.2).

From 1980 to 1991 the U.S. Geological Survey monitored rainwater collected at 33 sites around the United States. Concentrations of sulfates declined greatly at 26 of the 33 sites; however, nitrates were significantly decreased at only 3 of the 33 sites. The decreases in sulfate are directly related

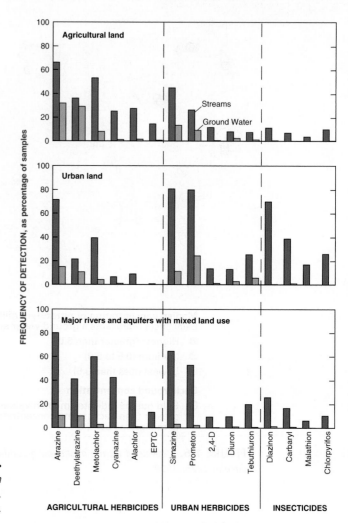

FIGURE 1.7.
Frequently detected pesticides in water of agricultural and urban areas. From U.S. Geological Survey, Circular 1225, 1999.

TABLE 1.2.   *Linkages between Emissions of SO_2 and NO_x and Important Environmental Issues[a]*

Problem	Linkage to acidic deposition	Reference
Coastal eutrophication	Atmospheric deposition is important in the supply of N to coastal waters.	Jaworski *et al.* (1997)
Mercury accumulation	Surface water acidification enhances mercury accumulation in fish.	Driscoll *et al.* (1994)
Decreased visibility	Sulfate aerosols are an important component of atmospheric particulates; they decrease visibility.	Malm *et al.* (1994)
Climate change	Sulfate aerosols increase atmospheric albedo, cooling the Earth and offsetting some of the warming potential of greenhouse gases; tropospheric O_3 and N_2O act as greenhouse gases.	Moore *et al.* (1997)
Tropospheric ozone	Emissions of NO_x contribute to the formation of ozone.	Seinfeld (1986)

[a] From Driscoll *et al.* (2001).

to the 30% decrease in sulfur dioxide emissions nationwide. The less dramatic decrease in nitrate concentrations may be due to emissions from more automobiles and new power plants and factories (*New York Times*, 1993). A 10-year $500 million study funded by the U.S. Government concluded that while there is some environmental damage from acid rain, the damage is much less than originally expected. In the United States, fewer than 1200 lakes have been extensively acidified, which is about 4% of the total number of lakes in the areas where acidification might be expected. However, in certain areas, such as the Adirondack Mountains, 41% of the lakes showed chronic or episodic acidification due to acidic deposition. The acidification of surface waters can result in decreases in survival, size, and density of aquatic life such as fish (Driscoll *et al.*, 2001). Except for red spruce at high elevations, little evidence was found that acid rain caused severe damage to forests in the United States. However, over time trees could suffer nutritionally because of depletion of nutrients leached from soils. Dramatic effects of acid rain have been observed in forests in Eastern Europe where sulfur and nitrogen oxide production is not being adequately controlled. More discussion on acid rain effects on soils is provided in Chapter 9.

Trace Elements

A trace element is an element present at a level <0.1% in natural materials such as the lithosphere; if the concentrations are high enough, they can be toxic to living organisms (Adriano, 1986). Trace elements include trace metals, heavy metals, metalloids (an element having both metallic and nonmetallic properties, e.g., As and B), micronutrients (chemical elements needed in small quantities for plant growth, i.e., <50 mg g^{-1} in the plant), and trace inorganics. Heavy metals are those elements having densities greater than 5.0 g cm^{-3}. Examples are Cd, Cr, Co, Cu, Pb, Hg, and Ni. Table 1.3 shows the occurrence and significance of trace elements in natural waters.

The sources of trace elements are soil parent materials (rocks), commercial fertilizers, liming materials, biosolids, irrigation waters, coal combustion residues, metal-smelting industries, auto emissions, and others. Table 1.4 shows the concentrations of trace elements in soil-forming rocks and other natural materials, while Table 1.5 illustrates trace element concentrations in biosolids from several countries.

One metalloid of increasing concern in the environment is arsenic (As). Arsenic is a known human carcinogen. Drinking water contaminated with As has been linked to cancer, diabetes, and cardiovascular problems. The source of the As in drinking water, particularly inorganic As, is often weathering of minerals in rocks and soils.

Total As in soils ranges from 0.1 to 97 ppm with a mean concentration of 7 ppm for surface soils in the United States (Dragun, 1991). Arsenic occurs in two major oxidation states, As(III) and As(V). As(III) is primarily present in anoxic environments while As(V) is found in oxic soils. Both As species primarily occur as oxyanions in the natural environment and strongly complex

TABLE 1.3.　Occurrence and Significance of Trace Elements in Natural Waters[a]

Element	Sources	Effects and significance	U.S. Public Health Service limit (mg liter^{-1})[b]	Occurrence: % of samples, highest and mean concentrations (μg liter^{-1})[c]
Arsenic	Mining by-product, pesticides, chemical waste	Toxic, possibly carcinogenic	0.05	5.5% (above 5 μg liter^{-1}), 336, 64
Beryllium	Coal, nuclear power, and space industries	Acute and chronic toxicity, carcinogenic	Not given	Not given
Boron	Coal, detergent formulations, industrial wastes	Toxic to some plants	1.0	98% (above 1 μg liter^{-1}), 5000, 101
Cadmium	Industrial discharge, mining waste, metal plating, water pipes	Replaces zinc biochemically, causes high blood pressure and kidney damage, destroys testicular tissue and red blood cells, toxic to aquatic biota	0.01	2.5%, not given, 9.5
Chromium	Metal plating, cooling-tower water additive (chromate), normally found as Cr(VI) in polluted water	Essential trace element (glucose tolerance factor), possibly carcinogenic as Cr(VI)	0.05	24.5%, 112, 9.7
Copper	Metal plating, industrial and domestic wastes, mining, mineral leaching	Essential trace element, not very toxic to animals, toxic to plants and algae at moderate levels	1.0	74.4%, 280, 15
Fluorine (fluoride ion)	Natural geological sources, industrial waste, water additive	Prevents tooth decay at above 1 mg liter^{-1}, causes mottled teeth and bone damage at around 5 mg liter^{-1} in water	0.8–1.7 depending on temperature	Not given
Iodine (iodide)	Industrial waste, natural brines, seawater intrusion	Prevents goiter	Not given	Rare in fresh water
Iron	Corroded metal, industrial wastes, acid mine drainage, low pE water in contact with iron minerals	Essential nutrient (component of hemoglobin), not very toxic, damages materials (bathroom fixtures and clothing)	0.05	75.6%, 4600, 52

TABLE 1.3. Occurrence and Significance of Trace Elements in Natural Waters[a] (contd)

Element	Sources	Effects and significance	U.S. Public Health Service limit (mg liter^{-1})[b]	Occurrence: % of samples, highest and mean concentrations (μg liter^{-1})[c]
Lead	Industry, mining, plumbing, coal, gasoline	Toxicity (anemia, kidney disease, nervous system), wildlife destruction	0.05	19.3% (above 2 μg liter^{-1}), 140, 23
Manganese	Mining, industrial waste, acid mine drainage, microbial action on manganese minerals at low pE	Relatively nontoxic to animals, toxic to plants at higher levels, stains materials (bathroom fixtures and clothing)	0.05	51.4% (above 0.3 μg liter^{-1}), 3230, 58
Mercury	Industrial waste, mining, pesticides, coal	Acute and chronic toxicity	Not given	Not given
Molybdenum	Industrial waste, natural sources, cooling-tower water additive	Possibly toxic to animals, essential for plants	Not given	32.7 (above 2 μg liter^{-1}), 5400, 120
Selenium	Natural geological sources, sulfur, coal	Essential at low levels, toxic at higher levels, causes "alkali disease" and "blind staggers" in cattle, possibly carcinogenic	0.01	Not given
Silver	Natural geological sources, mining, electroplating, film-processing wastes, disinfection of water	Causes blue-gray discoloration of skin, mucous membranes, eyes	0.05	6.6% (above 0.1 μg liter^{-1}), 38, 2.6
Zinc	Industrial waste, metal plating, plumbing	Essential element in many metalloenzymes, aids wound healing, toxic to plants at higher levels; major component of sewage sludge, limits land disposal of sludge	5.0	76.5% (above 2 μg liter^{-1}), 1180, 64

[a] Reprinted with permission from S. E. Manahan (1991). "Environmental Chemistry." Copyright Lewis Publishers, an imprint of CRC Press, Boca Raton, FL.
[b] U.S. Public Health Service (1962).
[c] Kopp and Kroner, "Trace Metals in Waters of the United States," U.S. EPA. The first figure is the percentage of samples showing the element; the second is the highest value found; the third is the mean value in positive samples (samples showing the presence of the metal at detectable levels).

TABLE 1.4. *Concentrations (mg kg⁻¹) of Trace Elements in Various Soil-Forming Rocks and Other Natural Materials[a,b]*

Element	Ultramafic igneous	Basaltic igneous	Granitic igneous	Shales and clays	Black shales	Deep-sea clays	Lime-stones	Sand-stones
Arsenic	0.3–16 / 3.0	0.2–10 / 2.0	0.2–13.8 / 2.0	— / 10	—	— / 13	0.1–8.1 / 1.7	0.6–9.7 / 2
Barium	0.2–40 / 1	20–400 / 300	300–1800 / 700	460–1700 / 700	70–1000 / 300	— / 2300	10 / —	— / 20
Beryllium	—	1.0	2–3	3	—	2.6	—	—
Cadmium	0–0.2 / 0.05	0.006–0.6 / 0.2	0.003–0.18 / 0.15	0–11 / 1.4	<0.3–8.4 / 1.0	0.1–1 / 0.5	0.05	0.05
Chromium	1000–3400 / 1800	40–600 / 200	2–90 / 20	30–590 / 120	26–1000 / 100	— / 90	— / 10	— / 35
Cobalt	90–270 / 150	24–90 / 50	1–15 / 5	5–25 / 20	7–100 / 10	— / 74	— / 0.1	— / 0.3
Copper	2–100 / 15	30–160 / 90	4–30 / 15	18–120 / 50	20–200 / 70	— / 250	— / 4	— / 2
Fluorine	—	20–1060 / 360	20–2700 / 870	10–7600 / 800	—	— / 1300	0–1200 / 220	10–880 / 180
Iron	94,000	86,500	14,000–30,000	47,200	20,000	65,000	3800	9800
Lead	— / 1	2–18 / 6	6–30 / 18	16–150 / 20	7–150 / 30	— / 80	— / 9	<1–31 / 12
Mercury	0.004–0.5 / 0.1?	0.002–0.5 / 0.05	0.005–0.4 / 0.06	0.005–0.51 / 0.09	0.03–2.8 / 0.5	0.02–2.0 / 0.4	0.01–0.22 / 0.04	0.001–0. / 0.05
Molybdenum	— / 0.3	0.9–7 / 1.5	1–6 / 1.4	— / 2.5	1–300 / 10	— / 27	— / 0.4	— / 0.2
Nickel	270–3600 / 2000	45–410 / 140	2–20 / 8	20–250 / 68	10–500 / 50	— / 225	— / 20	— / 2
Selenium	0.05	0.05	0.05	0.6	—	0.17	0.08	0.05
Vanadium	17–300 / 40	50–360 / 250	9–90 / 60	30–200 / 130	50–1000 / 150	— / 120	— / 20	— / 20
Zinc	— / 40	48–240 / 110	5–140 / 40	18–180 / 90	34–1500 / 100?	— / 165	— / 20	2–41 / 16

[a] From Tourtelot (1971).
[b] The upper values are the usually reported range, the lower values the average.

with metal oxides such as Al and Fe oxides as inner-sphere products. These oxides, and particularly Mn oxides, can effect the oxidation of As(III) to As(V) which reduces the toxicity of As. As can also occur as sulfide minerals such as arsenopyrite (FeAsS) and enargite (Cu_3AsS_4) at mining sites.

There has been much controversy over the MCL of As in drinking water. The current standard of 50 ppb in the United States was established in 1942 by the U.S. Public Health Service. The World Health Organization recommends a 10-ppb guideline. Since 1975, the U.S. EPA has been reassessing the 50-ppb level. In 1999, the National Research Council (NRC) published a report recommending that the 50-ppb level be lowered. In 2001 the NRC issued a new report estimating that humans who consume water with 3 ppb As daily (based on 1 liter consumption day^{-1}) have a 1 in 1000 risk of developing bladder or lung cancer during their lifetime. If the level of As in drinking water is 10 ppb, the risk is more than 3 in 1000 and at 20 ppb the risk is 7 in 1000 (NRC, 2001).

The U.S. Government has agreed to lower the MCL for As to 10 ppb. This new standard will affect 13 million people, primarily in the Western United States but also in parts of the Midwest and New England where As levels in well water exceed 10 ppb (Fig. 1.8). The 10-ppb standard will cost $181 million annually.

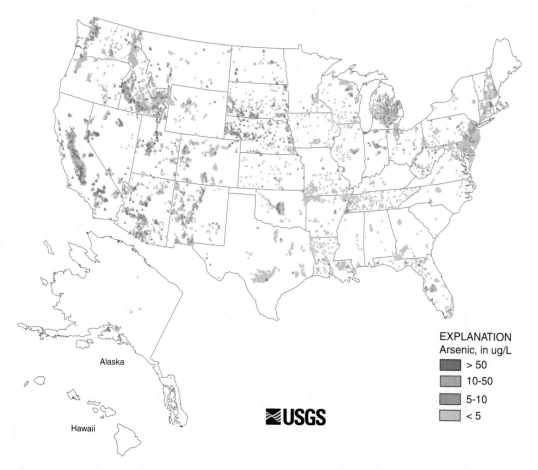

FIGURE 1.8.   *Arsenic concentrations in groundwater of the United States. From Welch et al. (2000), Arsenic in groundwater resources of the United States: U.S. Geological Survey Fact Sheet 063-00.*

TABLE 1.5.   Concentrations (mg kg^{-1}) of Trace Elements in Biosolids[a]

Element	United States[b] Mean	United States[b] Range	United Kingdom[c] Mean	United Kingdom[c] Range	Sweden[d] Mean	Sweden[d] Range	Canada[e] (mean)	New Zealand[f] (mean)
Ag	—	—	32	5–150	—	—	—	—
As	14.3	3–30	—	—	—	—	—	480
B	37.0	22–90	70	15–1,000	—	—	1,950	580
Ba	621	272–1,066	1,700	150–4,000	—	—	—	—
Be	<8.5	—	5	1–30	—	—	—	—
Bi	16.8	<1–56	34	<12–100	—	—	—	4.5
Cd	104	7–444	<200	<60–1,500	13	2–171	38	21
Co	9.6	4–18	24	2–260	15	2–113	19	—
Cr	1,441	169–14,000	980	40–8,800	872	20–40,615	1,960	850
Cu	1,346	458–2,890	970	200–8,000	791	52–3,300	1,600	720
F	167	370–739	—	—	—	—	—	—
Hg	8.6	3–18	—	—	6.0	<0.1–55	—	—
Mn	194	32–527	500	150–2,500	517	73–3,861	2,660	610
Mo	14.3	1–40	7	2–30	—	—	13	8
Ni	235	36–562	510	20–5,300	121	16–2,120	380	350
Pb	1,832	136–7,627	820	120–3,000	281	52–2,914	1,700	610
Sb	10.6	2–44	—	—	—	—	—	—
Se	3.1	1–5	—	—	—	—	—	—
Sn	216	111–492	160	40–700	—	—	—	80
Ti	2,331	1,080–4,580	2,000	<1,000–4,500	—	—	—	4,700
V	40.6	15–92	75	20–400	—	—	15	80
W	20.2	1–100	—	—	—	—	—	—
Zn	2,132	560–6,890	4,100	700–49,000	2,055	705–14,700	6,140	700

[a] From Adriano (1986), with permission from Springer-Verlag.
[b] Furr et al. (1976); includes Atlanta, Chicago, Denver, Houston, Los Angeles, Miami, Milwaukee, Philadelphia, San Francisco, Seattle, Washington, DC, and five cities in New York, with permission.
[c] Berrow and Webber (1972); includes 42 samples from different locations in England and Wales, with permission.
[d] Berrow and Odge (1972); from 93 treatment plants, with permission.

Hazardous Wastes

The disposal of hazardous wastes and their effects on the environment and human health are topics of worldwide importance. There is an array of potentially hazardous waste materials. These include mining waste, acid mine drainage, wastes from metal smelting and refining industries, pulp and paper industry wastes, petroleum refining wastes, wastes from paint and allied industries, pesticide applications, inorganic fertilizers, and municipal solid waste (Yong *et al.*, 1992).

According to the Resource Conservation and Recovery Act (RCRA) of 1976, solid waste is "any garbage, refuse, sludge, from waste treatment plants, water supply treatment plants or air pollution control facilities and other discarded material including solid, liquid semisolid, or contained gaseous materials resulting from industrial, commercial, mining and agricultural activities, and from community activities, but does not include solid or dissolved material in domestic sewage or irrigation return flows." This definition includes nearly all kinds of industrial and consumer waste discharge—solid, semiliquid, and liquid.

Hazardous waste is defined as "a solid waste, or combination of solid wastes, which because of its quantity, concentration, or physical, chemical, or infectious characteristics may: (a) cause, or significantly contribute to an increase in serious irreversible, or incapacitating reversible, illness; or (b) pose a substantial present or potential hazard to human health or the environment when improperly treated, stored, transported, or disposed of, or otherwise managed" (Resource Conservation and Recovery Act, 1976, Public Law 94-580).

Case Study of Pollution of Soils and Waters

An extensive case study that illustrates pollution of Department of Energy (DOE) sites and military bases in the United States has recently been conducted (Table 1.6). These are located around the United States and were sites for weapons production. Substantial radioactive wastes were produced. At some military bases toxic chemicals were disposed of in water supplies and other areas that are now leaking.

A report (Riley *et al.*, 1992) documented contamination of soils/sediments and groundwater at 91 waste sites on 18 DOE facilities. These facilities occupy 7280 km^2 in the 48 contiguous states. Most of the wastes were disposed of on the ground or in ponds, pits, injection wells, and landfills, and are contaminating the subsurface environment. Contamination is also resulting from leaking underground storage tanks and buried chemicals and wastes. The results of the survey found 100 individual chemicals or mixtures in the groundwater and soil/sediments of these sites.

TABLE 1.6.   *Contamination at U.S. Energy Department Sites and Costs of Cleanup[a]*

Site and location	Type of contamination	Spent as of Sept. 1990 on cleanup[b]	Estimated or approved cost of cleanup[b]
1. Hanford Nuclear Reservation[c] (Richland, WA)	Plutonium and other radioactive nuclides, toxic chemicals, heavy metals, leaking radioactive-waste tanks, groundwater and soil contamination, seepage into the Columbia River	$1,000 plus	$30,000–50,000
2. McClellan Air Force Base[d] (Sacramento, CA)	Solvents, metal-plating wastes, degreasers, paints, lubricants, acids, PCBs in groundwater	61.0	170.5
3. Hunters Point Naval Air Station[e] (San Francisco, CA)	Chemical spills in soil, heavy metals, solvents	21.5	114.0
4. Lawrence Livermore National Laboratory[c] (Livermore, CA)	Chemical and radioactive contamination of buildings and soils	N.A.	1,000 plus
5. Castle Air Force Base[d] (Merced, CA)	Solvents, fuels, oils, pesticides, cyanide, cadmium in soil, landfills, and disposal pits	12.7	90.0
6. Edwards Air Force Base[e] (Kern County, CA)	Oil, solvents, petroleum by-products in abandoned sites and drum storage area	21.0	53.4
7. Nevada Test Site[c] (near Las Vegas, NV)	Radioactive groundwater contamination	100 plus	1,000 plus
8. Idaho National Engineering Laboratory[c] (near Idaho Falls, ID)	Radioactive wastes, contamination of Snake River aquifer, chemical-waste lagoons	N.A.	5,000 plus
9. Tooele Army Depot[f] (Tooele County, UT)	Heavy metals, lubricants, paint primers, PCBs, plating and explosives wastes in groundwater and ponds	10.0	64.4
10. Rocky Mountain Arsenal[f] (Denver, CO)	Pesticides, nerve gas, toxic solvents, and fuel oil in shallow, leaking pits	315.5	2,037.1
11. Rocky Flats Plant[c] (Golden, CO)	Plutonium, americium, chemicals, other radioactive wastes in groundwater, lagoons and dump sites	200 plus	1,000 plus
12. Los Alamos National Laboratory[c] (Los Alamos, NM)	Millions of gallons of radioactive and toxic chemical wastes poured into ravines and canyons across hundreds of sites	N.A.	1,000 plus
13. Tinker Air Force Base[d] (Oklahoma City, OK)	Trichloroethylene and chromium in underground water	20.1	69.7
14. Twin Cities Army Ammunition Plant[f] (New Brighton, MN)	Chemical by-products and solvents from ammunition manufacturing	28.0	59.9
15. Lake City Army Ammunition Plant[f] (Independence, MO)	Toxic metals and chemicals in groundwater	25.9	55.1

TABLE 1.6. Contamination at U.S. Energy Department Sites and Costs of Cleanup[a] (contd)

Site and location	Type of contamination	Spent as of Sept. 1990 on cleanup[b]	Estimated or approved cost of cleanup[b]
16. Louisiana Army Ammunition Plant[f] (Doyline, LA)	Hazardous wastes, groundwater contamination	38.0	66.9
17. Oak Ridge National Laboratory[c] (Oak Ridge, TN)	Mercury, radioactive sediments in streams, lakes, and groundwater	1,000 plus	4,000–8,000
18. Griffiss Air Force Base[d] (Rome, NY)	Heavy metals, greases, solvents, caustic cleaners, dyes in tank farms and groundwater and disposal sites	7.3	100.0
19. Letterkenny Army Depot[f] (Franklin County, PA)	Oil, pesticides, solvents, metal-plating wastes, phenolics, painting wastes in soil and water	11.9	56.2
20. Naval Weapons Station[e] (Colts Neck, NJ)	Heavy metals, lubricants, oil, corrosive acids in pits and disposal sites	1.1	33.8
21. Aberdeen Proving Ground[f] (Aberdeen, MD)	Arsenic, napalm, nitrates, and chemical warfare agents contaminating soil and groundwater	19.8	579.4
22. Camp Lejeune Military Reservation[e] (Jacksonville, NC)	Lithium batteries, paints, thinners, pesticides, PCBs in soil and potentially draining into New River	3.0	59.0
23. Cherry Point Marine Air Corps Station[e] (Cherry Point, NC)	Untreated wastes soaking creek sediments with heavy metals, industrial wastes, and electroplating wastes	1.6	51.6
24. Savannah River Site[c] (Aiken, SC)	Radioactive waste burial grounds, toxic chemical pollution, contamination of groundwater	N.A.	5,000 plus
25. Mound Laboratory[c] (Miamisburg, OH)	Plutonium in soil and toxic chemical wastes	150 plus	500 plus
26. Feed Materials Production Center[c] (Fernald, OH)	Uranium and chemicals in ponds and soil	600 plus	1,000–3,000

[a] From Energy Department, Air Force, Army, Navy, General Accounting Office; *New York Times*, August 15, 1991.
[b] Army, Navy, and Air Force figures represent costs for committed or approved cleanup activities. Energy Department figures represent total estimated costs. All costs are in millions of dollars.
[c] Energy Department site.
[d] Air Force site.
[e] Navy site.
[f] Army site.

The most prevalent metals were Pb, Cr, As, and Zn while the major anion found was NO_3^- (Fig. 1.9). Greater than 50% of all DOE facilities contained 9 of the 12 metals and anions shown in Fig. 1.9 in the groundwater. The sources of the metals and anions are associated with reactor operations (Cr and Pb), irradiated fuel processing (NO_3^-, Cr, CN^-, and F^-), uranium recovery (NO_3^-), fuel fabrication (Cr, NO_3^-, and Cu), fuel production (Hg), and isotope separation (Hg) (Stenner *et al.*, 1988; Rogers *et al.*, 1989; Evans *et al.*, 1990). The most prevalent inorganic species in soils/sediments at the DOE sites were Cu, Cr, Zn, Hg, As, Cd, and NO_3^- (Fig. 1.9). Radionuclides that were most common in groundwater were tritium, U, and Sr. In soils/sediments, U, Pu, and Cs were the most prevalent.

Figure 1.10 shows that 19 chlorinated hydrocarbons were found in the groundwaters. The most common ones were trichloroethylene, 1,1,1-trichloroethane and 1,2-dichloroethylene, tetrachloroethylene, 1,1-dichloroethane, and chloroform. In soils/sediments, trichloroethylene, 1,1,1-trichloroethane, tetrachloroethylene, and dichloromethane were found at 50% or more of the sites. Fuel hydrocarbons most often found in groundwaters were toluene, xylene, benzene, and ethylbenzene. In soils/sediments the same fuel hydrocarbons were most often found but some polyaromatic hydrocarbons, such as phenanthrene, anthracene, and fluoranthene, also were detected. These latter compounds are not very soluble, which explains why they were not detected in the groundwaters. Sources of the high-molecular-weight hydrocarbons were coal and coal wastes (fly ash) from coal-fired electric power and steam-generating facilities located at many of the DOE sites. Sources of low-

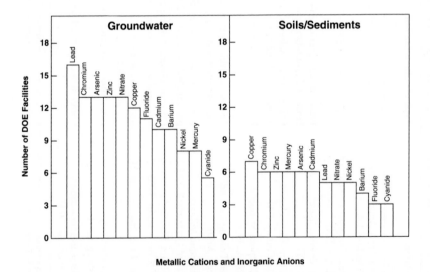

FIGURE 1.9. *Frequency of occurrence of selected metals and inorganic anions in groundwater and soils/sediments at DOE facilities. From Riley et al. (1992), with permission. This research or report was supported by the Subsurface Science program, Office of Health and Environmental Research, U.S. Department of Energy (DOE).*

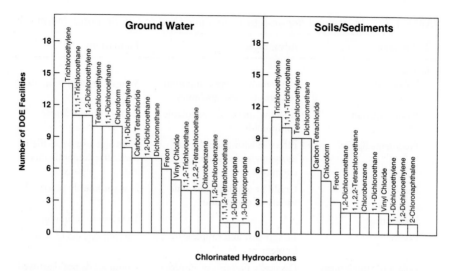

FIGURE 1.10. *Frequency of occurrence of chlorinated hydrocarbons in groundwater and soils/sediments at DOE facilities. From Riley et al. (1992), with permission. This research or report was supported by the Subsurface Science program, Office of Health and Environmental Research, U.S. Department of Energy (DOE).*

molecular-weight hydrocarbons were gasoline- and petroleum-derived fuels from leaking above- and underground tanks. Ketones, primarily acetone, methyl ethyl ketone, and methyl isobutyl ketone, were found in the groundwater, while acetone was the most prevalent ketone found in soils/sediments. Ketones are employed in nuclear fuels processing.

Other chemicals and compounds detected less frequently at the DOE sites included phthalates, pesticides, and chelating agents (e.g., EDTA, ethylenediaminetetraacetic acid), and organic acids such as oxalic and citric acids.

Soil Decontamination

Numerous attempts to decontaminate polluted soils with the use of an array of both *in situ* and non-*in-situ* techniques are being made (Table 1.7). None of these is a panacea for remediating contaminated soils and often more than one of the techniques may be necessary to optimize the cleanup effort. The complexity of soils and the presence of multiple contaminants also makes most remediation efforts arduous and costly (Sparks, 1993).

In Situ Methods

In situ methods are used at the contamination site. Soil does not need to be excavated, and therefore exposure pathways are minimized.

TABLE 1.7.   *In situ and Non-in-Situ Techniques Used in Soil Decontamination[a]*

Technology	Advantages	Limitations	Relative costs
In situ			
Volatilization	Can remove some compounds resistant to biodegradation	Volatile organic compounds only	Low
Biodegradation	Effective on some nonvolatile compounds	Long-term timeframe	Moderate
Phytoremediation	Effective with a number of inorganic and organic chemicals	Plants are often specific for particular contaminants	Low to medium
Leaching	Could be applicable to wide variety of compounds	Not commonly practiced	Moderate
Vitrification		Developing technology	High
Passive	Lowest cost and simplest to implement	Varying degrees of removal	Low
Isolation/containment	Physically prevents or impedes migration	Compounds not destroyed	Low to moderate
Non-*in-situ*			
Land treatment	Uses natural degradation processes	Some residuals remain	Moderate
Thermal treatment	Complete destruction possible	Usually requires special features	High
Asphalt incorporation	Use of existing facilities	Incomplete removal of heavier compounds	Moderate
Solidification	Immobilizes compounds	Not commonly practiced for soils	Moderate
Groundwater extraction and treatment	Product recovery, groundwater restoration		Moderate
Chemical extraction		Not commonly practiced	High
Excavation	Removal of soils from site	Long-term liability	Moderate

[a] Adapted from Preslo *et al.* (1988). Copyright Lewis Publishers, an imprint of CRC Press, Boca Raton, FL.

VOLATILIZATION

In situ volatilization causes mechanical drawing or air venting through the soil. A draft fan is injected or induced, which causes an air flow through the soil, via a slotted or screened pipe, so that air can flow but entrainment of soil particles is restricted. Some treatment, e.g., activated carbon, is used to recover the volatilized contaminant. This technique is limited to volatile organic carbon materials (Sparks, 1993).

BIODEGRADATION

In situ biodegradation involves the enhancement of naturally occurring microorganisms by stimulating their numbers and activity. The microorganisms then assist in degrading the soil contaminants. A number of soil, environmental, chemical, and management factors affect biodegradation of soil pollutants including moisture content, pH, temperature, the microbial community present, and the availability of nutrients. Biodegradation is facilitated by aerobic soil conditions and soil pH in the range of 5.5–8.0, with an optimal pH of about 7 and temperature in the range of 293–313 K. It is important to realize that a microbe may be effective in degrading one pollutant, but not another. Moreover, microbes may be effective in degrading one form of a specific pollutant but not another.

PHYTOREMEDIATION

The use of plants to decontaminate soils and water (phytoremediation) can be quite effective (Fig. 1.11). There are hundreds of plant species that can detoxify pollutants. For example, sunflowers can absorb uranium, certain ferns have high affinity for As, alpine herbs absorb Zn, mustards can absorb Pb, clovers take up oil, and poplar trees destroy dry-cleaning solvents (*New York Times*, 2001).

Recently the brake fern *(Pteris vittata)* was found to be an As hyper-accumulator (Brooks, 1998) and very effective in remediation of a Central Florida soil contaminated with chromated copper arsenate (Ma *et al.,* 2001). Brake ferns extracted 1,442–7,526 mg kg^{-1} As from the contaminated soils. The uptake of As into the fern fronds was rapid, increasing from 29.4 to 15,861 in two weeks. Almost all of the As present in the plant was inorganic, and there were indications that As(V) was converted to As(III) during translocation from roots to fronds.

LEACHING

This method involves leaching the in-place soil with water and often with a surfactant (a surface-active substance that consists of hydrophobic and hydrophilic regions; surfactants lower the surface tension) to remove the contaminants. The leachate is then collected, downstream of the site, using a collection system for treatment and/or disposal. The use of this method has been limited since large quantities of water are often used to remove the pollutants and, consequently, the waste stream is large and disposal costs can be high.

The effectiveness of a leaching technique also depends on the permeability, porosity, homogeneity, texture, and mineralogy of the soil, which all affect the desorbability (release) of the contaminant from the soil and the leaching rate of contaminants through the soil (Sparks, 1993).

VITRIFICATION

In *in situ* vitrification the contaminants are solidified with an electric current, resulting in their immobilization. Vitrification may immobilize pollutants for as long as 10,000 years. Since a large amount of electricity is necessary, the technique is costly.

ISOLATION/CONTAINMENT

With this method, contaminants are held in place by installing subsurface physical barriers such as clay liners and slurry walls to minimize lateral migration. Scientists and engineers have also added surfactants to clay minerals (organo-clays) to enhance retention of organic pollutants (Xu *et al.,* 1997) and used organo-clays in liners to minimize the mobility of pollutants and in waste-water treatment (Soundararajan *et al.,* 1990). Further discussion of organo-clays is provided in Chapter 2.

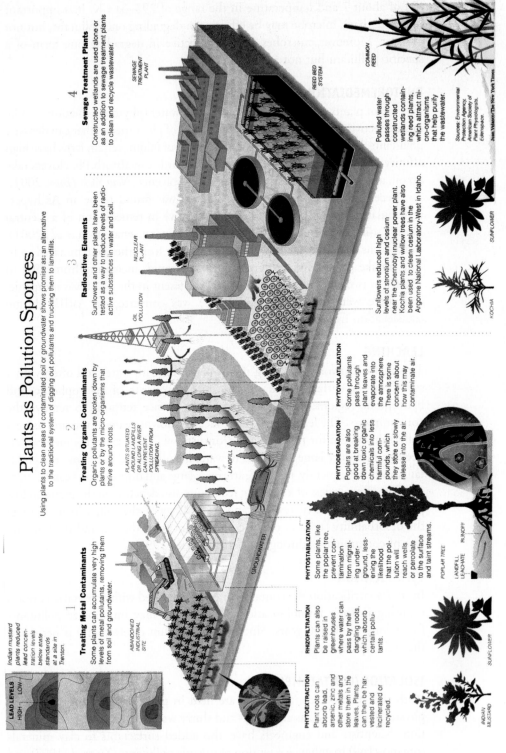

FIGURE 1.11. Schematic showing multiple use of plants in environmental remediation. From *New York Times* (2001), with permission.

PASSIVE REMEDIATION

With this method, natural processes such as volatilization, aeration, biodegradation, and photolysis are allowed to occur; these processes may cause decontamination. Passive remediation is simple and inexpensive and requires only monitoring of the site. Factors that affect this type of remediation include biodegradation, adsorption, volatilization, leaching, photolysis, soil permeability, groundwater depth, infiltration, and the nature of the contaminant.

Non-in-Situ Methods

Non-*in-situ* methods involve removal of the contaminated soil, usually by excavation, and the soil is then treated on-site, or transported to another location and then treated. With these methods there are obviously concerns about exposure of the contaminants in the moving and hauling process.

LAND TREATMENT

With this technique, the contaminated soil is excavated and spread over land so that natural processes such as biodegradation or photodegradation can occur to decontaminate the soil. The land area is prepared by grading to remove rocks and other debris and the area is surrounded by berms to lessen runoff. The soil pH is adjusted to 7.0 to immobilize heavy metals and to enhance the activity and effectiveness of soil microbes. Nutrients are also added for microbial stimulation. The contaminated soil is then spread on the site and mixed with the other soil to enhance the contact between the contaminant and microbes and to promote aerobic conditions (Sparks, 1993).

THERMAL TREATMENT

With thermal treatment, the excavated soil is exposed to high heats using a thermal incinerator. The high temperature breaks down the pollutants, and the released volatiles are then collected and moved through an afterburner and combusted or recovered with solvents.

ASPHALT INCORPORATION

With this method, contaminated soils are put into hot asphalt mixes. These mixtures are then used in paving. The asphalt and soil are heated while they are mixed. This causes volatilization or decomposition of some of the contaminants. The remaining pollutants are then immobilized in the asphalt.

SOLIDIFICATION/STABILIZATION

This technique involves the addition of an additive to excavated, contaminated soil so that the contaminants are encapsulated. The mixture is then landfilled. Thus, the contaminants are not free to move alone; however, they are not destroyed. This method has been employed to minimize inorganic pollutant contamination.

CHEMICAL EXTRACTION

In this treatment the excavated soil is mixed with a solvent, surfactant, or solvent/surfactant mixture to remove the contaminants. The solvent/surfactant and released contaminants are then separated from the soil. The soil is then washed or aerated to remove the solvent/surfactant and the latter is then filtered for fine particles and treated to remove the contaminants. This technique is expensive and is not often used.

EXCAVATION

With this method, the contaminated soil is removed and disposed elsewhere (e.g., a landfill). Landfills usually contain liners, such as clay, that diminish the mobility of the contaminants, or the landfills should be located on sites where the soil permeability is low. Landfills require large land areas and often pose hazards for humans. Excavation and disposal costs are high, and there are also liability problems, safety concerns, odor production, and potential runoff and groundwater contamination problems.

Molecular Environmental Soil Chemistry

It has become increasingly recognized that if we are going to predict and model fate/transport, toxicity, speciation (form of), bioavailability, and risk assessment of plant nutrients, toxic metals, oxyanions, radionuclides, and organic chemicals at the landscape scale, we must have fundamental information at multiple scales, and our research efforts must be multi- and interdisciplinary (Fig. 1.12).

With the advent of state-of-the-art analytical techniques, some of which are synchrotron-based (see discussions that follow), one can elucidate reaction mechanisms at small scale. This has been one of the major advances in the environmental sciences over the past decade. The use of small-scale techniques in environmental research has resulted in a new multidisciplinary field of study that environmental soil chemists are actively involved in—molecular environmental science. Molecular environmental science can be defined as the study of the chemical and physical forms and distribution of contaminants in soils, sediments, waste materials, natural waters, and the atmosphere at the molecular level.

There are a number of areas in environmental soil chemistry where the application of molecular environmental science is resulting in major frontiers. These include speciation of contaminants, which is essential for understanding release mechanisms, spatial resolution, chemical transformations, toxicity, bioavailability, and ultimate impacts on human health; mechanisms of microbial transformations, e.g., bioremediation; phytoremediation; development of predictive models; effective remediation and waste management strategies; and risk assessment. The application of molecular environmental science will be illustrated throughout the following chapters.

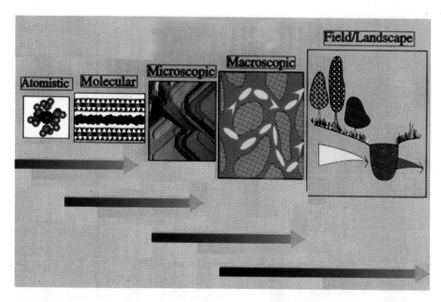

FIGURE 1.12. *Illustration of the various spatial scales that environmental scientists are interested in. From Bertsch and Hunter (1998), with permission.*

Electromagnetic Spectrum of Light

The use of intense light to understand mechanisms of soil chemical reactions and processes has revolutionized the field of environmental soil chemistry, or more appropriately, molecular environmental soil chemistry.

The electromagnetic spectrum of light is shown in Fig. 1.13. Electromagnetic radiation has both particle and wave properties such that light at a particular wavelength corresponds to a particular scale of detection (O'Day, 1999). For example, longer wave radiation detects larger objects while shorter wave radiation detects smaller objects. Light employed to see an object must have a wavelength similar to the object's size. Light has wavelengths longer or shorter than visible light. On the longer side are radio waves, microwaves, and infrared radiation. Shorter wavelength light includes ultraviolet, X-rays, and gamma rays. The shorter the wavelength, the higher the frequency and the more energetic or intense is the light. Light generated at shorter wavelengths such as X-rays is not visible by the human eye and must be detected via special means.

Each region of the spectrum is characterized by a range of wavelengths and photon energies that will determine the degree to which light will penetrate and interact with matter. At wavelengths from 10^{-7} to 10^{-10} m, one can explore the atomic structure of solids, molecules, and biological structures. Atoms, molecules, proteins, chemical bond lengths, and minimum distances between atomic planes in crystals fall within this wavelength range and can be detected. The binding energies of many electrons in atoms, molecules, and biological systems fall in the range of photon energies between 10 and

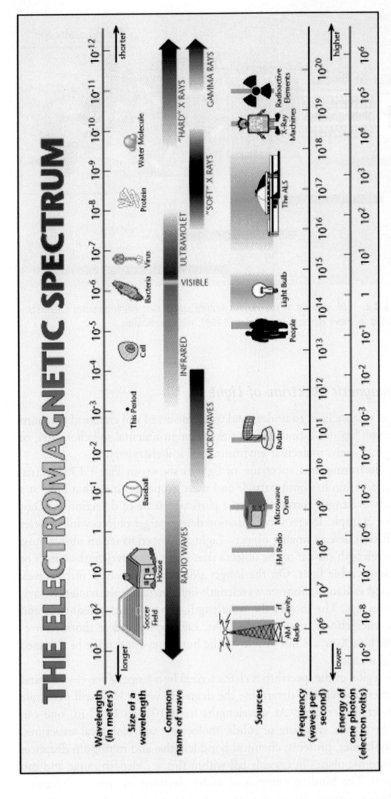

FIGURE 1.13. *Electromagnetic spectrum of light covering a wide range of wavelengths and photon energies. Courtesy of the Advanced Light Source, Lawrence Berkeley National Laboratory.*

10,000 eV. When absorbed by an atom, a photon causes an electron to separate from the atom or can cause the release or emission of other photons. By detecting and analyzing such e⁻ or photon emissions, scientists can better understand the properties of a sample.

Synchrotron Radiation

Intense light can be produced at a synchrotron facility. Synchrotron radiation is produced over a wide range of energies from the infrared region with energies <1 eV to the hard X-ray region with energies of 100 keV or more. There are a number of synchrotron facilities throughout the world (Table 1.8). In the United States major facilities are found at National Laboratories. These include the National Synchrotron Light Source (NSLS) at Brookhaven National Laboratory, the Advanced Photon Source (APS) at Argonne National Laboratory (Fig. 1.14), the Advanced Light Source (ALS) at Lawrence Berkeley National Laboratory, and the Stanford Synchrotron Radiation Laboratory (SSRL) at Stanford University. The NSLS and SSRL are second- and first-generation hard X-ray light sources, respectively, the APS is a third-generation hard X-ray light source, and the ALS offers a third-generation soft X-ray source.

TABLE 1.8. *Selected First-, Second-, and Third-Generation Synchrotron Research Facilities[a]*

Acronym	Facility	Location
	First-Generation Sources	
SSRL	Stanford Synchrotron Radiation Laboratory	Stanford, CA
CHESS	Cornell High Energy Synchrotron Source	Ithaca, NY
LURE	Laboratoire pour l'Utilisation de Rayonnement Electromagnétique	Orsay, France
HASYLAB	Hamburger Synchrotronstrahlungs Labor	Hamburg, Germany
	Second-Generation Sources	
SRS	Synchrotron Radiation Source	Daresbury, United Kingdom
KEK	Photon Factory	Tsukuba, Japan
NSLS	National Synchrotron Light Source	Upton, NY
BESSY	Berliner Elektronenspeicherring-Gesellschaft für Synchrotronstrahlung	Berlin, Germany
	Third-Generation Sources	
APS	Advanced Photon Source	Argonne, IL
ALS	Advanced Light Source	Berkeley, CA
ESRF	European Synchrotron Radiation Facility	Grenoble, France
SPring-8	Super Photon ring—8 GeV	Nishi Harima, Japan

[a] See Winick and Williams (1991) for a complete list worldwide. From Schulze and Bertsch (1999), with permission.

FIGURE 1.14.   *The Advanced Photon Source (APS), Argonne National Laboratory.*
Courtesy of the Advanced Photon Source, Argonne National Laboratory.

Synchrotrons are large machines (Fig. 1.15). The APS has a storage ring that is 1,104 m in circumference (Fig. 1.14) while the NSLS storage ring is 170 m in circumference. In the synchrotron, charged particles, either e^- or positrons, are injected into a ring-shaped vacuum chamber maintained at an ultra-high vacuum ($\sim 10^{-9}$ Torr). The particles enter the ring by way of an injection magnet and then travel around the ring at or near the speed of light, steered by bending magnets. Additional magnets focus and shape the particle beam as it travels around the ring. Synchrotron radiation or light is emitted when the charged particles go through the bending magnets, or through insertion devices, which are additional magnetic devices called wigglers or undulators inserted into straight sections of the ring. Beamlines allow the X-rays to enter experimental stations, which are shielded rooms that contain instrumentation for conducting experiments (Schultze and Bertsch, 1999).

Synchrotron radiation has enabled soil and environmental scientists to employ a number of spectroscopic and microscopic analytical techniques to understand chemical reactions and processes at molecular and smaller scales. Spectroscopies (Table 1.9) reveal chemical information and deal with the interaction of electromagnetic radiation with matter. A large number of spectroscopic techniques are a function of both large frequency or energy ranges of electromagnetic radiation involved and the approach used for probing the interaction over a given frequency range (Bertsch and Hunter, 1998). Microscopic techniques (Table 1.9) provide spatial information and arise from the interaction of energy with matter that either focuses or rasters radiation in some way to produce an image (O'Day, 1999).

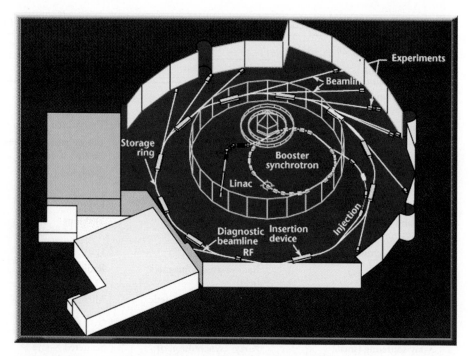

FIGURE 1.15. *Schematic diagram of a synchrotron X-ray source. Courtesy of the Advanced Light Source, Lawrence Berkeley National Laboratory.*

X-Ray Absorption Spectroscopy

One of the most widely used synchrotron-based spectroscopic techniques is X-ray absorption spectroscopy (XAS). XAS can be used to study most elements in crystalline or non-crystalline solid, liquid, or gaseous states over a concentration range of a few milligrams per liter to the pure element. It is also an *in situ* technique, which means that one can study reactions in the presence of water. This is a major advantage over many molecular scale techniques, which are *ex situ*, often requiring drying of the sample material, placing it in an ultra-high vacuum (UHV), heating the sample, or employing particle bombardment. Such conditions can alter the sample, creating artifacts, and do not simulate most natural soil conditions. XAS is an element-specific, bulk method that yields information about the local structural and compositional environment of an absorbing atom. It "sees" only the 2 or 3 closest shells of neighbors around an absorbing atom (<6 Å; note that Å will be used rather than nm to describe XAS analyses since Å is the standard unit used in the XAS scientific literature) due to the short electron mean free path in most substances. Using XAS one can ascertain important soil chemical information such as the oxidation state, information on next nearest neighbors, bond distances (accurate to ±0.02 Å), and coordination numbers (accurate to ±15–20%) (Brown *et al.*, 1995). Application of XAS to various soil chemical processes will be provided in later chapters.

TABLE 1.9. *Summary of Selected Analytical Methods for Molecular Environmental Soil Chemistry*[a]

Analytical method	Type of energy	
	Source	Signal
Absorption, emission, and relaxation spectroscopies		
IR[b] and FTIR	Infrared radiation	Transmitted infrared radiation
Synchrotron XAS (XANES and EXAFS)	Synchrotron X-rays	Transmitted or fluorescent X-rays; electron yield
Synchrotron microanalysis (XRF, XANES)	Synchrotron X-rays	Fluorescent X-rays
EELS (also called PEELS)	Electrons	Electrons
XPS and Auger spectroscopy	X-rays	Electrons
Resonance spectroscopies		
NMR	Radio waves (+ magnetic field)	Radio waves
ESR (also called EPR)	Microwaves (+ magnetic field)	Microwaves
Scattering and ablation		
X-ray scattering (small angle, SAXS; wide angle, WAXS)	X-rays (synchrotron or laboratory)	Scattered X-rays
SIMS	Charged ion beam	Atomic mass
LA-ICP-MS	Laser	Atomic mass
Microscopies		
STM	Tunneling electrons	Electronic perturbations
AFM (also called SFM)	Electronic force	Force perturbation
HR-TEM and STEM	Electrons	Transmitted or secondary electrons
SEM/EM with EDS or WDS chemical analysis	Electrons	Secondary, or backscattered electrons; fluorescent X-rays

[a] From O'Day (1999), with permission.
[b] Abbreviations are IR, infrared; FTIR, Fourier transform infrared; XAS, X-ray absorption spectroscopy; XANES, X-ray absorption near-edge structure; EXAFS, extended X-ray absorption fine structure; XRF, X-ray fluorescence; EELS, electron energy loss spectroscopy; PEELS, parallel electron energy loss spectroscopy; XPS, X-ray photoelectron spectroscopy; NMR, nuclear magnetic resonance; ESR, electron spin resonance (also known as EPR); EPR, electron paramagnetic resonance (also known as ESR); SAXS, small-angle X-ray scattering; WAXS, wide-angle X-ray scattering; SIMS, secondary ion mass spectrometry; LA-ICP-MS, laser ablation inductively coupled plasma mass spectrometry; STM, scanning tunneling microscopy; AFM, atomic force microscopy (also known as scanning force microscopy, SFM); HR-TEM, high-resolution transmission electron microscopy; STEM, scanning transmission electron microscopy; SEM, scanning electron microscopy; EM, electron microscopy; EDS, energy dispersive spectrometry; WDS, wavelength dispersive spectrometry.

An XAS experiment, which results in a spectrum (Fig. 1.16), consists of exposing a sample to an incident monochromatic beam of synchrotron X-rays, which is scanned over a range of energies below and above the absorption edge (K, L, M) of the element of interest. When X-rays interact with matter a number of processes can occur: X-ray scattering production of optical photons, production of photoelectrons and Auger electrons, production of fluorescence X-ray photons, and positron–electron pair production.

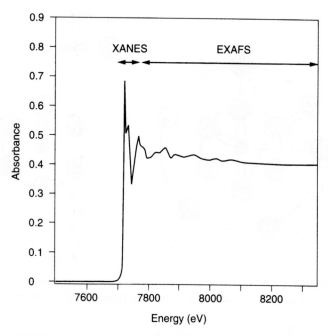

FIGURE 1.16. *Co K-edge X-ray absorption spectrum of $CoCO_3$ recorded in the transmission mode showing the XANES and EXAFS regions. The preedge region is from a few eV to ≈10 eV below the main absorption edge and shows a small preedge feature due to a 1s → 3d bound-state electron transition. From Xu (1993), with permission.*

In the X-ray energy range of 0.5 to 100 keV, photoelectron production dominates and causes X-ray attenuation by matter. When the energy of the incident X-ray beam ($h\nu$) < binding energy (E_b) of a core electron on the element of interest, absorption is minimal. However, when $h\nu \approx E_b$, electron transitions to unoccupied bound energy levels arise, contributing the main absorption edge and causing features below the main edge, referred to as the preedge portion of the spectrum (Fig. 1.16). As $h\nu$ increases beyond E_b, electrons can be ejected to unbound levels and stay in the vicinity of the absorber for a short time with excess kinetic energy. In the energy region extending from just above to about 50 eV above E_b and the absorption edge, electrons are multiply scattered among neighboring atoms (Fig. 1.17a), which produces the XANES (X-ray absorption near edge structure) portion of the spectrum (Fig. 1.16). Fingerprint information, such as oxidation states, can be gleaned from this portion of the XAS spectrum. When $h\nu$ is about 50 to 1000 eV above E_b and the absorption edge, electrons are ejected from the absorber, singly or multiply scattered from first- or second-neighbor atoms back to the absorber, and then leave the vicinity of the absorber (Fig. 1.17b), creating the EXAFS (extended X-ray absorption fine structure) portion (Fig. 1.16) of the spectrum (Brown *et al.*, 1995). The EXAFS spectrum is caused by interference between outgoing and backscattered photoelectrons,

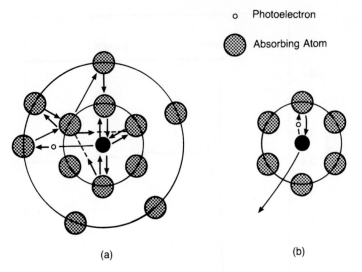

FIGURE 1.17. *Electron scattering process leading to (a) XANES and (b) EXAFS. From Brown et al. (1988), with permission. Copyright 1988 Springer-Verlag GmbH.*

which modulates the atomic absorption coefficient (Fig. 1.16). Analyses of the EXAFS spectrum provide information on bond distances, coordination number, and next nearest neighbors (Brown *et al.*, 1995).

XAS experiments can be conducted in several modes that differ in the type of detected particle: transmission (X-rays transmitted through the sample), fluorescence (fluorescent X-rays emitted due to absorption of the incident X-ray beam), or electron-yield (emitted photons). In a transmission experiment, the incident (I_0) and transmitted (I_1) X-ray intensities are recorded as a function of increasing incident X-ray energy (E) to yield an absorption spectrum, which is plotted as ln (I_0/I_1) vs E (in eV) (Fig. 1.16). The relationship between these intensities and the linear absorption coefficient μ (in cm^{-1}) of a sample of thickness x (in cm) is ln (I_0/I_1) = μx. In a fluorescence experiment, the X-ray fluorescence from the sample, I_f, can be measured and ratioed with I_0 as I_f/I_0, which is proportional to μ for dilute samples. Fluorescence methods are preferred for elements that may be contained in low concentrations on mineral surfaces (Brown *et al.*, 1995).

The Lytle detector, which is a solid angle, gas-filled ion chamber detector, is frequently used in fluorescence experiments. Figure 1.18 shows the experimental apparatus for fluorescence XAS measurements using a Lytle detector. Samples are loaded into a mylar-windowed sample holder made of low Z materials (aluminum or Teflon).

The XANES region of the spectrum, while not providing as much quantitative information as the EXAFS region, is often more intense and can provide qualitative or semi-quantitative information on the oxidation state of the measured element (Brown *et al.*, 1995). Such information can be obtained by comparing the features of the XANES spectrum of the sample with features of XANES spectra for well-characterized reference compounds

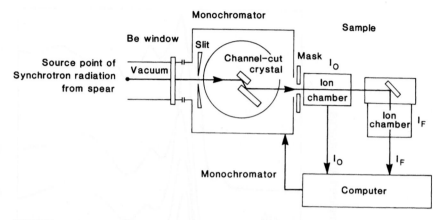

FIGURE 1.18. *Schematic illustration of the experimental setup used for fluorescence-yield XAS measurements. From Wong (1986), with permission.*

(Fig. 1.19). Some species, such as Cr, yield remarkably different, easily recognizable XANES spectra (Fig. 1.20). In Fig. 1.20 it is easy to differentiate Cr(III) from Cr(VI) as there is a prominent preedge feature for Cr(VI) that is absent for Cr(III).

Analysis of an EXAFS spectrum is illustrated in Fig. 1.21 and involves extracting structural parameters including interatomic distances (*R*), coordination numbers (CN), and identity of first, second, and more distant shells of neighbors around an absorber (Brown *et al.*, 1995).

To derive accurate structural parameters it is also necessary to obtain experimental EXAFS data for model or reference compounds that have known structures and contain the absorber and nearest-neighbor backscatters of interest. More detail on XAS methodology, sample preparation, and data analyses can be found in a number of excellent sources (Brown *et al.*, 1988, 1995; Brown, 1990; Fendorf and Sparks, 1996; Bertsch and Hunter, 1998; Fendorf, 1999; O'Day, 1999; Schulze and Bertsch, 1999).

Other Molecular-Scale Spectroscopic and Microscopic Techniques

As shown in Table 1.9, there are a number of *in situ* and *ex situ* analytical methods that are used in molecular environmental science.

The principal invasive *ex situ* techniques used for soil and aquatic systems are XPS, Auger electron spectroscopy (AES), and SIMS. Each of these techniques yields detailed information about the structure and bonding of minerals and the chemical species present on the mineral surfaces. XPS is the most widely used non-*in-situ* surface-sensitive technique. It has been used to study sorption mechanisms of inorganic cations and anions in soil and aquatic systems.

Examples of *in situ* techniques are ESR, FTIR, NMR, XAS, and Mössbauer spectroscopies.

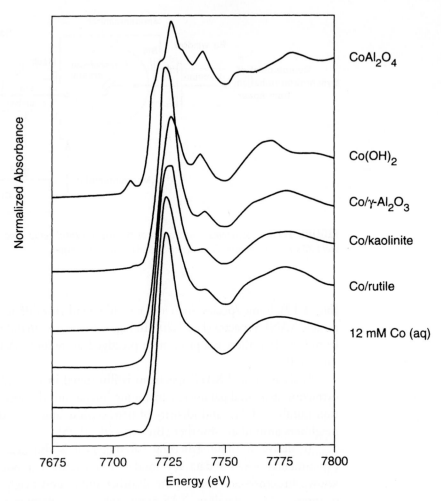

FIGURE 1.19.   *Co K-edge XANES spectra of CoAl$_2$O$_4$ spinel, crystalline Co(OH)$_2$, three samples with aqueous Co(II) sorbed on γ-Al$_2$O$_3$, kaolinite [Si$_4$Al$_4$O$_{10}$(OH)$_8$], and rutile (TiO$_2$), and a 12 mM aqueous Co(NO$_3$)$_2$ solution. From Chisholm-Brause et al. (1990b). Reprinted with permission from Nature. Copyright 1990 Macmillan Magazines Ltd.*

ESR spectroscopy is a technique for detecting paramagnetism. Electron paramagnetism occurs in all atoms, ions, organic free radicals, and molecules with an odd number of electrons. ESR is based upon the resonant absorption of microwaves by paramagnetic substances and describes the interaction between an electronic spin subjected to the influence of a crystal field and an external magnetic field (Calas, 1988). The method is applicable to transition metals of Fe^{3+}, Cu^{2+}, Mn^{2+}, V^{4+}, and molybdenum (V) and has been used widely to study metal ion sorption on soil mineral components (McBride, 1982; McBride *et al.*, 1984; Bleam and McBride, 1986) and soil organic matter (Senesi and Sposito, 1984; Senesi *et al.*, 1985).

Application of IR spectroscopy to the study of soil chemical processes and reactions has a long history. The introduction of Fourier transform techniques has made a significant contribution to the development of new

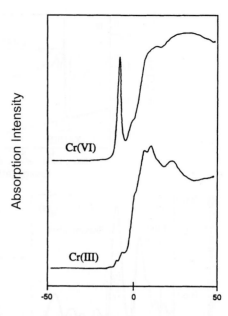

FIGURE 1.20.
Comparison of XANES spectra for Cr(III) and Cr(VI).
From Fendorf et al. (1994c), with permission.

investigation techniques such as diffuse reflectance infrared Fourier transform (DRIFT) and attenuated total reflectance (ATR) spectroscopy. IR spectroscopy now extends far beyond classical chemical analysis and is applied successfully to study sorption processes of inorganic and organic soil components. These techniques, and other vibrational spectroscopies, such as Raman, are the subject of numerous reviews (Hair, 1967; Bell, 1980; McMillan and Hofmeister, 1988; Johnston *et al.*, 1993; Piccolo, 1994).

The use of NMR spectroscopy to study surfaces has a shorter history, and fewer applications than vibrational spectroscopies. The primary reason is that the sensitivity of NMR is much lower than that of IR. Properties that might be exploited are the chemical shift, NMR relaxation times, and magnetic couplings to nuclei that are characteristic of a surface (Wilson, 1987). Most NMR studies in the field of soil science concentrate on the characterization of soil organic matter and soil humification processes and, therefore, involve ^1H, ^{13}C, and ^{15}N NMR. Reviews on these and related topics are available (Wershaw and Mikita, 1987; Wilson, 1987; Johnston *et al.*, 1993; Hatcher *et al.*, 1994). Because it is virtually impossible to obtain any useful molecular information by observing the nucleus of a paramagnetic metal directly, studies of cation exchange have focused on diamagnetic metals, such as Cd^{2+}, which have a spin of $^1/_2$ and an acceptable natural abundance (e.g., 12 and 13%, respectively, for the two NMR-active isotopes, ^{113}Cd and ^{111}Cd). NMR is essentially a bulk spectroscopic technique. The advent of high-resolution, solid-state NMR techniques, such as magic angle sample spinning (MAS) and cross polarization (CP), along with more sensitive, high-magnetic-field, user-friendly, pulsed NMR spectrometers has brought increased applications

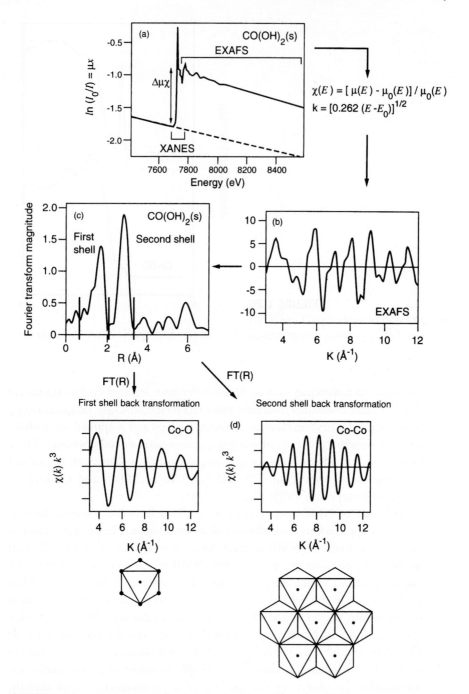

FIGURE 1.21. *Illustration of the steps involved in analysis of an EXAFS spectrum, using the Co K-edge XAS of solid Co(OH)$_2$ as an example. Subtraction of background (dashed line in the EXAFS spectrum) yields the EXAFS function ($\chi(k) \times k^3$) vs k ($\AA^{-1}$). Fourier transformation of the background-subtracted EXAFS function yields a radial structure function (Fourier transform magnitude vs R ($\AA$)), which contains peaks at different distances associated with various shells of neighboring atoms around Co. Fourier filtering of the different shells and back Fourier transformation yields the frequency contribution of each neighboring shell of atoms around Co. In this example, the first peak in the radial structure function is caused by backscattering of the photoelectron from six nearest-neighbor O atoms at 2.08 $\AA$ from Co, and the second peak by backscattering from six second-neighbor Co atoms at 3.17 $\AA$ from the central Co atom (after O'Day, 1992).*

to heterogeneous aqueous systems (Johnston *et al.,* 1993). In particular, ^{27}Al and ^{29}Si NMR in zeolites and other minerals have proven valuable for the structural elucidation of samples whose disorder has prevented diffraction techniques from being very useful (Altaner *et al.,* 1988; Herrero *et al.,* 1989; Woessner, 1989).

Spatial information on soil chemical processes can be gleaned from microscopic analyses. Scanning electron microscopy, TEM, and HRTEM are well-established methods for acquiring both chemical and micromorphological data on soils and soil materials. TEM can provide spatial resolution of surface alterations and the amorphous nature or degree of crystallinity of sorbed species (ordering). It can also be combined with electron spectroscopies to determine elemental analysis.

From the very inception of the STM in 1981, it was apparent that this technique would revolutionize the study of mineral surfaces and surface-related phenomena. Indeed, by the end of the 1980s, applications of STM were beginning to appear in the earth sciences literature (Hochella *et al.,* 1989; Eggleston and Hochella, 1990). However, the major event for the environmental science community came with the development of SFM, also known as AFM. SFM allows imaging of mineral surfaces in air or immersed in solution, at subnanometer scale resolution (Maurice, 1996). SFM employs an atomic-sized tip positioned by a microcantilever and rastered over a surface (O'Day, 1999). Applications to date include determining the molecular-to-atomic scale structure of mineral surfaces (Johnsson *et al.,* 1991); probing forces at the mineral/water interface (Ducker *et al.,* 1992); visualizing sorption of hemimicelles and macromolecular organic substances such as humic and fulvic acid (Manne *et al.,* 1994; Maurice, 1996); determining clay particle thicknesses and morphology of clay-sized particles (Hartman *et al.,* 1990; Maurice *et al.,* 1995); imaging soil bacteria (Grantham and Dove, 1996); and measuring directly the kinetics of growth, dissolution, heterogeneous nucleation, and redox processes (Dove and Hochella, 1993; Junta and Hochella, 1994; Fendorf *et al.,* 1996). Excellent references on SFM methodology and applications can be found in a number of reviews (Hochella *et al.,* 1998; Maurice, 1998).

Suggested Reading

Adriano, D. C. (1986). "Trace Elements in the Terrestrial Environment." Springer-Verlag, New York.

Bowen, H. J. M. (1979). "Environmental Chemistry of the Elements." Academic Press, London.

Brown, G. E., Jr. (1990). Spectroscopic studies of chemisorption reaction mechanisms at oxide–water interfaces. *In* "Mineral–Water Interface Geochemistry" (M. F. Hochella and A. F. White, Eds.), Rev. in Mineralogy 23, pp. 309–353. Mineral. Soc. Am., Washington, DC.

Brown, G. E., Jr., Parks, G. A., and O'Day, P. A. (1995). Sorption at mineral–water interfaces: Macroscopic and microscopic perspectives.

In "Mineral Surfaces" (D. J. Vaughan and R. A. D. Pattrick, Eds.), pp. 129–183. Chapman and Hall, London.

Dragun, J. (1988). "The Soil Chemistry of Hazardous Materials." Hazardous Materials Control Research Institute, Silver Spring, MD.

Fendorf, S. E., Sparks, D. L., Lamble, G. M., and Kelley, M. J. (1994). Applications of X-ray absorption fine structure spectroscopy to soils. *Soil Sci. Soc. Am. J.* **58**, 1585–1595.

Hassett, J. J., and Banwart, W. L. (1992). "Soils and Their Environment." Prentice Hall, Englewood Cliffs, NJ.

Hochella, M. F., Jr. (1990). Atomic structure, microtopography, composition, and reactivity of mineral surfaces. *In* "Mineral–Water Interface Geochemistry" (M. F. Hochella, Jr., and A. F. White, Eds.), Rev. in Mineralogy 23, pp. 87–132. Mineral. Soc. Am., Washington, DC.

Hochella, M. F., Jr., Rakovan, J. F., Rosso, K. M., Bickmore, B. R., and Rufe, E. (1998). New directions in mineral surface geochemical research using scanning probe microscopies. *In* "Mineral–Water Interfacial Reactions" (D. L. Sparks and T. J. Grundl, Eds.), ACS Symp. Ser. 715, pp. 37–56. Am. Chem. Soc., Washington, DC.

Manahan, S. E. (1991). "Environmental Chemistry," 5th ed. Lewis, Chelsea, MI.

Maurice, P. A. (1998). Atomic force microscopy as a tool for studying the reactivities of environmental particles. *In* "Mineral–Water Interfacial Reactions" (D. L. Sparks and T. J. Grundl, Eds.), ACS Symp. Ser. 715, pp. 57–66. Am. Chem. Soc., Washington, DC.

O'Day, P. A. (1999). Molecular environmental geochemistry. *Rev. Geophys.* **37**, 249–274.

Sawhney, B. L., and Brown, K., Eds. (1989). "Reactions and Movement of Organic Chemicals in Soils," SSSA Spec. Publ. 22. Soil Sci. Soc. Am., Madison, WI.

Schulze, D. G., and Bertsch, P. M. (1999). Overview of synchrotron X-ray sources and synchrotron X-rays. *In* "Synchrotron X-ray Methods in Clay Science" (D. G. Schulze, J. W. Stucki, and P. M. Bertsch, Eds.), pp. 1–18. Clay Minerals Soc., Boulder, CO.

Sparks, D. L. (1993). Soil decontamination. *In* "Handbook of Hazardous Materials" (M. Corn, Ed.), pp. 671–680. Academic Press, San Diego, CA.

Sparks, D. L. (1994). Soil chemistry. *In* "Encyclopedia of Agricultural Science" (C. J. Arntzen, Ed.), pp. 75–81. Academic Press, San Diego, CA.

Thomas, G. W. (1977). Historical developments in soil chemistry: Ion exchange. *Soil Sci. Soc. Am. J.* **41**, 230–238.

Thornton, I., Ed. (1983). "Applied Environmental Geochemistry." Academic Press, London.

Yong, R. N., Mohamed, A. M. O., and Warkentin, B. P. (1992). "Principles of Contaminant Transport in Soils," Dev. Geotech. Eng. 73. Elsevier, Amsterdam.

2 Inorganic Soil Components

Introduction

Soils are complex assemblies of solids, liquids, and gases. For example, in a typical silt loam soil ideal for plant growth the solid component in the surface horizon represents about 50% of the volume (about 45% mineral and 5% organic matter), gases (air) comprise about 20–30%, and water typically makes up the remaining 20–30%. Of course, the distribution of gases and water in the pore space component can change quickly depending on weather conditions and a host of other factors.

The median and range of elemental content of soils from around the world are given in Table 2.1. The elements that are found in the highest quantities are O, Si, Al, Fe, C, Ca, K, Na, and Mg. These are also the major elements found in the Earth's crust and in sediments (Table 2.1). Oxygen is the most prevalent element in the Earth's crust and in soils. It comprises about 47% of the Earth's crust by weight and greater than 90% by volume (Berry and Mason, 1959).

The inorganic components of soils represent more than 90% of the solid components. Their properties such as size, surface area, and charge behavior greatly affect many important equilibrium and kinetic reactions and processes in soils.

The inorganic components of soils include both primary and secondary minerals (defined below), which range in size (particle diameter) from clay-sized colloids (<2 μm or 0.002 mm) to gravel (>2 mm) and rocks. Table 2.2 lists the major primary and secondary minerals that are found in soils. A mineral can be defined as a natural inorganic compound with definite physical, chemical, and crystalline properties. A primary mineral is one that has not been altered chemically since its deposition and crystallization from molten lava. Examples of common primary minerals in soils include quartz and feldspar. Other primary minerals found in soils in smaller quantities include pyroxenes, micas, amphiboles, and olivines. Primary minerals occur primarily in the sand (2–0.05 mm particle diameter) and silt (0.05–0.002 mm particle diameter) fractions of soils but may be found in slightly weathered clay-sized fractions. A secondary mineral is one resulting from the weathering of a primary mineral, either by an alteration in the structure or from reprecipitation of the products of weathering (dissolution) of a primary mineral. Common secondary minerals in soils are aluminosilicate minerals such as kaolinite and montmorillonite, oxides such as gibbsite, goethite, and birnessite, amorphous materials such as imogolite and allophane, and sulfur and carbonate minerals. The secondary minerals are primarily found in the clay fraction of the soil but can also be located in the silt fraction.

Pauling's Rules

Most of the mineral structures in soils are ionically bonded and their structures can be predicted based on Pauling's Rules (Pauling, 1929). Ionic bonds are formed when an ion interacts with another ion of opposite charge in the mineral structure to form a chemical bond. Covalent bonds are those which result from a sharing of electrons. Most chemical bonds have a combination of ionic and covalent character. For example, the Si–O bond is equally ionic and covalent. The Al–O bond is approximately 40% covalent and 60% ionic (Sposito, 1989). Pauling's Rules (1929) are provided below with a brief description of their meaning in soil mineral structures.

RULE 1

A coordinated polyhedron of anions is formed about each cation, the cation–anion distance being determined by the radius sum (sum of cation and anion radii) and the coordination number of the cation by the radius ratio.

Ionic radii (IR) of cations and anions commonly found in inorganic soil minerals and the coordination number and radius ratio (assuming oxygen is the dominant anion) of the common cations are given in Table 2.3. The coordination number (CN), which is a function of the radius ratio, is the number of nearest anions surrounding the cation in a mineral. In soils, cations in mineral structures have coordination numbers of 4, 6, 8, or 12. The radius ratio is the ratio of the

TABLE 2.1. *Contents of Elements in Soils, the Earth's Crust, and Sediments*

Element	Soils (mg kg^{-1}) Median[a]	Range[a]	Median[b]	Range[b]	Earth's crust (mean)[c]	Sediments (mean)[c]
Al	72,000	700–<10,000	71,000	10,000–300,000	82,000	72,000
As	7.2	<0.1–97	6	0.10–40	1.5	7.7
B	33	<20–300	20	2–270	10	100
Ba	580	10–5,000	500	100–3,000	500	460
Be	0.92	<1–15	0.30	0.01–40	2.6	2
Br	0.85	<0.5–11	10	1–110	0.37	19
C, total	25,000	600–370,000	20,000	7,000–500,000	480	29,400
Ca	24,000	100–320,000	15,000	700–500,000	41,000	66,000
Cd	—	—	0.35	0.01–2	0.11	0.17
Cl	—	—	100	8–1,800	130	190
Co	9.1	<3–70	8	0.05–65	20	14
Cr	54	1–2,000	70	5–1,500	100	72
Cs	—	—	4	0.3–20	3	4.2
Cu	25	<1–700	30	2–250	50	33
F	430	<10–3,700	200	20–700	950	640
Fe	26,000	100–>100,000	40,000	2,000–550,000	41,000	41,000
Ga	17	<5–70	20	2–100	18	18
Ge	1.2	<0.1–2.5	1	0.10–50	1.8	1.7
Hg	0.09	<0.01–4.6	0.06	0.01–0.50	0.05	0.19
I	1.2	<0.5–9.6	5	0.10–25	0.14	16
K	15,000	50–63,000	14,000	80–37,000	21,000	20,000
La	37	<30–200	40	2–180	32	41
Li	24	<5–140	25	3–350	20	56
Mg	9,000	50–>100,000	5,000	400–9,000	23,000	14,000
Mn	550	<2–7,000	1,000	20–10,000	950	770
Mo	0.97	<3–15	1.2	0.1–40	1.5	2
N	—	—	2,000	200–5,000	25	470
Na	12,000	<500–100,000	5,000	150–25,000	23,000	5,700
Nb	11	<10–100	10	6–300	20	13
Nd	46	<70–300	35	4–63	38	32
Ni	19	<5–700	50	2–750	80	52
O	—	—	490,000	—	474,000	486,000
P	430	<20–6,800	800	35–5,300	1,000	670
Pb	19	<10–700	35	2–300	14	19
Rb	67	<20–210	150	20–1,000	90	135
S, total	1,600	<800–48,000	700	30–1,600	260	2,200
Sb	0.66	<1–8.8	1	0.20–10	0.2	1.2
Sc	8.9	<5–50	7	0.50–55	16	10
Se	0.39	<0.1–4.3	0.4	0.011	0.05	0.42
Si	310,000	16,000–450,000	330,000	250,000–410,000	277,000	245,000
Sn	1.3	<0.1–10	4	1–200	2.2	4.6
Sr	240	<5–3,000	250	4–2,000	370	320

continued

TABLE 2.1.   *Contents of Elements in Soils, the Earth's Crust, and Sediments (contd)*

Element	Soils (mg kg^{-1})				Earth's crust (mean)[c]	Sediments (mean)[c]
	Median[a]	Range[a]	Median[b]	Range[b]		
Th	9.4	2.2–31	9	1–35	12	9.6
Ti	2,900	70–20,000	5,000	150–25,000	5,600	3,800
U	2.7	0.29–11	2	0.70–9	2.4	3.1
V	80	<7–500	90	3–500	160	105
Y	25	<10–200	40	10–250	30	40
Yb	3.1	<1–50	3	0.04–12	3.3	3.6
Zn	60	<5–2,900	90	1–900	75	95
Zr	230	<20–2,000	400	60–2,000	190	150

[a] From U.S. Geological Survey Professional Paper 1270 (1984), with permission. Represents analyses from soils and other surficial materials from throughout the continental United States (regoliths including desert sands, sand dunes, loess deposits, and beach and alluvial deposits containing little or no organic matter).
[b] From Bowen (1979) and references therein, with permission. Represents soil analyses from throughout the world.
[c] From Bowen (1979) and references therein, with permission.

TABLE 2.2.   *Common Primary and Secondary Minerals in Soils[a]*

Name	Chemical formula[b]
	Primary Minerals
Quartz	SiO_2
Muscovite	$KAl_2(AlSi_3O_{10})\ (OH)_2$
Biotite	$K(Mg,Fe)_3(AlSi_3O_{10})\ (OH)_2$
Feldspars	
Orthoclase	$KAlSi_3O_8$
Microcline	$KAlSi_3O_8$
Albite	$NaAlSi_3O_8$
Amphiboles	
Tremolite	$Ca_2Mg_5Si_8O_{22}(OH)_2$
Pyroxenes	
Enstatite	$MgSiO_3$
Diopside	$CaMg(Si_2O_6)$
Rhodonite	$MnSiO_3$
Olivine	$(Mg,Fe)_2SiO_4$
Epidote	$Ca_2(Al,Fe)_3Si_3O_{12}(OH)$
Tourmaline	$(Na,Ca)\ (Al,Fe^{3+},\ Li,\ Mg)_3Al_6(BO_3)_3(Si_6O_{18})\ (OH)_4$
Zircon	$ZrSiO_4$
Rutile	TiO_2
	Secondary Minerals
Clay minerals[c]	
Kaolinite	$Si_4Al_4O_{10}(OH)_8$
Montmorillonite	$M_x\ (Al,\ Fe^{2+},\ Mg)_4Si_8O_{20}(OH)_4$ (M = interlayer metal cation)
Vermiculite	$(Al,\ Mg,\ Fe^{3+})_4(Si,\ Al)_8O_{20}(OH)_4$
Chlorite	$[M\ Al\ (OH)_6](Al,\ Mg)_4(Si,\ Al)_8\ O_{20}(OH,\ F)_4$
Allophane	$Si_3Al_4O_{12} \cdot nH_2O$
Imogolite	$Si_2Al_4O_{10} \cdot 5H_2O$
Goethite	$\alpha\text{-FeOOH}$
Hematite	$\alpha\text{-Fe}_2O_3$

TABLE 2.2. *Common Primary and Secondary Minerals in Soils[a] (contd)*

Name	Chemical formula[b]
	Secondary Minerals
Maghemite	$\gamma\text{-}Fe_2O_3$
Ferrihydrite	$Fe_5HO_8 \cdot 4H_2O$
Bohemite	$\gamma\text{-}AlOOH$
Gibbsite	$\gamma\text{-}Al\,(OH)_3$
Pyrolusite	$\beta\text{-}MnO_2$
Birnessite	$\delta\text{-}MnO_2$
Dolomite	$Ca\,Mg(CO_3)_2$
Calcite	$CaCO_3$
Gypsum	$CaSO_4 \cdot 2H_2O$
Jarosite	$KFe_3(SO_4)_2(OH)_6$

[a] Adapted from "Mineralogy: Concepts, Descriptions, Determinations" by L. G. Berry and B. Mason. Copyright © 1959 by W. H. Freeman and Company. Also adapted from C. S. Hurlbut, Jr., and C. Klein, "Manual of Mineralogy," 19th ed. Copyright © 1977 John Wiley & Sons, Inc. Reprinted by permission of John Wiley & Sons, Inc.
[b] An explanation for the chemical formula can be found in the text.
[c] Formulas are for the full-cell chemical formula unit.

cation radius to the anion radius. Cations having ionic radii less than a critical minimum radius ratio can fit between closely packed anions having different configurations (Fig. 2.1). For elements in the same group of the periodic table the IR increase as the atomic number increases. For positive ions of the same electronic structure the IR decrease with increasing valence. For elements such as Mn that exist in multiple valence states, the IR decrease with increasing positive valence.

Figure 2.1 shows the relationship of radius ratio, coordination number, and the geometrical arrangements of nearest anions around a central cation. The IR of the oxygen ion in minerals is assumed to be 0.140 nm. Based on Fig. 2.1 one can see that the Si^{4+} cation would occur in fourfold or tetrahedral coordination with O^{2-} (the radius ratio would be $0.039/0.140 = 0.279$); i.e., four oxygen anions can surround the cation and result in a tetrahedral coordination configuration such as that shown in Fig. 2.1. Aluminum (Al^{3+}) could also occur in fourfold coordination with O^{2-} since the radius ratio is $0.051/0.140 = 0.364$. In fact, Al^{3+} can occur in either four- or sixfold coordination, depending on the temperature during crystallization of the mineral. High temperatures cause low coordination numbers, i.e., fourfold coordination, while at low temperatures sixfold coordination is favored.

Based on the information given in Fig. 2.1, Fe^{2+} (0.074 nm), Fe^{3+} (0.064 nm), and Mg^{2+} (0.066 nm) would occur in octahedral coordination. As noted above, Al^{3+} could also occur in octahedral coordination. The coordination number of these cations is 6, so that six O groups arrange themselves around the cation, as shown in Fig. 2.1.

RULE 2

In a stable coordination structure the total strength of the valency bonds which reach an anion from all neighboring cations is equal to the charge of the anion.

TABLE 2.3. *Ionic Radius (IR), Radius Ratio, and Coordination Number (CN) of Common Cations and Anions in Inorganic Soil Minerals[a]*

	IR (nm)	Radius ratio[b]	CN
O^{2-}	0.140	—	—
F	0.133	—	—
Cl^-	0.181	—	—
Si^{4+}	0.039	0.278	4
Al^{3+}	0.051	0.364	4,6
Fe^{3+}	0.064	0.457	6
Mg^{2+}	0.066	0.471	6
Ti^{4+}	0.068	0.486	6
Fe^{2+}	0.074	0.529	6
Mn^{2+}	0.080	0.571	6
Na^+	0.097	0.693	8
Ca^{2+}	0.099	0.707	8
K^+	0.133	0.950	8,12
Ba^{2+}	0.134	0.957	8,12
Rb^+	0.147	1.050	8,12

[a] Adapted from C. S. Hurlbut, Jr., and C. Klein, "Manual of Mineralogy," 19th ed. Copyright 1977 John Wiley & Sons, Inc. Reprinted by permission of John Wiley & Sons, Inc.
[b] Ratio of cation radius to O^{2-} radius (O^{2-} radius = 0.140 nm).

This rule is known as the Electrostatic Valency Principle. This can be expressed as $s = Z/CN$, where s is the electrostatic bond strength to each coordinated anion, Z is the valence of the cation, and CN is the coordination number (Pauling, 1929). Thus, for Si in tetrahedral coordination, the electrostatic bond strength would be $Z(4^+)$ divided by CN(4), which equals 1. For Al in octahedral coordination the electrostatic bond strength would be $Z(3^+)$ divided by CN(6) or 0.5. If Al substitutes for Si in the tetrahedral layer, the electrostatic bond strength would be $Z(3^+)$ divided by CN(4) or 0.75, not 1. On the other hand, if Mg^{2+} substitutes for Al^{3+} in the octahedral layer then the electrostatic bond strength is $2^+/6$ or 0.33, not 0.5.

RULE 3

The existence of edges, and particularly of faces, common to the anion polyhedra in a coordinated structure decreases its stability; this effect is large for cations with high valency and small coordination number, and is especially large when the radius ratio approaches the lower limit of stability of the polyhedron.

Rule 3 is a statement of Coulomb's Law for cations and indicates that there are three ways for tetrahedra and octahedra polyhedra (Figs. 2.2A and 2.2B, respectively) to bond: point-to-point, the most stable configuration, edge-to-edge, and face-to-face, the least stable configuration (Fig. 2.3). With

Radius Ratio	Coordination Number	Geometrical Arrangements of Nearest Anions around a Central Cation	
0.15 - 0.22	3	Corners of an equilateral triangle	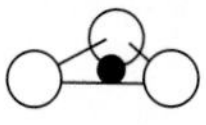
0.22 - 0.41	4	Corners of a tetrahedron	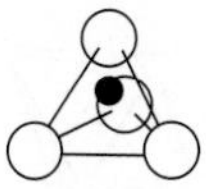
0.41 - 0.73	6	Corners of an octahedron	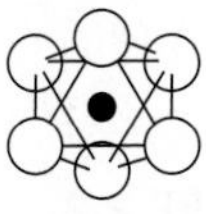
0.73 - 1.00	8	Corners of a cube	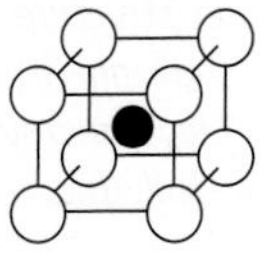
1.00	12	Corners of a cubo-octahedron	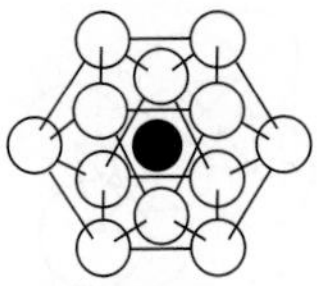

FIGURE 2.1. *Relationship of radius ratio, coordination number, and geometrical arrangement of nearest anions around a central cation. Adapted from Dennen (1960), with permission.*

cations with high valence such as Si^{4+} the tetrahedra are bonded point-to-point and with cations of lower valence such as Al^{3+} the octahedra are bonded edge-to-edge. Polyhedra are not bonded face-to-face.

RULE 4

In a crystal containing different cations those of high valency and small coordination number tend not to share polyhedron elements with each other.

This rule is saying that highly charged cations stay as far from each other as possible to lessen their contribution to the crystal's Coulomb energy (Pauling, 1929).

RULE 5

The number of essentially different kinds of constituents in a crystal tends to be small.

This is because all substances tend to form the lowest possible potential energy. Many kinds of constituents would result in a complex structure characterized by strains which would cause a high potential energy and instability. The different kinds of constituents refer to crystallographic configurations, tetrahedra and octahedra.

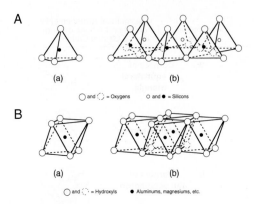

FIGURE 2.2. *(A) Diagrammatic sketch showing (a) single SiO₄ tetrahedron and (b) sheet structure of tetrahedra arranged in a hexagonal network. (B) Diagrammatic sketch showing (a) single octahedra unit and (b) sheet of octahedral units. From R. E. Grim, "Clay Mineralogy." Copyright © 1968 McGraw–Hill. Reproduced with permission of McGraw–Hill, Inc.*

FIGURE 2.3. *The sharing of a corner, an edge, and a face by a pair of tetrahedra and by a pair of octahedra. From "The Nature of the Chemical Bond" by L. Pauling, 3rd ed. Copyright © 1960 Cornell University. Used by permission of the publisher, Cornell University Press.*

Primary Soil Minerals

Some of the most important and prevalent primary minerals in soils are the feldspars (Table 2.2). They are common in the sand and silt fractions of soils and can also be found in the clay fraction. They comprise 59.5, 30.0, and 11.5% by weight of igneous rock, shale, and sandstone, respectively. Metamorphic rocks also contain feldspars (Huang, 1989). The K feldspars are important sources of K in soils and often compose a major component of the mineral form of soil K (Sparks, 1987). Feldspars are anhydrous three-dimensional aluminosilicates of linked SiO_4 and AlO_4 tetrahedra that contain cavities that can hold Ca^{2+}, Na^+, K^+, or Ba^{2+}

to maintain electroneutrality (Huang, 1989). Feldspars can be divided into two main groups, alkali feldspars, ranging in composition from $KAlSi_3O_8$ to $NaAlSi_3O_8$, and the plagioclases, ranging from $NaAlSi_3O_8$ to $CaAl_2Si_2O_8$.

Olivines, pyroxenes, and amphiboles are known as accessory minerals in soils and are found in the heavy specific gravity fractions. Pyroxenes and amphiboles are ferromagnesian minerals with single- and double-chain structures, respectively, of linked silica tetrahedra. They make up 16.8% by weight of igneous rocks (Huang, 1989). Olivines are green neosilicates in which Mg^{2+} and Fe^{2+} are octahedrally coordinated by O atoms. They are prevalent in igneous rocks, are sources of soil micronutrients, and are generally present in quantities smaller than those of pyroxenes and amphiboles (Huang, 1989).

Secondary Soil Minerals

Phyllosilicates

Clay is a general term for inorganic material that is <2 mm in size, whereas clay mineral refers to a specific mineral that mainly occurs in the clay-sized fraction of the soil (Moore and Reynolds, 1989). Without question, the secondary clay minerals (phyllosilicates) in soils play a profound role in affecting numerous soil chemical reactions and processes as we shall see in this chapter and in following chapters.

Clay minerals are assemblages of tetrahedral and octahedral sheets (Fig. 2.4) The silica tetrahedral sheet is characterized by a number of properties. The Si–O bond distance is about 0.162 nm, the O–O distance is about 0.264 nm, the tetrahedra are arranged so that all tips are pointing in the same direction and the bases of all tetrahedra are in the same plane, and tetrahedra are bonded point-to-point.

The aluminum octahedral sheet has an O–O distance of 0.267 nm and the OH–OH distance is 0.294 nm. Bonding of Al octahedra occurs via edges.

When one tetrahedral sheet is bonded to one octahedral sheet a 1:1 clay mineral results. Thus, the full-cell chemical formula for an ideal 1:1 clay would be $Si_4^{IV}Al_4^{VI}O_{10}(OH)_8$, where the superscripts represent four- and sixfold coordination in the tetrahedral and octahedral sheets, respectively. When two tetrahedral sheets are coordinated to one octahedral sheet, a 2:1 clay mineral results. The ideal full-cell chemical formula for a 2:1 clay mineral would be $Si_8^{IV}Al_4^{VI}O_{20}(OH)_4$, e.g., pyrophyllite. Between the sheets (i.e., interlayer space) cations may be octahedrally coordinated with hydroxyls, such as chlorites, and they may be present as individual cations, which may or may not be hydrated, as in micas, vermiculites, and smectites.

Isomorphous substitution, which occurs when the mineral forms, is the "substitution of one atom by another of similar size in the crystal lattice without disrupting the crystal structure of the mineral" (Glossary of Soil Science Terms, 1987). Thus the size of the cationic radius determines which cations can substitute in the tetrahedral and octahedral sheets. In the tetrahedral sheet Al^{3+} usually substitutes for Si^{4+} but P can also substitute. In the octahedral sheet Fe^{2+}, Fe^{3+}, Mg^{2+}, Ni^{2+}, Zn^{2+}, or Cu^{2+} can substitute for Al^{3+}. Thus, a cation with coordination number of 4 could substitute for Si^{4+} in the tetrahedral sheet and a cation of coordination number 6 could substitute for Al^{3+} in the octahedral sheet. As a result of this isomorphous substitution, a net negative charge associated with the 6 oxygens or hydroxyls of the octahedrons and with the 4 oxygens of the tetrahedrons develops.

As an example of this suppose that one Al^{3+} substitutes for one Si^{4+} in the tetrahedral sheet of an ideal 1:1 clay, $Si_4^{IV}Al_4^{VI}O_{10}(OH)_8$. After substitution the clay now has the formula $(Si_3Al_1)^{IV}Al_4^{VI}O_{10}(OH)_8$. The total negative charge is -28 and the total positive charge is $+27$. The net charge on the clay is -1, which is balanced by the presence of cations near the outer surface of the 1:1 clay.

Clays can be classified as dioctahedral or trioctahedral, depending on the number of cation positions in the octahedral sheet that are occupied (Table 2.4). If two of the three positions are filled, then the clay is dioctahedral. If all three positions are filled, the clay is trioctahedral. For example, if Al^{3+} is present in the octahedral sheet, only two-thirds of the cation positions are filled (dioctahedral); for every 6 OH^- anions, only two Al^{3+} satisfy the anionic charge and a formula of $Al_2(OH)_6$ results. When Mg^{2+} is present, all three cation sites are filled since Mg is divalent and three atoms of Mg^{2+} would be necessary to satisfy the 6 OH^- ions (trioctahedral). The formula for this sheet would be $Mg_3(OH)_6$.

Trioctahedral minerals, which often contain Mg^{2+}, are found in areas with drier climates, while dioctahedral clays, which usually contain Al^{3+} in the octahedral sheet, are found in wet climates. Thus, in the United States, east of the Mississippi River the soils have predominately dioctahedral minerals, while the clay fraction of soils west of the Mississippi is dominated by trioctahedral minerals.

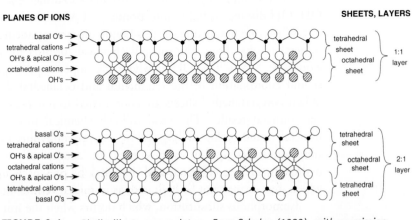

FIGURE 2.4. *Phyllosilicate nomenclature. From Schulze (1989), with permission.*

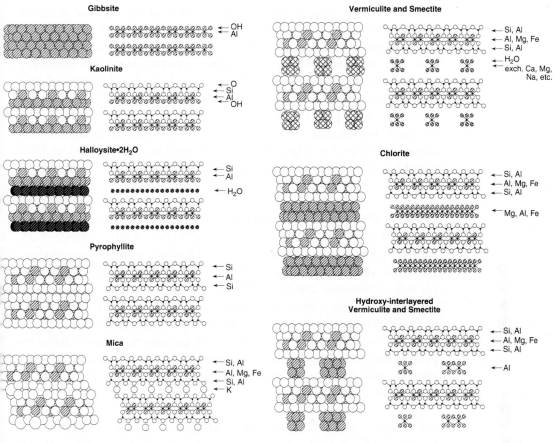

FIGURE 2.5. *Structural scheme of soil minerals based on octahedral and tetrahedral sheets. From Schulze (1989), with permission.*

1:1 Clays (Kaolin–Serpentine Group). This group can be divided into dioctahedral kaolins and trioctahedral serpentines. The most common dioctahedral kaolin is kaolinite (Table 2.4, Fig. 2.5). Looking at the structure of kaolinite (Fig. 2.5), one sees that the basic structure consists of a silica tetrahedral sheet bonded to an aluminum octahedral sheet. Layers of kaolinite stack by hydrogen bonding (an electrostatic bond between a positively charged H^+ ion and a negatively charged ion such as O^{2-}; Klein and Hurlbut, 1985) and thus there are no interlayer spaces present, unlike what is found in 2:1 clay minerals. The ideal full-cell chemical formula for kaolinite is $Si_4^{IV}Al_4^{VI}O_{10}(OH)_8$ and there is little, if any, isomorphic substitution in kaolinite.

Other dioctahedral kaolins are dickite, nacrite, and halloysite (Table 2.4). The ideal full-cell chemical formula for these clays is the same as that for kaolinite, with the major difference being the stacking sequence of layers. Halloysite (Fig. 2.5) has water molecules between each 1:1 layer and the ability to adsorb large quantities of monovalent cations such as NH_4^+. Drying will cause the water molecules to be removed and the clay layers to collapse together (Newman and Brown, 1987). Halloysite is also characterized by a tubular morphology, whereas kaolinite, when examined with microscopic techniques, has a platy structure.

Two trioctahedral serpentines that are 1:1 clays are antigorite and chrysotile (Table 2.4). These clays have all three octahedral positions filled with Mg. There is also no substitution in either the tetrahedral or octahedral sheets, and thus they have no permanent negative charge.

2:1 Clays (Pyrophyllite–Talc Group).   The ideal full-cell chemical formula for the dioctahedral 2:1 type clays is $Si_8^{IV}Al_4^{VI}O_{20}(OH)_4$ (pyrophyllite). The two representative clays of the pyrophyllite–talc group are pyrophyllite, a dioctahedral clay, and talc, a trioctahedral clay (Table 2.4). The layer charge per half-cell formula unit (full-cell formula unit divided by 2), x, on these clays is 0 due to no apparent isomorphous substitution. In addition, there is no interlayer space in pyrophyllite and talc (Newman and Brown, 1987). However, even though there is little if any permanent negative charge on pyrophyllite, the edge sites that are present can have a significant influence on the retention of metals and on various physical properties of the clay (Keren and Sparks, 1994).

2:1 Clays (Smectite–Saponite Group).   Clay minerals in the smectite–saponite group are characterized by a layer charge of 0.2–0.6 per half-cell formula unit (Table 2.4). The group includes the subgroups dioctahedral smectites and trioctahedral saponites. The dioctahedral smectites are represented by montmorillonite, beidellite, and nontronite. The ideal half-cell chemical formula for montmorillonite is $M_{0.33}$, $H_2OAl_{1.67}(Fe^{2+}, Mg^{2+})_{0.33} Si_4O_{10}(OH)_2$, where M refers to a metal cation in the interlayer space between sheets. The tetrahedral cations are Si^{4+} and the octahedral cations are Al^{3+}, Fe^{2+}, and Mg^{2+}. One can calculate the negative charge in the tetrahedral sheet as 0 and in the octahedral sheet as −0.33. Thus, the net negative charge is −0.33, which is balanced by exchangeable cations represented by $M_{0.33}$. The other feature that characterizes montmorillonite is the presence of water molecules in the interlayer space (Fig. 2.5). This causes montmorillonite to take on shrink–swell characteristics.

The major difference between montmorillonite and the other two dioctahedral smectites, beidellite and nontronite, is that isomorphous substitution in these minerals occurs in the tetrahedral sheet (i.e., Al^{3+} substitutes for Si^{4+}) rather than in the octahedral sheet. Nontronite also is an Fe-bearing mineral with Fe^{3+} being the predominant element in the octahedral sheet.

The trioctahedral members of the smectite–saponite group are saponite and hectorite. In saponite, substitution occurs in the tetrahedral sheet. In hectorite, a Li-bearing mineral, substitution occurs in the octahedral sheet (Table 2.4).

2:1 Clays (Vermiculite Group). Dioctahedral and trioctahedral vermiculites have a layer charge of 0.6–0.9 per half-cell formula unit. Dioctahedral vermiculite is characterized by substitution in both the tetrahedral and octahedral sheets, while trioctahedral vermiculite has substitution only in the tetrahedral sheet and all three of the octahedral cation positions are filled with Mg (Table 2.4).

While vermiculites are characterized by a platy morphology, similar to that for micas, they contain interlayer water (Fig. 2.5). Vermiculites form mainly from weathering of micas, particularly phlogopites and biotites, and interlayer K^+ is replaced by other interlayer cations. However, vermiculites have a layer charge lower than that of micas, and the Fe^{2+} in the octahedral sheet is oxidized to Fe^{3+} (Newman and Brown, 1987).

Let us take the half-cell chemical formula for dioctahedral vermiculite given in Table 2.4, since there is substitution in both the tetrahedral and octahedral sheets, and see how one would calculate the net layer charge on the clay:

$$M_{0.74} \, H_2O \, (Si^{4+}_{3.56}Al^{3+}_{0.44}) \, (Al^{3+}_{1.4}Mg^{2+}_{0.3}Fe^{3+}_{0.3}) \, O_{10} \, (OH)_2$$

The total positive charge on the clay is Si^{4+} ($3.56 \times 4^+ = 14.24$) + Al^{3+} ($1.4 + 0.44 = 1.84 \times 3^+ = 5.52$) + Mg^{2+} ($0.30 \times 2^+ = 0.60$) + Fe^{3+} ($0.30 \times 3^+ = 0.90$) = +21.26.

The total negative charge is O^{2-} ($10 \times 2^- = -20$) + OH^{-1} ($2 \times 1^- = -2$) = −22.

Thus, the net layer charge on dioctahedral vermiculite is + 21.26 + (−22) = −0.74.

This negative charge on the clay is balanced out by metal cations in the interlayer space represented as $M_{0.74}$ in Table 2.4. Please note that since H_2O is in the interlayer space and is not structural water, it is not considered in calculating the net layer charge.

2:1 Clays (Illite Group). Illite has a layer charge of about 0.8 per half-cell formula unit, intermediate between smectite and mica. Grim *et al.* (1937) developed the term illite to describe clay-size mica that was found in argillaceous rocks. Other terms that have been used in lieu of illite are hydromica, hydromuscovite, hydrous illite, hydrous mica, K-mica, micaceous clay, and sericite. Illite has more Si^{4+}, Mg^{2+}, and H_2O but less tetrahedral Al^{3+} and K^+ and water than muscovite. While K^+ is the predominant interlayer ion along with divalent ions such as Ca^{2+} and Mg^{2+}, NH_4^+ can also occur in illite.

2:1 Clays (Mica). Micas have a layer charge of about 1.0 per half-cell formula unit and are both dioctahedral, e.g., muscovite and paragonite, and trioctahedral, e.g., biotite, phlogopite, and lepidolite (Table 2.4). With the

TABLE 2.4.　*Partial Classification of Phyllosilicate Clay Minerals*

Type	Group	Subgroup	Species	Composition (half-cell chemical formula unit)			
				Interlayer space	Cations		Anions
					Tetrahedral sheet	Octahedral sheet	
1:1	Kaolin–serpentine $(x \sim 0)^a$	Kaolins (dioctahedral)	Kaolinite	—	Si_2	Al_2	$O_5(OH)_4$
			Dickite	—	Si_2	Al_2	$O_5(OH)_4$
			Nacrite	—	Si_2	Al_2	$O_5(OH)_4$
			Halloysite	$2H_2O$	Si_2	Al_2	$O_5(OH)_4$
		Serpentines (trioctahedral)	Antigorite	—	Si_2	Mg_3	$O_5(OH)_4$
			Chrysotile	—	Si_2	Mg_3	$O_5(OH)_4$
2:1	Pyrophyllite–talc $(x \sim 0)$	Pyrophyllite (dioctahedral)	Pyrophyllite	—	Si_4	Al_2	$O_{10}(OH)_2$
		Talc (trioctahedral)	Talc	—	Si_4	Mg_3	$O_{10}(OH)_2$
	Smectite–saponite $(x \sim 0.2\text{–}0.6)$	Smectites (dioctahedral)	Montmorillonite	$(M_{0.33}, H_2O)^b$	Si_4	$Al_{1.67}{}^c(Fe^{2+},Mg)_{0.33}$	$O_{10}(OH)_2$
			Beidellite	$(M_{0.33}, H_2O)$	$(Si_{3.67}Al_{0.33})$	Al_2	$O_{10}(OH)_2$
			Nontronite	$(M_{0.33}, H_2O)$	$(Si_{3.67}Al_{0.33})$	Fe_2^{3+}	$O_{10}(OH)_2$
		Saponite (trioctahedral)	Saponite	$(M_{0.33}, H_2O)$	$(Si_{3.67}Al_{0.33})$	Mg_3	$O_{10}(OH)_2$
			Hectorite	$(M_{0.33}, H_2O)$	Si_4	$(Mg_{2.67}Li_{0.33})$	$O_{10}(OH)_2$
	Vermiculite $(x \sim 0.6\text{–}0.9)$	Dioctahedral vermiculite	Dioctahedral vermiculite	$(M_{0.74}, H_2O)$	$(Si_{3.56}Al_{0.44})$	$(Al_{1.4}Mg_{0.3}Fe_{0.3}^{3+})$	$O_{10}(OH)_2$
		Trioctahedral vermiculite	Trioctahedral vermiculite	$(M_{0.70}, H_2O)$	$(Si_{3.3}Al_{0.7})$	Mg_3	$O_{10}(OH)_2$
	Illite $(x > 0.6, x < 0.9)$	Illite	Illite	$M_{0.74}$ (predominately K)	$(Si_{3.4}Al_{0.6})$	$(Al_{1.53}Fe_{0.22}^{3+}Fe_{0.03}^{2+}Mg_{0.28}^{2+})$	$O_{10}(OH)_2$
	Mica $(x \sim 1)$	Dioctahedral mica	Muscovite	K	(Si_3Al_1)	Al_2	$O_{10}(OH,F)_2$
			Paragonite	Na	(Si_3Al_1)	Al_2	$O_{10}(OH,F)_2$
		Trioctahedral mica	Biotite	K	$(Si,Al)_4$	$(Mg,Fe,Al)_3$	$O_{10}(OH,F)_2$
			Phlogopite	K	(Si_3Al_1)	Mg_3	$O_{10}(OH,F)_2$
			Lepidolite	K	$(Si,Al)_4$	$(Li,Al)_3$	$O_{10}(OH,F)_2$
	Brittle mica $(x \sim 2)$	Dioctahedral brittle mica	Margarite	Ca	(Si_2Al_2)	Al_2	$O_{10}(OH,F)_2$
		Trioctahedral brittle mica	Clintonite	Ca	$(Si_{1.3}Al_{2.7})$	$(Mg_{2.3}Al_{0.7})$	$O_{10}(OH,F)_2$

TABLE 2.4. Partial Classification of Phyllosilicate Clay Minerals (contd)

Type	Group	Subgroup	Species	Interlayer space	Tetrahedral sheet	Octahedral sheet	Anions
					Composition (half-cell chemical formula unit)		
					Cations		
2:1:1	Chlorite (x is variable)	Di, dioctahedral chlorite	Donbassite	$(Mg_{0.3}Al_{1.9}OH_6)$	$(Si_{3.9}Al_{0.1})$	$(Al_{1.8}Mg_{0.2})$	$O_{10}(OH,F)_2$
		Di, trioctahedral chlorite	Cookeite	$(Li_1Al_{1.93}OH_6)$	$(Si_{3.51}Al_{0.49})$	$(Al_{1.78}Li_{0.22})$	$O_{10}(OH)_2$
		Tri, trioctahedral chlorite	Clinochlore	$(Mg_2Al_1OH_6)$	(Si_3Al)	Mg_3	$O_{10}(OH)_2$

[a] x = Layer charge per half-cell formula unit.
[b] M = Metal cation.
[c] Represents a statistical average for isomorphic substitution.

exception of paragonite, a Na-bearing mica, the other micas have K^+ in the interlayer positions to satisfy the negative charge resulting from isomorphous substitution. Thus, micas are major K-bearing minerals in soils and as they weather, the nonexchangeable K is released for plant uptake. Weathering converts micas to partially expansible 2:1 clay minerals such as illite and vermiculite. The released K^+ from layer and edge weathering results in frayed edges and "wedge zones" (see Chapter 9 for a discussion of these) that play profound roles in K selectivity and K fixation (Sparks, 1987).

2:1:1 Clays (Chlorite Group).   Chlorites are often referred to as 2:1:1 clays since they are 2:1 clays with a hydroxide interlayer, either gibbsite-like $[Al(OH)_x]$ or brucite-like $[Mg(OH)_x]$ where x is <3, that is continuous across the interlayer sheet and is octahedrally coordinated. This sheet is positively charged because there are fewer than 3 OH^- per Al^{3+} in the sheet. The interlayer sheet is bound to the 2:1 clay electrostatically, and the tetrahedral layer is bonded to the interlayer sheet by hydrogen bonding.

Chlorites can be trioctahedral in both octahedral sheets, i.e., the octahedral sheet of the 2:1 layer and the interlayer octahedral hydroxide sheet, and are referred to as tri,trioctahedral chlorites. Chlorites that are dioctahedral in the 2:1 layer and trioctahedral in the interlayer hydroxide sheet are referred to as di,trioctahedral chlorites.

2:1 Clays (Intergrade Clay Minerals).   Intergrade clay minerals are intermediates between smectites and vermiculites. The interlayer space is composed of exchangeable cations and gibbsite-like or brucite-like islands that are not continuous as in chlorite. Accordingly, the interlayer space does not collapse on heating as readily as with smectites and vermiculites but it does collapse more easily than the complete interlayer hydroxy sheets in chlorite. A prominent intergrade clay mineral in many southeastern and mid-Atlantic United States soils is hydroxy-interlayered vermiculite (HIV), which is characterized by hydroxy–Al interlayers. The hydroxy–Al in these acid soils is an important source of nonexchangeable Al. In alkaline soils the interlayer material is hydroxy–Mg.

INTERSTRATIFIED CLAY MINERALS

Since the 2:1 and 1:1 layers of clays are strongly bonded internally but are weakly bonded to each other, layers can stack together to form interstratified clays. Examples include interstratified smectite with talc-type units and smectite and mica units such as 1:1 mica–dioctahedral smectite or hectorite. Another example would be interstratification of smectite or vermiculite with chlorite. For example, a 1:1 regularly interstratified chlorite–smectite structure would contain four tetrahedral sheets, two in each of the 2:1 layers, three octahedral sheets, one in each of the 2:1 layers and the other in the hydroxidic interlayer, and one expanding interlayer space containing the exchangeable cations. The anion content is $O_{40}(OH)_{20}$. Interstratification of smectite with kaolinite can also occur (Newman and Brown, 1987).

ALLOPHANE AND IMOGOLITE

Allophanes form from volcanic ash materials and are major components of volcanic-derived soils. They may also be found in the clay fraction of many nonvolcanically derived soils. Volcanic soils containing allophane usually contain significant organic matter and have low bulk densities. The SiO_2/Al_2O_3 ratio of allophanes varies from 0.84 to nearly 2, the characteristic ratio for kaolinite. Aluminum is in both tetrahedral and octahedral coordination. Imogolite has an almost constant SiO_2/Al_2O_3 ratio of 1 and Al is only in octahedral coordination; while it has little charge resulting from isomorphic substitution, imogolite can adsorb substantial quantities of monovalent cations, like halloysite (Newman and Brown, 1987). Microscopic analyses of imogolite reveal thread-like particles that are bundles of parallel tubes about 20 nm in diameter. Allophane exhibits spherical particles 30–50 nm in diameter (Brown, 1980).

FIBROUS CLAY MINERALS

Palygorskite (previously referred to as attapulgite) and sepiolite are fibrous minerals that do not have continuous octahedral sheets. They contain ribbons of 2:1 phyllosilicates. One ribbon is linked to the next by inversion of SiO_4 tetrahedra along a set of Si–O–Si bonds.

Oxides, Hydroxides, and Oxyhydroxides

Aluminum, iron, and manganese oxides play extremely important roles in the chemistry of soils. While they may not be found in large quantities, they have significant effects on many soil chemical processes such as sorption and redox because of their high specific surface areas and reactivity.

The general term "oxides" refers to metal hydroxides, oxyhydroxides, and hydrous oxides (where nonstoichiometric water is in the structure). Some of the oxides found in soils are listed in Table 2.5. Oxides are ubiquitous in soils. They may exist as discrete crystals, as coatings on phyllosilicates and humic substances, and as mixed gels.

ALUMINUM OXIDES

A number of crystalline Al hydroxides, oxyhydroxides, and oxides are found in natural settings. The oxyhydroxides are less common than the hydroxides. Only two oxides are naturally formed in soils, gibbsite ($Al(OH)_3$, an Al hydroxide) and boehmite (γ-AlOOH, an oxyhydroxide). Diaspore (α-AlOOH) and corundum (α-Al_2O_3, an anhydrous Al oxide) can also be found, but are much less common (Table 2.5). The anhydrous Al oxides may be found in some igneous and metamorphic rocks.

Gibbsite is the most common Al oxide mineral. It is often found in highly weathered soils such as Oxisols in tropical areas and in Ultisols. The structure of gibbsite is shown in Fig. 2.6 (Hsu, 1989). There are two planes of closely packed OH^- ions with Al^{3+} ions between the planes. The Al^{3+} atoms are found in two of the three octahedral positions

TABLE 2.5.   *Oxides, Oxyhydroxides, and Hydroxides Found in Soils[a]*

Aluminum oxides
    Bayerite α-Al(OH)$_3$
    Boehmite γ-AlOOH
    Corundum Al$_2$O$_3$
    Diaspore α-AlOOH
    Gibbsite γ-Al(OH)$_3$
Iron oxides
    Akaganeite β-FeOOH
    Ferrihydrite Fe$_5$HO$_8$·4H$_2$O
    Feroxyhyte δ-FeOOH
    Goethite α-FeOOH
    Hematite α-Fe$_2$O$_3$
    Lepidocrocite γ-FeOOH
    Maghemite γ-Fe$_2$O$_3$
    Magnetite Fe$_3$O$_4$
Manganese oxides
    Birnessite δ-MnO$_2$
    Manganite γ-MnOOH
    Pyrolusite β-MnO$_2$
Titanium oxides
    Anatase TiO$_2$
    Ilmenite FeTiO$_3$
    Rutile TiO$_2$

---

[a] Adapted from Taylor (1987), Hsu (1989), McKenzie (1989), Schulze (1989), and Schwertmann and Cornell (1991).

(dioctahedral) and are in hexagonal rings. In the interior of gibbsite, each Al^{3+} shares six OH$^-$ with three other Al^{3+} ions and the OH$^-$ ion is bridged between Al^{3+} (Fig. 2.6). On the edge of gibbsite, each Al^{3+} shares only four OH$^-$ with two other Al^{3+}, and the other two coordination sites are filled by one OH$^-$ and one H$_2$O; neither is bridged between Al^{3+} ions. Due to H bonding, the OH$^-$ ions in a particular unit are directly above the OH$^-$ ions in another unit (Hsu, 1989).

IRON OXIDES
The major Fe oxides and oxyhydroxides are listed in Table 2.5. The octahedron is the basic structural unit for all Fe oxides. Each Fe atom is surrounded by six oxygens or both O^{2-} and OH$^-$ ions. The various Fe oxides differ primarily in the octahedra arrangements and how they are linked (Kämpf *et al.*, 2000). Goethite and hematite are called the α phases (structures based on hexagonal close packing (hcp) of the anions), while lepidocrocite and maghemite are the γ phases (structures based on cubic close packing (ccp) of the anions). The α phases are more stable than the γ phases. The Fe^{3+} can be replaced by Al^{3+}, Mn^{3+}, and Cr^{3+} via isomorphous substitution. Other cations such as Ni, Ti, Co, Cu, and Zn can also be found in the Fe oxide structure (Schwertmann and Cornell, 1991).

Goethite is the most common and thermodynamically stable Fe oxide in soils. It has double bands of FeO(OH) octahedra sharing edges and corners, the bands bonded partially by H bonds. Goethite exhibits needle-shaped crystals with grooves and edges. Hematite is the second most common Fe oxide in soils. It is common in highly weathered soils and gives many red soils their color. Its structure consists of FeO_6 octahedra connected by edge- and face-sharing. Other Fe oxides are maghemite, which is common in highly weathered soils, and ferrihydrite, a hydrated semicrystalline material (Schwertmann and Cornell, 1991).

MANGANESE OXIDES

Manganese oxides (Table 2.5) are quite common in soils. They provide a source of Mn, an essential element for plants, they can adsorb heavy metals, and they are a natural oxidant of certain metals such as As(III) and Cr(III). Manganese oxides occur in soils as coatings on soil particles, in cracks and veins, and as nodules as large as 2 cm in diameter. Most Mn oxides are amorphous. The most stable form of Mn oxide is pyrolusite but it is uncommon in soils. Birnessite is the most prevalent Mn oxide in soils (McKenzie, 1989).

Carbonate and Sulfate Minerals

The major carbonates found in soils are calcite ($CaCO_3$), magnesite ($MgCO_3$, which is very unstable and is transformed to $Mg(OH)_2$), dolomite ($CaMg(CO_3)_2$), ankerite (($Ca,Fe,Mg)_2(CO_3)_2$), siderite ($FeCO_3$), and rhodochrosite

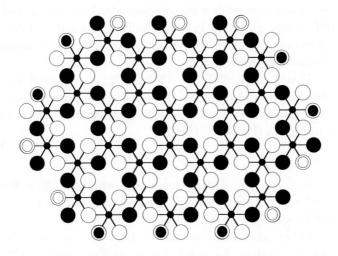

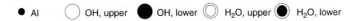

FIGURE 2.6. *Schematic representation of the unit layer of Al(OH)₃. Edge, unshared OH may protonate as a function of pH to form H₂0. From Hsu (1989), with permission.*

($MnCO_3$). The major sulfate mineral is gypsum, $CaSO_4 \cdot 2H_2O$. The carbonate and sulfate minerals are highly soluble compared to silica soil minerals and are most prevalent in arid and semiarid regions (Doner and Lynn, 1989).

Specific Surface of Soil Minerals

An important physical property of secondary soil minerals is their high surface areas (Table 2.6). The total surface area of a secondary mineral depends on both internal and external surface areas.

External Surface Area Measurement

External surface area is normally determined by measuring the adsorption of N_2 gas (the adsorbate) at temperatures near the boiling point of liquid N_2 (77 K). Surface area analysis is based on the Brunauer, Emmett, and Teller (1938) theory. The common linearized BET equation is (Carter *et al.*, 1986)

$$\frac{P}{V(P_0 - P)} = \left(\frac{1}{V_m c}\right) + \left[\left(\frac{(c-1)}{V_m c}\right)\left(\frac{P}{P_0}\right)\right] , \qquad (2.1)$$

where P = pressure (Pa) of a gas at equilibrium with a solid, P_0 = gas pressure (Pa) required for saturation at the temperature of the experiment, V = volume (m^3) of gas adsorbed at pressure P, V_m = volume (m^3) of gas required for monolayer coverage over the complete adsorbent surface, and $c = \exp[E_1 - E_2]$ /RT, where E_2 is the heat of liquification of the gas, E_1 is the heat of adsorption of the first layer of adsorbate, R is the gas constant, and T is the absolute temperature.

If one plots $P/V(P_0 - P)$ vs P/P_0, a linearized plot should result. The slope would equal $(c - 1)/V_m c$ and the intercept would equal $1/V_m c$. Once the value of V_m is determined, the specific surface can be calculated according to (Hiemenz, 1986)

$$E = N_a \sigma V_m/V_0, \qquad (2.2)$$

where E = specific surface (m^2) of the adsorbent, N_a = Avogadro's number (6.02×10^{23} molecules mol^{-1}), σ = cross-sectional area (m^2) of the adsorbate, and V_0 = molar volume (m^3).

Other adsorbates can also be used in external surface area analysis, such as ethylene for solids having surface areas < 1000 cm^2 g^{-1} and krypton for solids with intermediate surface areas. Nitrogen is the most frequently used adsorbate, particularly with solids that have surface areas > 10 m^2 g^{-1}. It is not good to use N_2 with a clay mineral like montmorillonite since its surface area is primarily internal and nonpolar N_2 molecules do not penetrate the interlayer regions between the layer sheets.

TABLE 2.6. *Specific Surface of Selected Soil Minerals*

Mineral	Specific surface $(m^2\ g^{-1})$
Kaolinite	7–30
Halloysite	10–45
Pyrophyllite	65–80
Talc	65–80
Montmorillonite	600–800
Dioctahedral vermiculite	50–800
Trioctahedral vermiculite	600–800
Muscovite	60—100
Biotite	40—100
Chlorite	25—150
Allophane	100—800
Aluminum oxides	100—220[a]
Iron oxides	70—250[a]
Manganese oxides	5–360[a]

[a] From Kämpf *et al.* (2000).

Total Surface Area Measurement

Ethylene glycol monoethyl ether (EGME) is a common material used to measure the total surface area of a mineral. One applies excess EGME to the adsorbent that has been dried over a P_2O_5 desiccant in a vacuum desiccator and then removes all but a monolayer of the EGME from the mineral surface using a vacuum desiccator containing $CaCl_2$. Then, by plotting the weight of EGME retained on the solid as a function of time, one can determine when the weight of EGME becomes constant. At this point, one assumes that monolayer coverage has occurred. To calculate the specific surface, one can use (Carter *et al.*, 1986)

$$A = W_g / [(W_s)(0.000286)], \tag{2.3}$$

where A = specific surface $(m^2\ g^{-1})$, W_g = weight (g) of EGME retained by the sample after monolayer equilibration, and W_s = weight (g) of P_2O_5 (desiccant) dried sample; 2.86×10^{-4} g EGME are required to form a monolayer on 1 m^2 of surface. To obtain internal surface area (S_I),

$$S_T = S_E + S_I$$

or

$$S_I = S_T - S_E, \tag{2.4}$$

where S_T = total surface area $(m^2\ g^{-1})$, S_E = external surface area $(m^2\ g^{-1})$, and S_I = internal surface area $(m^2\ g^{-1})$.

Surface Charge of Soil Minerals

Types of Charge

Soil minerals can exhibit two types of charge, permanent or constant charge and variable or pH-dependent charge. In the remainder of the book, we will refer to these as constant and variable charge. In most soils there is a combination of constant and variable charge. Constant charge is invariant with soil pH and results from isomorphous substitution. Therefore, this component of charge was developed when the mineral was formed. Examples of inorganic soil components that exhibit constant charge are smectite, vermiculite, mica, and chlorite. The constant charge terminology is attributed to Schofield (1949). He found that the cation exchange capacity (CEC) of a clay subsoil at Rothamsted Experimental Station in England was relatively constant between pH 2.5 and 5, but increased between pH 5 and 7. He also showed that the CEC of montmorillonite did not vary below pH 6, but increased as pH increased from 6 to 9. He ascribed the constant charge component to isomorphous substitution and the variable component to ionization of H^+ from SiOH groups on the clay surfaces. His conclusion about constant charge is still valid, if the clays are relatively pure. However, we now know that the contribution of SiOH groups to the creation of negative charge is not significant in most soils since the pK_a (see Box 2.1 for derivation and description of K_a and pK_a) of silicic acid is so high (about 9.5) that ionization of the SiOH group would occur only at very alkaline pH.

The variable charge component in soils changes with pH due to protonation and deprotonation of functional groups (Chapter 5) on inorganic soil minerals such as kaolinite, amorphous materials, metal oxides, oxyhydroxides, and hydroxides, and layer silicates coated with metal oxides and soil organic matter.

Cation Exchange Capacities of Secondary Soil Minerals

The negative charge on constant charge minerals that results from isomorphous substitution and on variable charge minerals that results from deprotonation of functional groups is balanced by positive charge in the form of exchangeable cations. While cation and anion exchange phenomena will be discussed in detail later (Chapter 6), it would be instructive at this point to briefly describe the CEC of some of the important secondary minerals in soils. A major component of a soil's CEC is attributable to the secondary clay minerals. The other major component is the organic matter fraction that will be discussed in Chapter 3. The CEC of secondary minerals is very important in affecting the retention of inorganic and organic species.

Table 2.7 shows the CEC for some important secondary minerals. The measured CEC of kaolinite is variable, depending on the degree of

BOX 2.1 *Calculation of Acid Dissociation Constants (K_a) and Derivation of the Henderson–Hasselbalch Equation*

(I)   Calculation of K_a and pK_a values.

Suppose that a 0.1 M CH$_3$COOH (acetic acid) solution is ionized 1.3% at 298 K. What is the K_a for CH$_3$COOH?

$$\underset{\text{acid}}{CH_3COOH} + H_2O \overset{K_a}{\rightleftharpoons} H_3O^+ + \underset{\text{base}}{CH_3COO^-} \qquad (2.1a)$$

or

$$\underset{\text{acid}}{CH_3COOH} \overset{K_a}{\rightleftharpoons} H^+ + \underset{\text{base}}{CH_3COO^-}.$$

Brönsted and Lowry defined an acid as a proton donor and a base as a proton acceptor. According to Lewis an acid is an electron–pair acceptor and a base is an electron–pair donor. The definitions of Lewis are general and also include species that do not have a reactive H$^+$ (Harris, 1987). In Eq. (2.1a) CH$_3$COOH is an acid and CH$_3$COO$^-$ is a base because the latter can accept a proton to become CH$_3$COOH. Acetic acid and CH$_3$COO$^-$ are a conjugate acid–base pair. The K_a for the second reaction in Eq. (2.1a) is

$$K_a = \frac{[H^+][CH_3COO^-]}{[CH_3COOH]}, \qquad (2.1b)$$

where brackets indicate concentration in mol liter^{-1}, [H$^+$] = 0.1 mol liter^{-1} × 0.013 = 0.0013 mol liter^{-1}, [CH$_3$COO$^-$] = 0.1 mol liter^{-1} × 0.013 = 0.0013 mol liter^{-1}, and [CH$_3$COOH] = (0.1000 mol liter^{-1}) − (0.0013 mol liter^{-1}) = 0.0987 mol liter^{-1}. Substituting these values in Eq. (2.1b) gives

$$K_a = \frac{[0.0013 \text{ mol liter}^{-1}][0.0013 \text{ mol liter}^{-1}]}{[0.0987 \text{ mol liter}^{-1}]}$$

$$= 1.712 \times 10^{-5} \text{ mol liter}^{-1}. \qquad (2.1c)$$

The higher the K_a, the more dissociation of the acid into products and the stronger the acid. Acetic acid has a low K_a and thus is slightly dissociated. It is a weak acid. One can also calculate a pK_a for CH$_3$COOH where

$$pK_a = -\log K_a = -\log 1.712 \times 10^{-5} \text{ mol liter}^{-1}, \qquad (2.1d)$$

$$pK_a = 4.77. \qquad (2.1e)$$

The lower the pK_a, the stronger the acid. For example, hydrochloric acid (HCl), a strong acid, has a $pK_a = -3$.

(II)   Derivation of Henderson–Hasselbalch Equation

One can derive a relationship, known as the Henderson–Hasselbalch equation, between pK_a and pH that is very useful in studying weak acids and in preparing buffers. A buffer is a solution whose pH is relatively constant when a small amount of acid or base is added. Since soils behave as weak acids, the Henderson–Hasselbalch equation is useful in understanding how, if pH and pK_a are known, the protonation and deprotonation of functional groups (formally defined in Chapter 5; an example would be the carboxyl (R—C=O—OH, where R is an aliphatic group) functional group of soil organic matter in soils) can be assessed:

$$-\log K_a = -\log \left[\frac{[H^+][CH_3COO^-]}{[CH_3COOH]} \right]. \tag{2.1f}$$

Rearranging,

$$-\log K_a = -\log[H^+] - \log\frac{[CH_3COO^-]}{[CH_3COOH]}. \tag{2.1g}$$

Simplifying,

$$pK_a = pH - \log \frac{[CH_3COO^-]}{[CH_3COOH]}. \tag{2.1h}$$

Rearrangement yields

$$pH = pK_a + \log \frac{[CH_3COO^-]}{[CH_3COOH]} \tag{2.1i}$$

or

$$pH = pK_a + \log \frac{[\text{conjugate base}]}{[\text{acid}]}. \tag{2.1j}$$

If $[CH_3COO^-] = [CH_3COOH]$, then

$$pH = pK_a + \log 1. \tag{2.1k}$$

Since $\log 1 = 0$, Eq. (2.1k) becomes

$$pH = pK_a. \tag{2.1l}$$

Therefore, when the pH = pK_a, 50% of the acid is dissociated (CH_3COO^-) and 50% is undissociated (CH_3COOH). If one is studying the dissociation of a carboxyl functional group associated with soil organic matter, and pK_a is 5, at pH 5, 50% would be in the undissociated carboxyl (R—C=O—OH) form and 50% would be in the dissociated, carboxylate (R—C=O—O⁻) form.

impurities in the clay and the pH at which the CEC is measured. If few impurities are present the CEC should be very low, <2 cmol kg^{-1}, since most scientists agree that the degree of isomorphic substitution in kaolinite is small. Impurities of 0.1 to 10% smectitic, micaceous, and vermiculitic layers in kaolinite (Moore and Reynolds, 1989) and anatase, rutile, feldspars, quartz, and iron oxides are often present. Selective dissolution analysis of kaolinite and separation into fine fractions to determine the degree of impurities and whether isomorphous substitution is present can be employed (Newman and Brown, 1987). Another explanation for the often unexpected CEC that is measured for kaolinite is negative charge resulting from surface functional edge groups such as AlOH which are deprotonated when neutral pH extracting solutions (e.g., ammonium acetate) are employed in CEC measurements.

Halloysite, which also has little if any ionic substitution, can have a higher measured CEC than expected. This is attributable to the association of halloysite with allophane (which often has a high CEC) in soils and to the large quantities of NH_4^+ and K^+ ions (which will be assumed to be exchangeable) often present between the clay layers. These ions result in an inflation of the CEC when it is measured. When they are accounted for, the CEC of halloysite is close to that of kaolinite.

Montmorillonite has a high CEC due to substantial isomorphic substitution and to the presence of fully expanded interlayers that promote exchange of ions. Even though the layer charge of dioctahedral vermiculite is higher than that for smectites, the CEC is less. This is because the interlayer space of vermiculite is not fully open due to K^+ ions satisfying some of the negative charge. Because K^+ has a small hydrated radius compared to other cations such as Ca^{2+} and Mg^{2+}, and because it has the proper coordination number, 12, it fits snugly into the interlayer space of vermiculites and micas. Since the coulombic attractive forces between the K^+ ion and the surfaces of the clay's interlayer are greater than the hydration forces between the individual K^+ ions, the interlayer space undergoes a partial collapse. The trapped K^+ is not always accessible for exchange, particularly when NH_4^+ is used as the index cation to measure CEC.

TABLE 2.7. *Cation Exchange Capacity (CEC) of Secondary Soil Minerals*

Mineral	CEC (cmol kg^{-1})
Kaolinite	2–15
Halloysite	10–40
Talc	<1
Montmorillonite	80–150
Dioctahedral vermiculite	10–150
Trioctahedral vermiculite	100–200
Muscovite	10–40
Biotite	10–40
Chlorite	10–40
Allophane	5–350

The measured CEC for dioctahedral vermiculite is also often affected by the presence of $Al(OH)_x$ material in the interlayer. This material often blocks the exchange of cations in the interlayer and consequently the measured CEC is lower than what would be expected based on the layer charge (see Chapter 9 for further discussion on effects of hydroxy–Al interlayer material on cation exchange). In trioctahedral vermiculite, the interlayer cations are usually Ca^{2+} or Mg^{2+}, and there are no restrictions to complete cation exchange.

While micas have a higher layer charge than vermiculites, the measured CEC is lower because the exchangeable K^+ in the interlayer, which satisfies the negative lattice charge, is usually inaccessible to index cations used in CEC analyses since it is "fixed" tightly by hydrogen bonding.

The $Al–(OH)_x$ or $Mg–(OH)_x$ interlayer sheet of chlorite is positively charged, satisfying some negative charge. The interlayer sheet also blocks the exchange of cations. Both of these factors cause the measured CEC of chlorite to be lower than one might predict based on layer charge. Allophane has Al^{3+} substituting for Si^{4+} and a lot of edge sites that can create a very high CEC. The high CEC can also be due to salt absorption, like what occurs with halloysite.

Identification of Minerals by X-ray Diffraction Analyses

Clay minerals and other secondary as well as primary minerals can be semiquantitatively identified by using X-ray diffraction analyses. If one wants to precisely quantify the mineral suites, then quantitative techniques including chemical, thermal, infrared, and surface area analyses must be employed. Structural formulas of clay minerals can be determined via chemical analyses (see Box 2.2).

To analyze and identify clay minerals by X-ray diffraction the patterns (diffractograms) are analyzed with particular emphasis on the peak's position, intensity, shape, and breadth. Values of 2θ are measured by employing Bragg's Law, and d-spacings (interplanar or diffraction spacing) are calculated or obtained from 2θ–d tabulations. Bragg's Law can be expressed as (Brindley and Brown, 1980)

$$n\lambda = 2d \sin\theta, \qquad (2.5)$$

where θ is the glancing angle of reflection or half the angle between the incident and diffracted beams, λ is the wavelength of the radiation, and n is the order of reflection which takes values 1, 2, 3, etc. For example, the reflection from the basal (001) plane is described as the (001) reflection or the diffraction is of first order. The experimentally measured parameter $d/n = \lambda/2\sin\theta$ is the d-spacing. More comprehensive details on X-ray diffraction theory are provided in several excellent references (Brindley and Brown, 1980; Newman, 1987; Moore and Reynolds, 1989).

Clay Separation and X-Ray Diffraction Analysis

Clays are usually separated by centrifugation after removal of carbonates with sodium acetate buffered to pH 5 and of organic matter with hydrogen peroxide or sodium hypochlorite. Iron oxides are then usually removed with a sodium dithionite ($Na_2S_2O_4$) solution, buffered at neutral pH with sodium bicarbonate ($NaHCO_3$) and sodium citrate which is added as a complexing agent to keep the iron in solution (Kunze and Dixon, 1986).

The clays are usually segregated into either the <2-μm fraction or the 2- to 0.2-μm and <0.2-μm fractions by centrifugation. They are then saturated with K^+ and Mg^{2+}, and excess salt is removed with deionized water. In the case of the Mg-saturated clays, a glycerol treatment is added. The glycerol solution aids in the identification of smectite. Clay is placed on a ceramic tile or on a glass slide or is transferred from a Millipore filter to a glass slide for analysis. The clay samples are then heated at several temperatures and analyzed by X-ray diffraction. Clays will give characteristic d-spacings at different temperatures that are diagnostic for their identification. Also, by heating K-saturated samples, certain clay minerals will collapse and be unstable at high temperatures. Moreover, K^+ is not very hydrated and heating will cause it to collapse. With the K-saturated samples, temperatures of 298, 383, 573, and 823 K should be used. For Mg-saturated samples, temperatures of 298 and 383 K are employed. Diagnostic d(001)-spacings for some common primary and secondary soil minerals are given in Table 2.8.

TABLE 2.8. *Diagnostic d(001)-Spacings (nm) of Soil Minerals as Determined from X-Ray Diffraction*

	K saturation (K)[a]				Mg saturation (K)[a]	
	298	383	573	823	298	383
Kaolinite	0.715–0.720	0.715–0.720	0.715–0.720	No peak[b]	0.715–0.720	0.715–0.720
Montmorillonite	1.4	1.4	1.4	1.0	1.82	1.42
Vermiculite	1.4	1.4	1.0	1.0	1.4	1.4
Mica	1.0	1.0	1.0	1.0	1.0	1.0
Chlorite	1.4	1.4	1.4	1.4	1.4	1.4
Hydroxy-interlayered vermiculite	1.4	*d*-spacing decreases with increased temperature to near 1.0 nm if vermiculitic and to 1.2 nm if more chloritic[c]			1.4	1.4
Quartz	0.426 (001 reflection) 0.334 (002 reflection)					
Feldspars	0.318–0.330					
Birnessite	0.72					
Gibbsite	0.484					

[a] Degrees Kelvin.
[b] Kaolinite is not structurally stable above 573 K.
[c] As temperature increases, the $Al(OH)_x$ interlayers become less stable and are removed.

BOX 2.2   *Calculation of Structural Formulas and CEC of Clay Minerals*

While X-ray diffraction can be used to semiquantitatively determine the mineral suites in soils and thermal and surface area analyses can be used to quantify soil minerals, these methods will not provide structural formulas. Such information can only be gleaned through chemical analyses. To obtain the necessary data, the mineral must first be broken down by acid dissolution. The procedure outlined below was first introduced by Ross and Hendricks (1945). In calculating the structural formulas it must be assumed that the SiO_2, Al_2O_3, etc., that are reported were a part of the clay mineral structure. While this may not be completely valid, it is usually true for clay minerals from natural deposits.

The analyses shown in Table 2.9 are for a 2:1 clay with an ideal formula of $Si_8Al_4O_{20}(OH)_4$. Thus, there are a total of 44 positive charges and 44 negative charges, assuming a full-cell chemical formula. To calculate the structural formula for this clay, the following assumptions are made: (1) a total of eight atoms can fit in the tetrahedral sheet, which can contain Si^{4+} and Al^{3+}; (2) a total of four atoms can fit in the octahedral sheet with Al^{3+}, Fe^{3+}, Fe^{2+}, Mg^{2+}, and Li^+ able to fit in this sheet; and (3) larger cations such as K^+, Na^+, and Ca^{2+} occur as exchangeable cations and satisfy negative charge resulting from isomorphic substitution.

The data given in Table 2.9 indicate that there are only 7.844 Si^{4+} atoms present in the clay mineral; a total of 8 atoms can fit in the tetrahedral sheet. Therefore, the remaining 0.156 atoms in the tetrahedral layer will be Al^{3+}. Therefore, this sheet has the cationic composition $(Si^{4+}_{7.844}, Al^{3+}_{0.156})^{IV}$. The remaining Al^{3+} atoms $(3.005 - 0.156 = 2.849)$ will go in the octahedral sheet plus enough Fe^{3+}, Fe^{2+}, and Mg^{2+} atoms, in that order, to yield 4 atoms.

TABLE 2.9.   *Structural Analysis for a 2:1 Clay Mineral*

Oxide[a]	Weight (%)	Atomic weight (g)	g eq cations[b]	g eq cations in total[c]	Atoms per unit cell[d]
SiO_2	50.95	60.06	3.39	31.376	7.844
Al_2O_3	16.54	101.94	0.974	9.015	3.005
Fe_2O_3	1.36	159.69	0.055	0.509	0.170
FeO	0.26	71.85	0.007	0.065	0.033
MgO	4.65	40.30	0.231	2.138	1.069
CaO	2.26	56.08	0.081	0.750	0.375
Na_2O	0.17	61.98	0.006	0.056	0.056
K_2O	0.47	94.20	0.010	0.093	0.093

[a] Total analyses of clay mineral expressed as oxides and based on a full-cell chemical formula unit.
[b] (Weight %/atomic weight) × (valence of cation) (numbers of atoms of the cation), e.g., for SiO_2 (50.95/60.06) (4).
[c] g eq cations × 44 (number of positive charges per unit cell)/(normalization factor). The normalization factor is Σ g eq cations, which for this example is 4.754.
[d] g eq cations in total/valence of cation.

Thus, the octahedral layer would have the cationic composition $(Al^{3+}_{2.849}, Fe^{3+}_{0.170}, Fe^{2+}_{0.033}, Mg^{2+}_{0.948})^{VI}$. The remaining Mg atoms, 0.121, plus the 0.375 Ca, 0.056 Na, and 0.093 K atoms will go in the interlayer of the clay as exchangeable cations to satisfy the negative charge created in the tetrahedral and octahedral sheets. The complete structural formula for the clay would be

$$(Si^{4+}_{7.844}, Al^{3+}_{0.156})^{IV} (Al^{3+}_{2.849}, Fe^{3+}_{0.170}, Fe^{2+}_{0.033}, Mg^{2+}_{0.948})^{VI}$$

$$O_{20}(OH)_4 (Mg^{2+}_{0.121}, Ca^{2+}_{0.375}, Na^+_{0.056}, K^+_{0.093}).$$

The overall positive charge is +31.844 in the tetrahedral sheet and +11.019 in the octahedral sheet. The positive charge is thus (+31.844) + (+11.019) = +42.863. The total negative charge is $[(20)(-2) + (4)(-1)]$ = −44. Therefore, the net charge on the clay is (+42.863) + (−44) = −1.137 or −1.14 equivalents (eq) unit cell^{-1}. This negative charge is balanced out by (0.121)(2) + (0.375)(2) + (0.056)(1) + (0.093)(1) = 1.141 or +1.14 eq unit cell^{-1} as exchangeable cations.

The cation exchange capacity of the clay can be calculated using the formula

$$\frac{\text{Net negative charge in eq unit cell}^{-1}}{\text{Molecular weight of the clay per unit cell}} (1000 \text{ meq eq}^{-1})(100 \text{ g})$$

$$= \frac{1.14 \text{ eq unit cell}^{-1}}{746.45 \text{ g}} (1000 \text{ meq eq}^{-1})(100 \text{ g}) = 152.72 \text{ meq } 100 \text{ g}^{-1}$$

$$= 152.72 \text{ cmol kg}^{-1}.$$

Use of Clay Minerals to Retain Organic Contaminants

There are many industrial uses of clays (for example, in oil and chemical industries). There has been much interest in using clays, particularly smectites, because of their high surface areas, for removal of organic pollutants from water. However, natural clays contain exchangeable metal ions, which makes their surfaces hydrophilic, and consequently they are quite ineffective in sorbing nonionic organic compounds (NOC) from aqueous solution. However, it has been found that if the metal ions are exchanged with large surfactant cations, such as long-chain alkylamine cations, the clay surfaces become hydrophobic or organophilic. Such clays are referred to as organoclays and are very effective in sorbing organic pollutants such as NOC (Mortland, 1970; Mortland et al., 1986; Xu et al., 1997). It also appears that the organoclay is a stable complex; i.e., the surfactants are not easily desorbed (Zhang et al., 1993). Zhang and Sparks

(1993) found that there was little adsorption of aniline and phenol on untreated montmorillonite, but that adsorption of both organics was significantly enhanced when the clay was modified with HDTMA$^+$ (hexadecyltrimethyl-ammonium).

Such organoclays could prove very advantageous in a number of settings. For example, clays employed as liners in landfills to diminish the transport of pollutants into water supplies could be modified, resulting in enhanced organic chemical retention and decreased mobility of the pollutants (Xu *et al.*, 1997). Also, the organoclays could be used in wastewater treatment and spill control situations (Soundarajan *et al.*, 1990).

Suggested Reading

Berry, L. G., and Mason, B. (1959). "Mineralogy—Concepts, Descriptions, Determinations." W. H. Freeman, San Francisco.

Brindley, G. W., and Brown, G., Eds. (1980). "Crystal Structure of Clay Minerals and Their X-Ray Identification," Monogr. 5. Mineral. Soc., London.

Dixon, J. B., and Weed, S. B., Eds. (1989). "Minerals in Soil Environments," SSSA Book Ser. 1. Soil Sci. Soc. Am., Madison, WI.

Gieseking, J. E., Ed. (1975). "Soil Components, Inorganic Components," Vol. 2. Springer-Verlag, New York.

Grim, R. E. (1968). "Clay Mineralogy," 2nd ed. McGraw–Hill, New York.

Jackson, M. L. (1964). Chemical composition of soils. *In* "Chemistry of the Soil" (F. E. Bear, Ed.), pp. 87–112. Reinhold, New York.

Kämpf, N., Scheinost, A. C., and Schulze, D. G. (2000). Oxide minerals. *In* "Handbook of Soil Science" (M. E. Sumner, Ed.), pp. F-125–168. CRC Press, Boca Raton, FL.

Marshall, C.E. (1964). "The Physical Chemistry and Mineralogy of Soils, Soil Materials," Vol. 1. Wiley, New York.

Moore, D. M., and Reynolds, Jr., R. C. (1989). "X-ray Diffraction and the Identification and Analysis of Clay Minerals." Oxford Univ. Press, New York.

Newman, A. C. D., Ed. (1987). "Chemistry of Clays and Clay Minerals," Mineral. Soc. Engl. Monogr. 6, Longman, Essex, England.

Rich, C. I., and Thomas, G. W. (1960). The clay fraction of soils. *Adv. Agron.* **12**, 1–39.

Schulze, D. G. (1989). An introduction to soil mineralogy. *In* "Minerals in Soil Environments" (J. B. Dixon and S. B. Weed, Eds.), pp. 1–34. SSSA Book Ser. 1. Soil Sci. Soc. Am., Madison, WI.

Schwertmann, U., and Cornell, R. M. (1991). "Iron Oxides in the Laboratory." VCH, Weinheim.

Weaver, C. E., and Pollard, L. D. (1973). "The Chemistry of Clay Minerals," Dev. Sedimentol. 15. Elsevier, Amsterdam.

Xu, S., Sheng, G., and Boyd, S. A. (1997). Use of organoclays in pollutant abatement. *Adv. Agron.* **59**, 25–62.

Yariv, S., and Cross, H. (1979). "Geochemistry of Colloid Systems." Springer-Verlag, New York.

3

Chemistry of Soil Organic Matter

Similar to the inorganic components of soil, soil organic matter (SOM) plays a significant role in affecting the chemistry of soils. Despite extensive and important studies, the molecular structure and chemistry of SOM is still not well understood. Moreover, because of its variability and close relationship with clay minerals and metal oxides the chemistry and reactions it undergoes with metals and organic chemicals are complex. In this chapter, background discussions on SOM content and function in soils and its composition, fractionation, structure, and intimate association with inorganic soil components will be covered. Additionally, environmentally important reactions between SOM and metals and organic contaminants will be discussed. For further in-depth discussions on these topics the reader is referred to the suggested readings at the end of this chapter.

Introduction

Humus and SOM can be thought of as synonyms, and include the total organic compounds in soils, excluding undecayed plant and animal tissues, their "partial decomposition" products, and the soil biomass (Stevenson, 1982). Schnitzer and Khan (1978) note that SOM is "a mixture of plant and animal residues in different stages of decomposition, substances synthesized microbiologically and/or chemically from the breakdown products, and the bodies of live and dead microorganisms and their decomposing remains." Humus includes humic substances (HS) plus resynthesis products of microorganisms which are stable and a part of the soil. Common definitions and terminology for these are given in Table 3.1.

Soil organic matter contents range from 0.5 to 5% on a weight basis in the surface horizon of mineral soils to 100% in organic soils (Histosols). In Mollisols of the prairie regions, SOM may be as high as 5% while in sandy soils, e.g., those of the Atlantic Coastal Plain of the United States, the content is often <1%. Even at these low levels, the reactivity of SOM is so high that it has a pronounced effect on soil chemical reactions.

Some of the general properties of SOM and its effects on soil chemical and physical properties are given in Table 3.2. It improves soil structure, water-holding capacity, aeration, and aggregation. It is an important source of macronutrients such as N, P, and S and of micronutrients such as B and Mo. It also contains large quantities of C, which provides an energy source for soil macroflora and microflora. The C/N ratio of soils is about 10–12:1.

Soil organic matter has a high specific surface (as great as 800–900 $m^2\,g^{-1}$) and a CEC that ranges from 150 to 300 cmol kg^{-1}. Thus, the majority of a surface soil's CEC is in fact attributable to SOM. Due to the high specific

TABLE 3.1. *Definitions of Soil Organic Matter (SOM) and Humic Substances[a]*

Term	Definition
Organic residues	Undecayed plant and animal tissues and their partial decomposition products.
Soil biomass	Organic matter present as live microbial tissue.
Humus	Total of the organic compounds in soil exclusive of undecayed plant and animal tissues, their "partial decomposition" products, and the soil biomass.
Soil organic matter	Same as humus.
Humic substances	A series of relatively high-molecular-weight, brown- to black-colored substances formed by secondary synthesis reactions. The term is used as a generic name to describe the colored material or its fractions obtained on the basis of solubility characteristics. These materials are distinctive to the soil (or sediment) environment in that they are dissimilar to the biopolymers of microorganisms and higher plants (including lignin).
Nonhumic substances	Compounds belonging to known classes of biochemistry, such as animo acids, carbohydrates, fats, waxes, resins, and organic acids. Humus probably contains most, if not all, of the biochemical compounds synthesized by living organisms.
Humin	The alkali insoluble fraction of soil organic matter or humus.
Humic acid	The dark-colored organic material that can be extracted from soil by various reagents and is insoluble in dilute acid.
Fulvic acid	The colored material that remains in solution after removal of humic acid by acidification.
Hymatomelanic acid	Alcohol soluble portion of humic acid.

[a] From F. J. Stevenson, "Humus Chemistry." Copyright © 1982 John Wiley & Sons, Inc. Reprinted by permission of John Wiley & Sons, Inc.

surface and CEC of SOM, it is an important sorbent of plant macronutrients and micronutrients, heavy metal cations, and organic materials such as pesticides. The uptake and availability of plant nutrients, particularly micronutrients such as Cu and Mn, and the effectiveness of herbicides are greatly affected by SOM. For example, manure additions can enhance micronutrient availability in alkaline soils where precipitation of the micronutrients at high pH reduces their availability. The complexation of low-molecular-weight SOM components such as fulvic acids (FA) with metals such as Al^{3+} and Cd^{2+} can decrease the uptake of metals by plants and their mobility in the soil profile.

Effect of Soil Formation Factors on SOM Contents

The quantity of soil organic matter in a soil depends on the five soil-forming factors first espoused by Jenny (1941) — time, climate, vegetation, parent material, and topography. These five factors determine the equilibrium level of SOM after a period of time. Of course, these factors vary for different soils, and thus SOM accumulates at different rates and, therefore, in varying quantities.

The accumulation rate of SOM is usually rapid initially, declines slowly, and reaches an equilibrium level varying from 110 years for fine-textured parent material to as high as 1500 years for sandy materials. The equilibrium level is attributed to organic acids that are produced which are resistant to microbial attack, the stability of humus due to its interactions with polyvalent cations and clays, and low amounts of one or more essential nutrients such as N, P, and S which limit the quantity of stable humus that can be synthesized by soil organisms (Stevenson, 1982).

Climate is an extremely important factor in controlling SOM contents because it affects the type of plant species, the amount of plant material produced, and the degree of microbial activity. A humid climate causes a forest association, while a semiarid climate creates grassland associations. Soils formed under grass usually have the highest SOM content, while desert, semidesert, and tropical soils have the lowest quantities of SOM. However, tropical soils often contain high quantities of HS, even though they are highly weathered. This is due to the formation of complexes between the HS and inorganic constituents such as quartz, oxides, and amorphous materials (organo–inorgano complexes) that are quite stable. In a complexed form the HS are less susceptible to microbial attack (Stevenson, 1982).

Vegetation also has a profound effect on SOM contents. Grassland soils, as mentioned above, are higher in SOM than forest soils. This is due to greater amounts of plants being produced in grassland settings, inhibition in nitrification that preserves N and C, higher humus synthesis which occurs in the rhizosphere, and the high base content of grassland soils which promotes NH_3 fixation by lignin (Stevenson, 1982).

TABLE 3.2.   *General Properties of Soil Organic Matter and Associated Effects in the Soil*[a]

Property	Remarks	Effect on soil
Color	The typical dark color of many soils is caused by organic matter.	May facilitate warming.
Water retention	Organic matter can hold up to 20 times its weight in water.	Helps prevent drying and shrinking. May significantly improve the moisture-retaining properties of sandy soils.
Combination with clay minerals	Cements soil particles into structural units called aggregates.	Permits exchange of gases. Stabilizes structure. Increases permeability.
Chelation	Forms stable complexes with Cu^{2+}, Mn^{2+}, Zn^{2+}, and other polyvalent cations.	May enhance the availability of micronutrients to higher plants.
Solubility in water	Insolubility of organic matter is because of its association with clay. Also, salts of divalent and trivalent cations with organic matter are insoluble. Isolated organic matter is partly soluble in water.	Little organic matter lost by leaching.
Buffer action	Organic matter exhibits buffering in slightly acid, neutral, and alkaline ranges.	Helps to maintain a uniform reaction in the soil.
Cation exchange	Total acidities of isolated fractions of humus range from 300 to 1400 cmol kg^{-1}.	May increase the CEC of the soil. From 20 to 70% of the CEC of many soils (e.g., Mollisols) is due to organic matter.
Mineralization	Decomposition of organic matter yields CO_2, NH_4^+, NO_3^-, PO_4^{3-} and SO_4^{2-}.	A source of nutrient elements for plant growth.
Combines with organic chemicals	Affects bioactivity, persistence and biodegradability of pesticides and other organic chemicals.	Modifies application rate of pesticides for effective control.

[a] From F. J. Stevenson, "Humus Chemistry." Copyright © 1982 John Wiley & Sons, Inc. Reprinted by permission of John Wiley & Sons, Inc.

The main effect of parent material on SOM content is the manner in which it affects soil texture. Clay soils have higher SOM contents than sandy soils. The type of clay mineral is also important. For example, montmorillonite, which has a high adsorption affinity for organic molecules, is very effective in protecting nitrogenous materials from microbial attack (Stevenson, 1982).

Topography, or the lay of the landscape, affects the content of SOM via climate, runoff, evaporation, and transpiration. Moist and poorly drained soils are high in SOM since organic matter degradation is lessened due to the anaerobic conditions of wet soil. Soils on north-facing slopes, which are wetter and have lower temperatures, are higher in SOM than soils on south-facing slopes, which are hotter and drier (Stevenson, 1982; Bohn *et al.*, 1985).

Cultivating soils also affects the content of SOM. When soils are first cultivated, SOM usually declines. In soils that were cultivated for corn production, it was found that about 25% of the N was lost in the first 20 years, 10% in the second 20 years, and 7% during the third 20 years (Jenny *et al.*,

1948). This decline is not only due to less plant residues, but also to improved aeration resulting from cultivation. The improved aeration results in increased microbial activity and lower amounts of humic materials. Wetting and drying of the soil also causes increased respiration, which reduces the amount of SOM (Stevenson, 1982).

Carbon Cycling and Sequestration

Atmospheric C was approximately 280 ppm in the preindustrial era and had increased to 370 ppm by 2000 (Lal, 2001). To stabilize future atmospheric C concentrations at 550 ppm (~ 2 times the preindustrial level) will require an annual reduction in worldwide CO_2 emissions from the projected level of 21 to 7 billion tons (measured as C) by the year 2100 (Hileman, 1999).

Over the past 150 years, the amount of C in the atmosphere, from greenhouse gases such as CO_2, CH_4, and N_2O, has increased by 30%. The increased levels of gases, particularly CO_2, are strongly linked to global warming. The increased levels of greenhouse gases are largely due to high levels of fossil fuel (oil, coal) combustion, deforestation, wildfires, and cultivation of land.

There are a number of ways to reduce atmospheric CO_2. These include the use of technology to develop energy-efficient fuels and the use of non-C energy sources such as solar, wind, water, and nuclear energy. Another way to reduce atmospheric CO_2 is by carbon sequestration.

Carbon sequestration is the long-term storage of C in oceans, soils, vegetation (especially forests), and geologic formations. The global carbon pools and the C cycle are shown in Fig. 3.1. The cycle is composed of both inputs (pools) and outputs (fluxes) in the environment. The ocean pool is estimated at 38,000 Pg (petagrams = 1×10^{15} g = 1 billion metric tons), the geologic pool at 5000 Pg, the soil organic C (SOC) pool, stored primarily in SOM, at 1500 (to 1 m depth)–2400 Pg (to 2 m depth), the atmospheric pool at 750 Pg, and the biotic pool (e.g., plants) at 560 Pg (Lal, 2001).

The atmospheric C pool has been increasing at the expense of the geological pool due to fossil fuel emission, the biotic pool due to deforestation and wildfires, and the soil pool due to cultivation and other anthropogenic disturbances (Lal, 2001). It has been estimated that land use changes and agriculture play an important role in emission of CO_2, CH_4, and N_2O and account for 20% of the increase in radioactive force (Lal, 2001).

The SOC pool, assuming an average content of 2400 Pg to 2 m depth, is 3.2 times the atmospheric pool and 4.4 times the biotic pool. Soils contain about 75% of the C pool on land, three times more than stored in living plants and animals. In addition to the SOC pool there is a soil inorganic carbon (SIC) pool that ranges from 695–748 Pg of CO_3^{2-} and is most important in subsurface horizons of arid and semiarid soils (Baties, 1996). The source of the SIC pool is primary (lithogenic) carbonates and pedogenic (secondary) carbonates, the latter being more important in C sequestration. The pedogenic carbonates are formed when H_2CO_3 chemically reacts with Ca^{2+} and/or Mg^{2+} in the soil solution in the upper portion of the profile and then is leached in lower soil horizons via irrigation. The rate of SIC sequestration by this mechanism may be 0.25 to 1 Mg C ha^{-1} year^{-1} (Wilding, 1999). Accordingly, the role that soils, particularly SOM, play in the global C cycle is immense, both in serving as a pool in sequestering C and also as a flux in releasing C (Fig. 3.1).

Land use and crop and soil management have drastic effects on the level of the SOC pool, and thus, C sequestration. Declines in the SOC pool are due to (a) mineralization of soil organic carbon, (b) transport by soil erosion processes, and (c) leaching into subsoil or groundwater (Fig. 3.2). The rate of SOC loss due to conversion from natural to agricultural ecosystems, particularly cultivation which enhances soil respiration and mineralization and decomposition of SOC, is more significant in tropical than in temperate-region soils, is higher from cropland than from pastureland, and is higher from soils with high SOC levels than with low initial levels (Mann, 1986). The loss of SOC due to cultivation may be as high as 60–80 Mg C ha^{-1}. Schlesinger (1984) estimated the loss of C from cultivated soils as large as 0.8×10^{15} g C year^{-1}. Some soils may lose the SOC pool at a rate of 2–12% year^{-1}, with a cumulative decrease of 50–70% of the original pool (Lal, 2001).

The use of limited cultivation (tillage) such as no-tillage can dramatically reduce C losses from soils by reducing mineralization and erosion, and promoting C sequestration. It has been estimated that extensive use of no-tillage in crop production could alone serve as a sink for 277 to 452×10^{12} g C, about 1% of the fossil fuel emissions during the next 30 years (Kern and Johnson, 1993). Cover crops such as legumes and crop rotation can also enhance C sequestration.

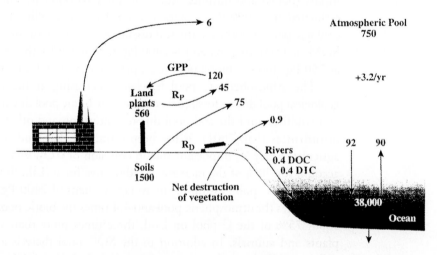

FIGURE 3.1. *The global carbon cycle. All pools are expressed in units of 10^{15} g C and all fluxes in units of 10^{15} g C/yr, averaged for the 1980s. Modified from Schlesinger (1997), with permission.*

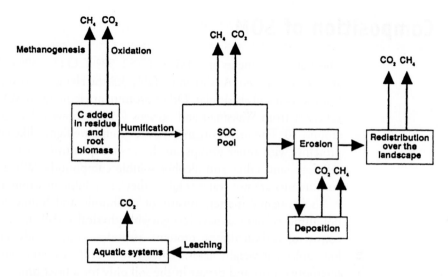

FIGURE 3.2. *Processes affecting soil carbon dynamics. From Lal (2001), with permission.*

Robertson *et al.* (2000) measured N_2O production, CH_4 oxidation, and soil C sequestration in cropped and unmanaged ecosystems in Midwestern USA soils. Except for a conventionally managed system (conventional tillage and chemical inputs), all cropping systems sequestered soil C. The no-till system (conventional chemical inputs) accumulated 30 g C m^{-2} $year^{-1}$ and the organic-based systems (reduced chemical inputs and organic with no chemical inputs), which included a winter legume cover crop, sequestered 8–11 g C m^{-2} $year^{-1}$.

Despite the gains in the soil C pool and thus, C sequestration, resulting from no-tillage agriculture and cover crops, these must be balanced by considering CO_2 fluxes due to manufacture of applied inorganic N fertilizers and irrigation of crops (Schlesinger, 1999). CO_2 is produced in N fertilizer production (0.375 moles of C per mole of N produced), and fossil fuels are used in pumping irrigation water. Additionally, the groundwaters of arid regions can contain as much as 1% dissolved Ca and CO_2 vs 0.036% in the atmosphere. When such waters are applied to arid soils CO_2 is released to the atmosphere $(Ca^{2+} + 2HCO_3^- \rightarrow CaCO_3\downarrow + H_2O + CO^2\uparrow)$.

Carbon sequestration can also be significantly enhanced by restoring soils degraded by erosion, desertification, salinity, and mining operations. Such practices as improving soil fertility by adding inorganic and organic fertilizer amendments, increasing biomass and decreasing erosion by establishing cover crops and crop rotations, implementing limited and no-tillage systems, and irrigating can increase C sequestration by 0.1 to 42 Mg ha^{-1} in terms of total SOC and from 0.1 to 4.5% in SOC content (Lal, 2001).

Composition of SOM

The main constituents of SOM are C (52–58%), O (34–39%), H (3.3–4.8%), and N (3.7–4.1%). As shown in Table 3.3 the elemental composition of HA from several soils is similar. Other prominent elements in SOM are P and S. Research from Waksman and Stevens (1930) showed that the C/N ratio is around 10. The major organic matter groups are lignin-like compounds and proteins, with other groups, in decreasing quantities, being hemicellulose, cellulose, and ether and alcohol soluble compounds. While most of these constituents are not water soluble, they are soluble in strong bases.

Soil organic matter consists of nonhumic and humic substances. The nonhumic substances have recognizable physical and chemical properties and consist of carbohydrates, proteins, peptides, amino acids, fats, waxes, and low-molecular-weight acids. These compounds are attacked easily by soil microorganisms and persist in the soil only for a brief time.

Humic substances can be defined as "a general category of naturally occurring, biogenic, heterogeneous organic substances that can generally be characterized as being yellow to black in color, of high molecular weight (MW), and refractory" (Aiken et al., 1985b). They are amorphous, partly aromatic, polyelectrolyte materials that no longer have specific chemical and physical characteristics associated with well-defined organic compounds (Schnitzer and Schulten, 1995). Humic substances can be subdivided into humic acid (HA), fulvic acid (FA), and humin. Definitions of HS are classically based on their solubility in acid or base (Schnitzer and Khan, 1972) as will be discussed later in the section Fractionation of SOM.

Several mechanisms for explaining the formation of soil HS have been proposed (Fig. 3.3). Selman Waksman's classical theory, the so-called lignin theory, was that HS are modified lignins that remain after microbial attack (pathway 4 of Fig. 3.3). The modified lignins are characterized by a loss of methoxyl (OCH_3) groups and the presence of o(ortho)-hydroxyphenols and oxidation of aliphatic side chains to form COOH groups. These lignins undergo more modifications and then result in first HA and then FA. Pathway 1, which is not considered significant, assumes that HS form from sugars (Stevenson, 1982).

The contemporary view of HS genesis is the polyphenol theory (pathways 2 and 3 in Fig. 3.3) that involves quinones (Fig. 3.4). In pathway 3 (Fig. 3.3) lignin is an important component of HS creation, but phenolic aldehydes and acids released from lignin during microbial attack enzymatically are altered to quinones, which polymerize in the absence or presence of amino compounds to form humic-like macromolecules. Pathway 2 (Fig. 3.3) is analogous to pathway 3 except the polyphenols are microbially synthesized from nonlignin C sources, e.g., cellulose, and oxidized by enzymes to quinones and then to HS (Stevenson, 1982).

While the polyphenol theory is currently in vogue to explain the creation of HS, all four pathways may occur in all soils. However, one pathway is usually prominent. For example, pathway 4, the lignin pathway, may be primary in poorly drained soils while the polyphenol pathways (2 and 3) may predominate in forest soils (Stevenson, 1982).

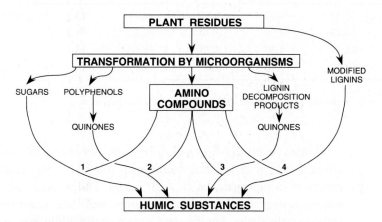

FIGURE 3.3. *Mechanisms for the formation of soil humic substances. Amino compounds synthesized by microorganisms are seen to react with modified lignins (pathway 4), quinones (pathways 2 and 3), and reducing sugars (pathway 1) to form complex dark-colored polymers. From F. J. Stevenson, "Humus Chemistry." Copyright © 1982 John Wiley & Sons, Inc. Reprinted by permission of John Wiley & Sons, Inc.*

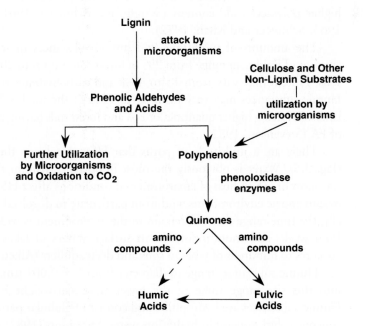

FIGURE 3.4. *Schematic representation of the polyphenol theory of humus formation. From F. J. Stevenson, "Humus Chemistry." Copyright © 1982 John Wiley & Sons, Inc. Reprinted by permission of John Wiley & Sons, Inc.*

TABLE 3.3. *The Elementary Composition of Humic Acids from Different Soils[a]*

Soil	Percentage				Ratio		
	C	H	N	O	C/N	C/H	O/H
A[b]	52.39	4.82	3.74	39.05	14.0	10.9	8.1
B	57.47	3.38	3.78	35.37	15.2	17.0	10.4
C	58.37	3.26	3.70	34.67	15.7	17.9	10.6
D	58.56	3.40	4.09	33.95	14.3	17.2	10.0

[a] From Kononova (1966).
[b] Soils A, B, C, and D represent soils of varying genesis, taxonomy, and physicochemical properties.

Humic acids are extremely common. According to Szalay (1964) the amount of C in the earth as humic acids (60×10^{11} Mg) exceeds that which occurs in living organisms (7×10^{11} Mg). Humic acids are found in soils, waters, sewage, compost heaps, marine and lake sediments, peat bogs, carbonaceous shales, lignites, and brown coals. While they are not harmful, they are not desirable in potable water (Stevenson, 1982).

One of the problems in studying humin is that it is not soluble. Thus methods that do not require solubilization are necessary. Carbon-13 (^{13}C) nuclear magnetic resonance (NMR) spectroscopy has greatly assisted in the study of humin since the high content of mineral matter in humin is not a factor. Humin is similar to HA. It is slightly less aromatic (organic compounds that behave like benzene; Aiken *et al.*, 1985b) than HA, but it contains a higher polysaccharide content (Wright and Schnitzer, 1961; Acton *et al.*, 1963; Schnitzer and Khan, 1972).

The amounts of nonhumic and humic substances in soils differ. The amount of lipids can range from 2% in forest soil humus to 20% in acid peat soils. Protein may vary from 15 to 45% and carbohydrates from 5 to 25%. Humic substances may vary from 33 to 75% of the total SOM with grassland soils having higher quantities of HA and forest soils having higher amounts of FA (Stevenson, 1982).

There are a multitude of paths that HS can take in the environment (Fig. 3.5). Water is obviously the most important medium that affects the transport of HS. A host of environmental conditions affect HS, ranging from oxic to anoxic environments, and from particulate to dissolved HS. Additionally, the time range that HS remain in the environment is wide. It can range from weeks and months for HS in surface waters of lakes, streams, and estuaries to hundreds of years in soils and deep aquifers (Aiken *et al.*, 1985b).

Humic substances range in diameter from 1 to 0.001 μm. While HA fit into this size range some of the lower-molecular-weight FA are smaller. Humic substances are hydrophilic and consist of globular particles, which in aqueous solution contain hydration water. Stevenson (1982) notes that HS are thought of as coiled, long-chain molecules or two- or three-dimensional cross-linked macromolecules whose negative charge is primarily derived from ionization of acidic functional groups, e.g., carboxyls.

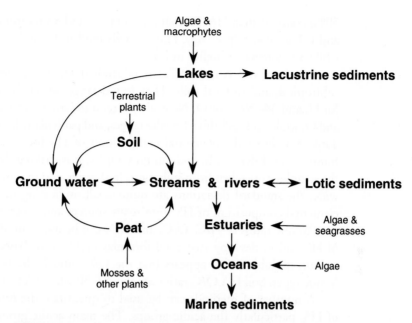

FIGURE 3.5. *Diagram of the many possible environmental flowpaths of humic substances. From G. R. Aiken, D. M. McKnight, R. L. Wershaw, and P. MacCarthy, in "Humic Substances in Soil, Sediments, and Water" (G. R. Aiken, D. M. McKnight, and R. L. Wershaw, Eds.), pp. 1–9. Copyright © 1985 John Wiley and Sons, Inc. Reprinted by permission of John Wiley and Sons, Inc.*

The average molecular weight of HS may range from 500 to 5000 Da for FA to 3,000 to 1,000,000 Da for HA (Stevenson, 1982). Soil HA have higher molecular weights than aquatic HA. The molecular weight measurements depend on pH, concentration, and ionic strength.

The lack of reproducibility in analytical methods makes the study of HS difficult and exacerbates the task of deriving a precise elemental composition. Table 3.4 shows the average elemental composition for HA and FA. Based on these data, mean formulas of $C_{10}H_{12}O_5N$ for HA and $C_{12}H_{12}O_9N$ for FA, disregarding S, could be derived. The major elements composing HA and FA are C and O. The C content varies from 41 to 59% and the O content varies from 33 to 50%. Fulvic acids have lower C (41 to 51%) but higher O (40 to

TABLE 3.4. *Average Values for Elemental Composition of Soil Humic Substances[a]*

	Humic acids (%)	Fulvic acids (%)
Carbon	53.8–58.7	40.7–50.6
Hydrogen	3.2–6.2	3.8–7.0
Oxygen	32.8–38.3	39.7–49.8
Nitrogen	0.8–4.3	0.9–3.3
Sulfur	0.1–1.5	0.1–3.6

[a] From C. Steelink, in "Humic Substances in Soil, Sediments, and Water" (G. R. Aiken, D. M. McKnight, and R. L. Wershaw, Eds.), pp. 457–476. Copyright © 1985 John Wiley & Sons, Inc. Reprinted by permission of John Wiley & Sons, Inc.

50%) contents than HA. Percentages of H, N, and S vary from 3 to 7, 1 to 4, and 0.1 to 4%, respectively. Humic acids tend to be higher in N than FA, while S is somewhat higher in FA.

Schnitzer and Khan (1972) have studied HS from arctic, temperate, subtropical, and tropical soils. They found ranges of 54–56% for C, 4–5% for H, and 34–36% for O. Neutral soils tend to have narrow ranges of C, H, and O, while acid soils show broader ranges, and particularly higher O contents. Table 3.5 shows the elemental composition of HS for a large number of humic acids, fulvic acids, and humin samples. It is striking that even though the samples were taken from soils, peat, freshwater, and marine sources worldwide, the standard deviations are quite small, indicating the similarities in elemental composition of HS from many sources and geographical areas.

Atomic ratios of H/C, O/C, and N/C can be useful in identifying types of HS and in devising structural formulas (Table 3.6). Based on the information in Table 3.6 it appears that the O/C ratio is the best indicator of humic types. Soil HA O/C ratios are about 0.50 while FA O/C ratios are 0.7.

A number of methods can be used to quantitate the functional groups of HS, particularly the acidic groups. The main acidic groups are carboxyl (R—C=O—OH) and acidic phenolic OH groups (presumed to be phenolic OH), with carboxyls being the most important group (Table 3.7). The total acidities of FA are higher than those for HA (Table 3.8). Smaller amounts of alcoholic OH, quinonic, and ketonic groups are also found. Fulvic acids are high in carboxyls, while alcoholic OH groups are higher in humin than in FA or HA, and carbonyls are highest in FA (Table 3.8).

TABLE 3.5. *Elemental Compositions for Humic Substances[a]*

	C	H	O	N
Group A				
Humic acid (410)	55.1 ± 5.0[b]	5.0 ± 1.1	35.6 ± 5.8	3.5 ± 1.5
Fulvic acid	46.2 ± 5.4	4.9 ± 1.0	45.6 ± 5.5	2.5 ± 1.6
Humin (26)[c]	56.1 ± 2.6	5.5 ± 1.0	34.7 ± 3.4	3.7 ± 1.3
Group B				
Soil humic acids (215)	55.4 ± 3.8	4.8 ± 1.0	36.0 ± 3.7	3.6 ± 1.3
Freshwater humic acids (56)	51.2 ± 3.0	4.7 ± 0.6	40.4 ± 3.8	2.6 ± 1.6
Peat humic acids (23)[d]	57.1 ± 2.5	5.0 ± 0.8	35.2 ± 2.7	2.8 ± 1.0
Group C				
Soil fulvic acids (127)	55.4 ± 3.8	4.8 ± 1.0	36.0 ± 3.7	3.6 ± 1.3
Freshwater fulvic acids (63)	51.2 ± 3.0	4.7 ± 0.6	40.4 ± 3.8	2.6 ± 1.6
Peat fulvic acids (12)	54.2 ± 4.3	5.0 ± 0.8	35.2 ± 2.7	2.8 ± 1.0

Note: Group A: average values for humic acid, fulvic acid, and humin from all over the world and not segregated by source. Group B: average values for humic acids from soil, freshwater, and peat sources. Group C: average values for fulvic acids from soil, freshwater, and peat sources.
[a] From MacCarthy (2001), with permission.
[b] The uncertainties with each value are absolute standard deviations; the numbers in parentheses following the name of each sample give the number of samples used in calculating the averages and standard deviations.
[c] Only 24 samples used for the nitrogen value in this row.
[d] Only 21 samples used for the nitrogen value in this row.

TABLE 3.6. *Atomic Ratios of Elements in Soil Humic and Fulvic Acids[a]*

Source	H/C	O/C	N/C	References
Soil fulvic acids				
Average of many samples	1.4	0.74	0.04	Schnitzer and Khan (1978)
Average of many samples	0.83	0.70	0.06	Ishiwatari (1975)
Average of many samples	0.93	0.64	0.03	Malcolm *et al.* (1981)
Soil humic acids				
Average of many samples	1.0	0.48	0.04	Schnitzer and Khan (1978)
Average of many samples	1.1	0.50	0.02	Ishiwatari (1975)
Neutral soils, average	1.1	0.47	0.06	Hatcher (1980)
Aldrich humic acid	0.8	0.46	0.01	Steelink *et al.* (1989)
Amazon HA/FA	0.97	0.57	0.04	Leenheer (1980)

[a] From C. Steelink, in "Humic Substances in Soil, Sediments, and Water" (G. R. Aiken, D. M. McKnight, and R. L. Wershaw, Eds.), pp. 457–476. Copyright © 1985 John Wiley & Sons, Inc. Reprinted by permission of John Wiley & Sons, Inc.

TABLE 3.7. *Some Important Functional Groups of SOM[a]*

Functional group	Structure
Acidic groups	
Carboxyl	$R—C{=}O(\text{–}OH)$[b]
Enol	$R—CH{=}CH—OH$
Phenolic OH	$Ar—OH$[b]
Quinone	$Ar{=}O$
Neutral groups	
Alcoholic OH	$R—CH_2—OH$
Ether	$R—CH_2—O—CH_2—R$
Ketone	$R—C{=}O(—R)$
Aldehyde	$R—C{=}O(—H)$
Ester	$R—C{=}O(—OR)$
Basic groups	
Amine	$R—CH_2—NH_2$
Amide	$R—C{=}O(—NH—R)$

[a] Adapted from F. J. Stevenson, "Humus Chemistry." Copyright © 1982 John Wiley & Sons, Inc. Reprinted by permission of John Wiley & Sons, Inc. and Thurman (1985) with permission of Kluwer Academic Publishers.
[b] R is an aliphatic (a broad category of carbon compounds having only a straight, or branched, open chain arrangement of the constituent carbon atoms; the carbon–carbon bonds may be saturated or unsaturated; Aiken *et al.*, 1985a) backbone and Ar is an aromatic ring.

TABLE 3.8. *Functional Groups in Humic Substances from 11 Florida Muck Samples (cmol kg^{-1}), with Standard Errors of the Means[a]*

	Total acidity	Carboxyls	Phenolic OH	Alcoholic OH	Carbonyls
Humins	510 ± 20	200 ± 20	310 ± 20	360 ± 30	260 ± 20
Humic acids	720 ± 40	310 ± 20	420 ± 30	130 ± 30	130 ± 10
Fulvic acids	860 ± 40	400 ± 20	460 ± 20	80 ± 20	430 ± 10

[a] From Zelazny and Carlisle (1974), with permission.

The chemical structures of some of the amino acids found in soils are shown in Table 3.9. The quantities of amino acids found in HS extracted from tropical soils are given in Table 3.10. High levels of amino acid nitrogen were found in HA, FA, and humin. There are high distributions of acidic and some neutral amino acids, particularly glycine, alanine, and valine.

Humic substances also contain small amounts of nucleic acids and their derivatives, chlorophyll and chlorophyll-degradation products, phospho-lipids, amines, and vitamins. The nucleic acids include DNA and RNA. They can be identified by the nature of the pentose sugar, i.e., deoxyribose or ribose, respectively (Stevenson, 1982).

Fractionation of SOM

Before one can suitably study SOM, it must be separated from the inorganic soil components. Fractionation of SOM lessens the heterogeneity of HS so that physical and chemical techniques can be used to study their structure and molecular properties (Hayes and Swift, 1978). The classical fractionation scheme (Fig. 3.6) involves precipitation of HS by adjustment of pH and salt concentrations, addition of organic solvents, or addition of metal ions.

Alkali extraction, usually with 0.1–0.5 M NaOH and Na_2CO_3 solutions, is based on solubility principles. Humic acid is soluble in alkali (base) and insoluble in acid while FA is soluble in both alkali and acid. Hymatomelanic acid is the alcohol soluble portion of HA, and humin is not soluble in alkali or acid. After extraction, the HA precipitate is usually frozen and thawed to remove water and then freeze-dried for subsequent analysis.

TABLE 3.9. *Chemical Structure of Some Protein Amino Acids Found in Soils[a]*

Neutral Amino Acids	Aromatic Amino Acids

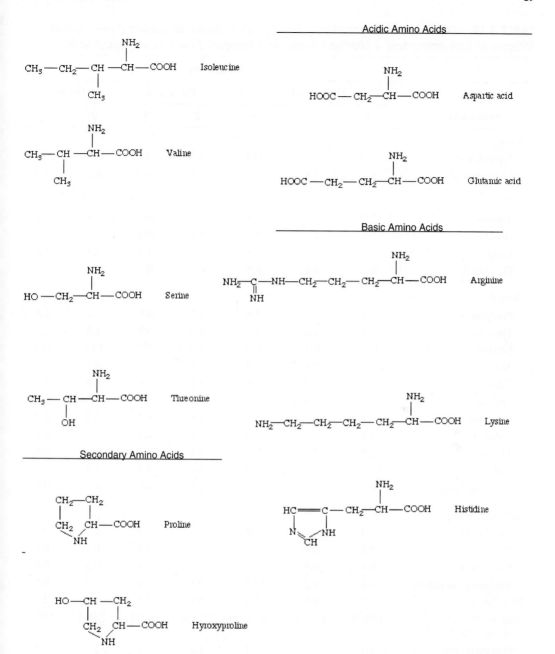

However, alkali extractions can dissolve silica, contaminating the humic fractions, and dissolve protoplasmic and structural components from organic tissues. Milder extractants, such as $Na_4P_2O_7$ and EDTA, dilute acid mixtures with HF, and organic solvents, can also be employed. However, less SOM is extracted (Stevenson, 1982).

TABLE 3.10. *Relative Molar Distribution of Amino Acids in Humic Substances (α-Amino Acid Nitrogen of Each Amino Acid × 100/Total Amino Acid Nitrogen) from Several Tropical Soils[a,b]*

	Soil number[c]						
	Humic acid			Fulvic acid		Humin	
Amino acid	2	3	5	2	5	2	5
Acidic							
Aspartic acid	13.0	11.7	11.8	26.1	23.1	11.8	23.2
Glutamic acid	8.5	8.8	8.6	15.0	20.9	8.2	10.1
Basic							
Arginine	2.0	2.2	2.3	0.5	0.3	1.9	1.9
Histidine	2.3	1.4	1.5	0.9	0.2	2.1	1.4
Lysine	3.3	3.1	3.5	1.9	1.4	2.7	2.6
Ornithine	0.7	0.8	0.7	0.9	0.7	0.6	1.1
Neutral							
Phenylalanine	3.2	3.3	2.9	1.3	0.9	2.9	1.7
Tyrosine	1.5	1.6	1.4	1.2	0.2	1.5	1.2
Glycine	11.2	10.9	11.1	12.6	13.5	13.1	11.2
Alanine	7.6	8.5	8.3	7.4	9.1	8.9	7.9
Valine	5.4	6.2	5.9	4.1	3.6	6.0	3.8
Leucine	5.1	5.8	5.1	3.0	1.7	5.3	4.4
Isoleucine	3.3	3.5	3.5	1.8	1.7	3.0	2.2
Serine	5.0	4.8	4.9	5.2	4.3	5.7	4.1
Threonine	4.9	4.7	5.2	4.4	3.9	5.3	3.8
Proline	4.2	6.2	4.9	4.4	3.3	4.2	2.7
Hydroxyproline	0.7	0.7	0.7	tr[d]	tr[d]	0.8	0.3
Sulfur-containing							
Methionine	0.5	1.0	0.8	0.4	0.3	0.7	0.3
Cystine	0.3	0.1	0.2	0.1	0.2	0.1	0.1
Cysteic acid	0.4	0.5	0.6	1.7	3.4	0.3	0.8
Methionine sulphoxide	0.6	0.2	0.3	0.1	1.0	0.2	0.3
Miscellaneous[e]	1.7	0.9	1.6	0.9	2.9	1.6	2.1
Total amino acid nitrogen (μM g^{-1})	1496.0	1476.0	1856.0	775.0	587.0	346.4	149.5
Amino sugar ratio[f]	1.3	1.3	1.3	1.2	1.4	3.5	2.5

[a] From M. Schnitzer, in "Humic Substances in Soil, Sediments, and Water" (G. R. Aiken, D. M. McKnight, and R. L. Wershaw, Eds.), pp. 303–325. Copyright © 1985 John Wiley & Sons, Inc. Reprinted by permission of John Wiley & Sons, Inc.
[b] Reprinted from *Soil Biology and Biochemistry*, Vol. **8**, F. J. Sowden, S. M. Griffith, and M. Schnitzer, The distribution of nitrogen in some highly organic tropic volcanic soils, pp. 55–60, copyright © 1976, with kind permission from Elsevier Science Ltd., The Boulevard, Langford Lane, Kidlington 0X5 1GB, UK.
[c] Numbers represent different soils.
[d] tr, Trace.
[e] Includes allo-isoleucine, α-NH$_2$-butyric acid; 2-4-diaminobutyric acid, diaminopimelic acid, β-alanine, ethanolamine, and unidentified compounds.
[f] Ratio of glucosamine/galactosamine.

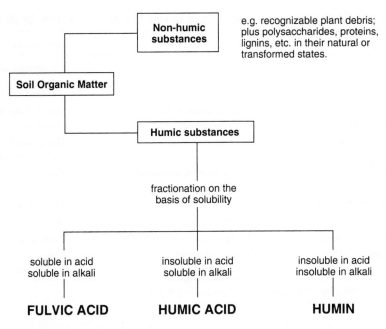

FIGURE 3.6. *Fractionation of soil organic matter and humic substances. From M.H.B. Hayes and R.S. Swift, The chemistry of soil organic colloids, in "The Chemistry of Soil Constituents" (D.J. Greenland and M.H.B. Hayes, Eds.), pp. 179–230. Copyright © 1978 John Wiley and Sons, Inc. Reprinted by permission of John Wiley and Sons, Inc.*

In addition to extraction and precipitation procedures, gel permeation chromatography, ultrafiltration membranes, adsorption on hydrophobic resins (XAD, nonionic methylmethacrylate polymer), adsorption on ion exchange resins, adsorption on charcoal and Al_2O_3, and centrifugation are also used for SOM fractionation (Buffle, 1984; Thurman, 1985). New electrophoretic methods including polyacrylamide gel electrophoresis, isoelectric focusing, and isotachophoresis are also promising techniques for SOM fractionation (Hayes and Swift, 1978). The use of XAD resins is considered by many researchers to be the best method to fractionate or isolate HS (Thurman *et al.*, 1978; Thurman, 1985). The advantages and limitations of various isolation procedures for HS are given in Table 3.11.

Molecular and Macromolecular Structure of SOM

While we know the elemental and functional group composition of HS, definitive knowledge of the basic "backbone structure" of SOM is still an enigma. Many structures have been proposed, and each of them is characterized by similar functional groups and the presence of aliphatic and aromatic components.

TABLE 3.11.   *Advantages and Limitations of Various Isolation Procedures for Humic Substances*[a]

Method	Advantages	Limitations
Precipitation	None	Fractionates sample, not specific for humus, slow with large volumes
Freeze concentration	All DOC (dissolved organic carbon) concentrated	Slow, tedious procedure, concentrates inorganics
Liquid extraction	Visual color removal	Not quantified by DOC, slow with large volumes
Ultrafiltration	Also separates by molecular weight	Slow
Strong anion-exchange	Efficient sorption	Does not desorb completely
Charcoal	Efficient sorption	Does not desorb completely
Weak anion-exchange	Adsorbs and desorbs efficiently	Resin bleeds DOC
XAD resin	Adsorbs and desorbs efficiently	Resin must be cleaned to keep DOC bleed low

[a] From Thorman (1985). Reprinted by permission of Kluwer Academic Publishers.

Use of advanced analytical techniques such as ^{13}C NMR spectroscopy has provided spectra of whole soils that show paraffinic C, OCH_3-C, amino acid-C, C in carbohydrates and aliphatic structures containing OH groups, aromatic C, phenolic C, and C in CO_2H groups (Arshad *et al.*, 1988). Based on the spectra, the aromaticity of SOM has been shown to seldom exceed 55% with the aliphaticity often being higher than aromaticity (Schnitzer and Preston, 1986). Similar to the ^{13}C NMR studies, pyrolysis-field ionization mass spectrometry (Py-FIMS) studies on SOM in whole soils show the presence of carbohydrates, phenols, lignin monomers, lignin dimers, alkanes, fatty acids, *n*-alkyl mono-, di-, and tri-esters, *n*-alkylbenzenes, methylnapthalenes, methylphenanthrenes, and N compounds (Schnitzer and Schulten, 1992). The carbohydrates, proteinaceous materials (amino acids, peptides, and proteins), and lipids (alkanes, alkenes, saturated and unsaturated fatty acids, alkyl mono-, di-, and tri-esters) in SOM are tightly bound by the aromatic SOM compounds (Schnitzer, 2000).

Using Py-FIMS and Curie-Point pyrolysis GC/MS (gas chromatography/ mass spectrometry), Schulten and Schnitzer (1993) proposed a 2-dimensional structure for HA (Fig. 3.7) in which aromatic rings are linked covalently by aliphatic chains. Oxygen is present as carboxyls, phenolic and alcoholic hydroxyls, esters, ethers, and ketones while nitrogen is present as heterocyclic structures and as nitriles. The elemental composition of the HA structure in Fig. 3.7 is $C_{308}H_{328}O_{90}N_5$; it has a molecular weight of 5.540 Da and an elemental analysis of 66.8% C, 6.0% H, 26.0% O, and 1.3% N. The C skeleton has high microporosity containing voids of different dimensions that can trap and bind other organics (e.g., carbohydrates and proteinaceous materials) and inorganic components (clay minerals and metal oxides) and water. Schnitzer and Schulten (1995) assumed that the carbohydrates and the proteinaceous materials were adsorbed on external surfaces as well in the voids and that H bonds significantly affected their immobilization.

Schulten and Schnitzer (1993), assuming that a molecular weight of HA interacted with 10% carbohydrates and 10% proteinaceous materials (values previously found by others for HA), proposed a HA with an elemental composition of $C_{342}H_{388}O_{124}N_{12}$, with a molecular weight of 6651 Da and an elementary analysis of 61.8% C, 5.9% H, 29.8% O, and 2.5% N. They compared these analyses and functional group content with those for HAs extracted from soils of three great soil groups (Table 3.12). The data for the soils compare favorably to the HAs extracted from the soils.

Schulten and co-workers (Schulten and Schnitzer, 1997; Schulten et al., 1998) expanded on their earlier two-dimensional structure of HA by using computational chemistry. Three-dimensional structures of HA, SOM, and whole soil have been proposed. A three-dimensional structure of a soil particle is shown in Fig. 3.8. Schnitzer (2000) assumed that the soil contained 3% SOM, 3% H_2O, and 94% inorganic components. Voids in the model SOM structure could contain organics, inorganics, and water. The functional groups could react with metals and inorganic minerals and also provide nutrients to plants and microbes. In Fig. 3.8 the SOM in the soil particle is bound to silicates via Fe^{3+} and Al^{3+} ions. It is surrounded by a model matrix of silica sheets. The modeled soil particle shows 23 H bonds, 13 of which are intramolecular, 9 are in the mineral matrix, and 1 is between the SOM and the silica sheet.

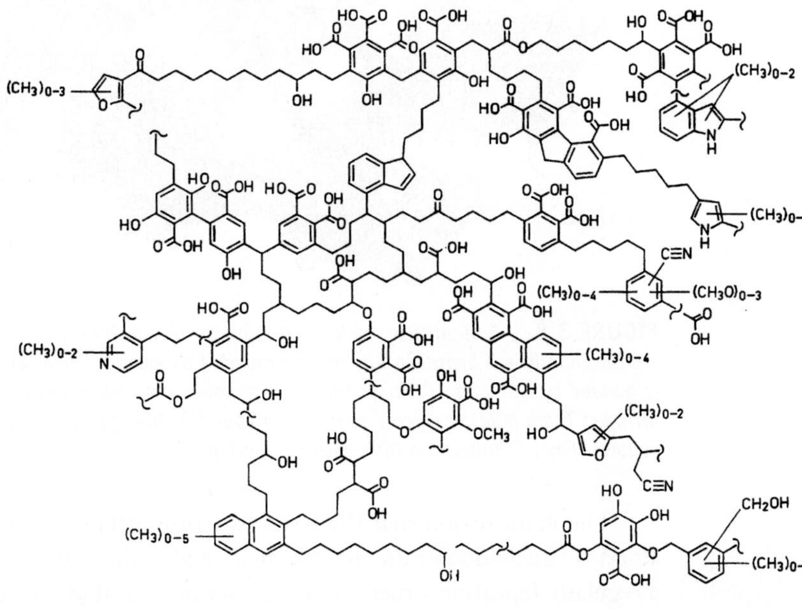

FIGURE 3.7. *Two-dimensional HA model structure. From Schulten and Schnitzer (1993), with permission. Copyright 1993 Springer-Verlag GmbH.*

TABLE 3.12. *Analytical Characteristics of HAs Extracted from Soils Belonging to Three Different Great Soil Groups and of the Proposed HA Structure[a]*

	Udic Boroll	Haplaquod	Haplaquoll	Proposed[b]
C (%)	56.4	58.2	54.2	61.8
H (%)	5.5	5.4	6.0	5.9
N (%)	4.1	3.1	6.0	2.5
S (%)	1.1	0.7	0.9	—
O (%)	32.9	32.6	32.9	29.8
Total acidity (cmol kg^{-1})	6.6	5.7	6.4	5.8
CO_2H (cmol kg^{-1}), phenolic	4.5	3.2	3.5	4.4
OH (cmol kg^{-1}), alcoholic	2.1	2.5	2.9	1.4
OH (cmol kg^{-1})	2.8	3.2	3.0	1.4
OCH_3 (cmol kg^{-1})	0.3	0.4	0.4	0.3

[a] From Schnitzer and Schulten (1995), with permission.
[b] MW = 6651 Da.

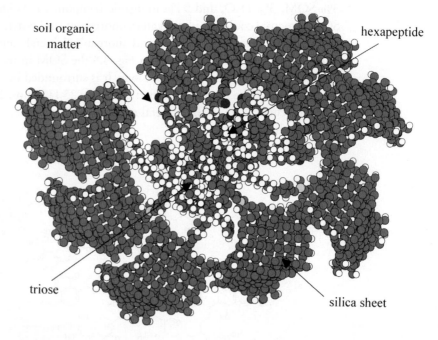

FIGURE 3.8. *Soil particle model consisting of HA (in center), containing in its voids a trisaccharide + a heptapeptide, and surrounded by eight silica sheets to which the HA is bonded by Fe^{3+} and Al^{3+} ions. The element colors are carbon (blue); hydrogen (white); nitrogen (dark blue); oxygen (red); silicon (purple); iron (green); and aluminum (light yellow). From Schnitzer (2000), with permission.*

The main reason that the basic structure of HS is still not fully understood is largely due to the heterogeneity and complexity of HS. There is not a regularly repeating structural unit or set of units that is characteristic of HS (MacCarthy, 2001). Consequently, no two molecules of HS are alike (Dubach and Mehta, 1963).

The macromolecular structure (size and shape) of HS is of great importance as it can affect the chemistry of SOM–mineral complexes, stability of organo-mineral aggregates, fate/transport of pollutants, biotransformations, and the C cycle in soils (Schnitzer, 1991). The conformation and macromolecular structure of HS are affected by pH, electrolyte concentration, ionic strength, and HA and FA concentrations (Chen and Schnitzer, 1976; Ghosh and Schnitzer, 1980).

Chen and Schnitzer (1976) used electron microscopy to study the macromolecular structure of FA (Fig. 3.9). At pH 2 (Fig. 3.9a) elongated fibers and bundles of fibers are evident. The fibers were either linear or curved, up to 6–7 μm in length, and 120–400 nm in thickness. At pH 4 (Fig. 3.9b) the fibers were thinner and fiber bundles more prevalent. The thickness was less, 120–200 nm. At pH 6 (Fig. 3.9c) there were less fibers, they were not as thick, and there were more bundles of closely knit fibers. At pH 7 (Fig. 3.9d) there was a fine network of tightly meshed fibers with parallel orientation. At pH 8 (Fig. 3.9e) the fine network had changed into a sheet-like-structure with varying thicknesses. At pH 9 (Fig. 3.9f) the sheets were thicker, and at pH 10 (Fig. 3.9g) homogeneous grain-like particles existed.

Ghosh and Schnitzer (1980) used viscosity surface pressure measurements to study the macromolecular structure of HAs and FAs (Fig. 3.10). From these data it was concluded that HAs and FAs are rigid uncharged spheroids at (1) high sample concentrations (>3.5 to 5.0 g liter^{-1}), (2) low pH (6.5 for HA and <3.5 for FA), and (3) electrolyte concentrations 0.05 M and higher. However, HAs and FAs are flexible, linear polyelectrolytes at (1) low sample

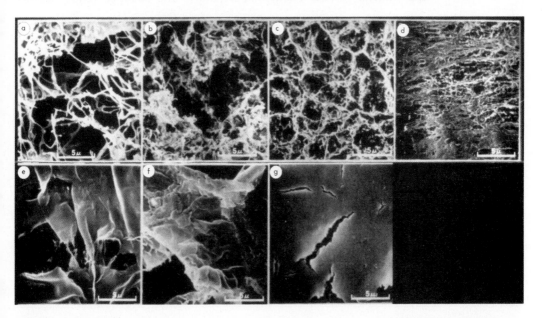

FIGURE 3.9. *Scanning electron micrograph of a freeze-dried FA at various pH values. Modified from Chen and Schnitzer (1976), with permission.*

concentrations (<3.5 g liter^{-1}), (2) pH >6.5 for HA and >3.5 for FA, and (3) electrolyte concentrations <0.05 M. Thus, in most soil solutions and fresh waters, where humic and salt concentrations are low, HA (at pH >6.5) and FA (at pH >3.5) should occur as flexible, linear polyelectrolytes.

With recent advances in the use of *in situ* molecular scale spectroscopies and microscopies, many of the conclusions of Schnitzer and his co-workers (Chen and Schnitzer, 1976; Ghosh and Schnitzer, 1980) about the macromolecular structure and conformation of SOM under varying environmental conditions have been directly confirmed.

In a beautiful study, Myneni *et al.* (1999) employed state-of-the-art *in situ* spectromicroscopy to study the macromolecular structure of HSs as a function of HS origin (soil vs fluvial), solution chemistry (pH, ionic strength), and mineralogy. As the concentration of HS increased above 1.3 g of C liter^{-1} aggregates of different sizes and shapes formed (Fig. 3.11).

In dilute, acidic, high-ionic-strength solutions (NaCl), HSs formed globular aggregates and ringlike structures (Fig. 3.11A). As the fulvic acid concentration increased, large sheet-like structures appeared and encapsulated smaller structures. Coiling was not prevalent and HS dispersed into small aggregates (<0.1 μm) in solutions of pH >8 (Fig. 3.11B). In concentrated, HS solutions formed globular aggregates bound together with thin films of HSs. In dilute, FA solutions of Ca^{2+}, Cu^{2+}, or Fe^{2+}, FA formed thread-and-netlike structures (Figs. 3.11C and 3.11D). As FA and cation concentration increased, the structures were larger and formed ringed sheets.

In situ atomic force microscopy (AFM) has been recently used to study the macromolecular structure as affected by pH, I, SOM concentration, and mineral surface. Namjesnik-Dejanovic and Maurice (2000) studied

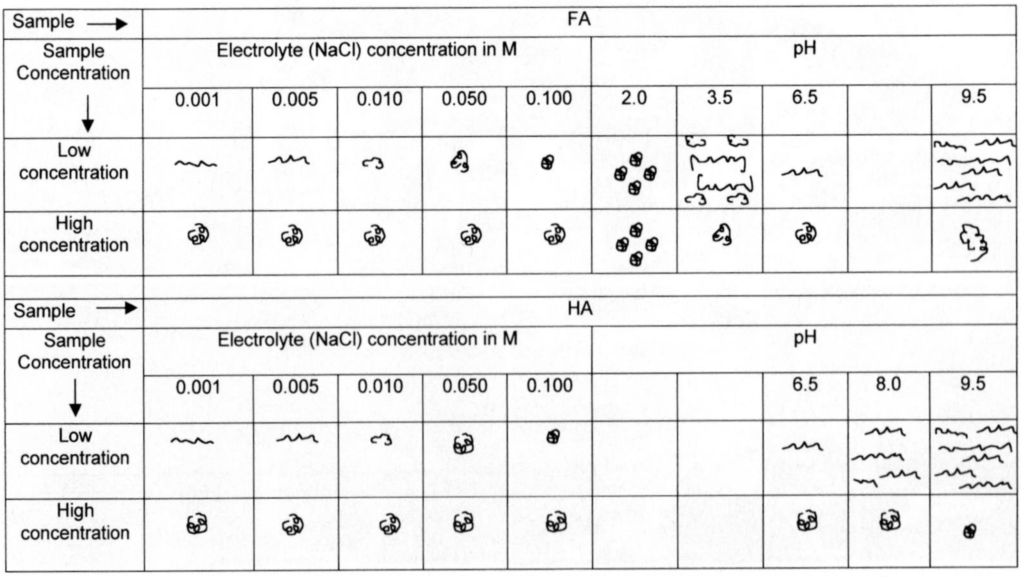

Sample →	FA									
Sample Concentration ↓	Electrolyte (NaCl) concentration in M					pH				
	0.001	0.005	0.010	0.050	0.100	2.0	3.5	6.5		9.5
Low concentration										
High concentration										

Sample →	HA									
Sample Concentration ↓	Electrolyte (NaCl) concentration in M					pH				
	0.001	0.005	0.010	0.050	0.100			6.5	8.0	9.5
Low concentration										
High concentration										

FIGURE 3.10. *Macromolecular HA and FA configurations at different pH values and electrolyte concentrations. From Ghosh and Schnitzer (1980), with permission of the publisher.*

SOM conformation on mica and hematite over a pH range of 3–11, I of 0.001–0.3 M, SOM concentrations of 20–100 mg C liter^{-1}, and type of electrolyte cation (Ca, Na, and L). At intermediate pH and low SOM concentrations on mica, larger spherical aggregates were observed while at higher SOM concentrations, ring structures with considerable porosity were seen. On hematite, at pH 4 and with dissolved Fe being present, globular structures with mean diameter of 0.3 μm were observed (Fig. 3.12). A similar macromolecular structure was observed by Myneni *et al.* (1999) using *in situ* spectromicroscopy (Fig. 3.11).

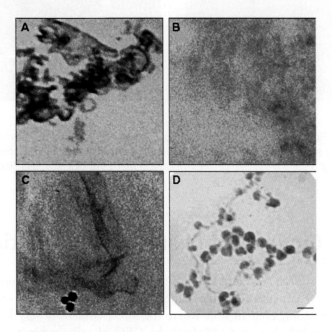

FIGURE 3.11. *Influence of pH, ionic strength, and complexing cations on the macromolecular structures of isolated fluvial fulvic acid. (A) pH is 3.0 ± 1.0, NaCl is 1.0 M, and C is ~1.5 g liter^{-1}. The average sizes of globular and ringlike structures were 0.3 μm (range of 0.2 to 0.45 μm) and 0.65 μm (range of 0.3 to 1.2 μm), respectively. The ratio of globular to ring structures was 70:30. In addition, sheetlike structures (average of 3 μm, range of 2 to 8 μm) were also noticed in concentrated HS solutions. (B) pH is 9.0 ± 1.0, NaCl is 0.5 M, and C is ~1.2 to 1.5 g liter^{-1}. Average aggregate size is <0.1 μm, with little deviation in size. In concentrated HS solutions, globular (average of 0.3 μm, range of 0.2 to 0.5 μm) and sheetlike structures (average of 1.5 μm, range of 1 to 5 μm) were also formed. (C) pH is ~4.0 ± 1.0, CaCl$_2$ is 0.018 M, and C is ~1.0 g liter^{-1}. The threadlike structures had an average length of ~3 μm (range of 2 to 6 μm) and a width of <0.15 μm. (D) pH is ~4.0 ± 1.0, Fe^{3+} is 1 mM, and C is ~0.1 g liter^{-1}. The average sizes of globular and threadlike structures were 0.3 μm (range of 0.25 to 0.4 μm) and 0.8 μm (range of 0.5 to 1.3 μm), respectively. Scale bar, 500 nm. From Myneni et al. (1999), with permission. Copyright 1999 American Association for the Advancement of Science.*

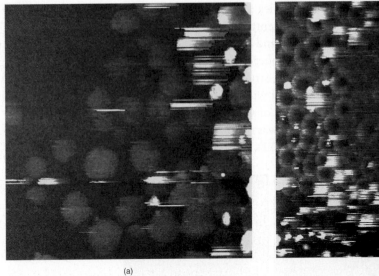

(a)   (b)

FIGURE 3.12. *(a) Flattened spherical aggregates of groundwater NOM imaged on hematite with light tapping; the underlying surface cannot be seen but is similar to (b), which contains flattened disk or ring-shaped aggregates (a) 3.5 µm and (b) 3.2 µm on a side. From Namjesnik-Dejanovic and Maurice (2000), with permission.*

Functional Groups and Charge Characteristics

The surface areas and CEC of SOM, as given earlier, are higher than those of clay minerals. The role that SOM plays in the retention of ions is indeed significant, even in soils where the SOM content is very low. It has been estimated that up to 80% of the CEC of soils is due to organic matter (Stevenson, 1982).

Organic matter is a variable charge soil component. Since its point of zero charge (PZC), defined as the pH at which the colloid particle has no net charge (see Chapter 5 for discussions on points of zero charge), is low, about 3, SOM is negatively charged at pH values greater than 3. As pH increases, the degree of negative charge increases due to the deprotonation or dissociation of H^+ from functional groups.

The major acidic functional groups are carboxyls ($pK_a < 5$), quinones that may dissociate as readily as carboxyl groups and have as low a pK_a, phenolic OH groups, and enols (Table 3.7). Since carboxyl and phenolic groups can deprotonate at pH's common in many soils, they are major contributors to the negative charge of soils. It has been estimated that up to 55% of the CEC from SOM is due to carboxyl groups (Broadbent and Bradfield, 1952) while about 30% of the CEC of SOM up to pH 7 is due to the quinonic, phenolic, and enolic groups. Neutral and basic functional groups important in SOM are also given in Table 3.7.

Two seminal papers (Helling *et al.*, 1964; Yuan *et al.*, 1967) in the scientific literature clearly established the importance of SOM in contributing to the CEC of a soil. Helling *et al.* (1964) measured the CEC of 60 Wisconsin soils at pH values between 2.5 and 8, and by multiple correlation analysis determined the contributions of OM and clay at each pH, and the variation in CEC of each as pH changed (Table 3.13). Based on these studies, Helling *et al.* (1964) concluded that the CEC of the clay fraction changed much less as pH increased compared to the CEC of the SOM fraction. They attributed this to the dominance of permanent charge of the clay fraction, primarily composed of montmorillonite and vermiculite, which would increase little with increase in pH. The small increases in CEC of the clay fraction were attributed to the dissociation of edge functional groups on the kaolinite surfaces as pH increased.

However, the CEC of the organic fraction of the soils increased dramatically with pH—from 36 cmol kg^{-1} at pH 2.5 to 215 cmol kg^{-1} at pH 8.0. This was due to the dissociation of H$^+$ from functional groups of the SOM. At pH 2.5, 19% of the CEC was due to SOM, while at pH 8.0 SOM accounted for 45% of the total CEC.

The importance of SOM in sandy soils to the cation exchange capacity is shown in Table 3.14. From 66.4 to 96.5% of the CEC of these soils was due to SOM.

Soil organic matter is a major contributor to the buffering capacity of soils. A typical titration curve of peat HA and soil HA is shown in Fig. 3.13. As base is added, pH increases, illustrating the large buffering capacity of HS that is apparent over a wide pH range and the weak acid characteristics of HS. Most buffer curves of HS from acid soils have a pK_a of about 6. The titration curve (Fig. 3.13) can be divided into three zones. Zone I, the most acid region, represents dissociation of carboxyl groups, while Zone III represents dissociation of phenolic OH and other very weak acid groups. Zone II is an intermediate range attributed to a combination of dissociation of weak carboxyl and very weak acid groups (Stevenson, 1982).

TABLE 3.13. *Contribution of Organic Matter and Clay Fractions to Soil Cation Exchange Capacity as Influenced by pH[a]*

Buffer pH	Clay fraction (cmol kg^{-1} clay)	Organic fraction (cmol kg^{-1} SOM)	% of CEC due to SOM
2.5	38	36	19
3.5	45	73	28
5.0	54	127	37
6.0	56	131	36
7.0	60	163	40
8.0	64	215	45

[a] Adapted from data of Helling *et al.* (1964), with permission.

TABLE 3.14. *Contribution of Organic Matter and Clay to the Cation Exchange Capacities of Sandy Soils[a]*

Soil group	Average CEC (cmol kg^{-1})	Relative contribution (%) SOM	Clay
Entisols			
Psamments	5.26	74.9	25.1
Aquipsamments	3.84	86.8	15.2
Quartipsamments	5.63	75.7	24.3
Acid family	3.83	78.7	21.3
Nonacid family	4.21	95.4	4.6
Phosphatic family	10.58	77.4	22.6
Inceptisols			
Aquepts and Umbrepts	8.17	69.2	30.8
Mollisols			
Aqualls	12.93	66.4	33.6
Spodosols			
Aquods	5.53	96.5	3.5
All soils	6.77	76.1	23.9

[a] Adapted from Yuan, T. L., Gammon, N., Jr., and Leighty, R. G. (1967), Relative contribution of organic and clay fractions to cation-exchange capacity of sandy soils from several groups. *Soil Sci.* **104**, 123–128. Used with permission of Williams and Wilkins.

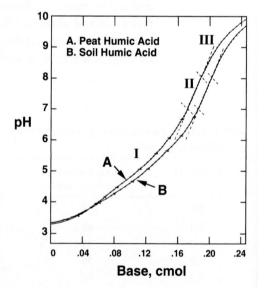

FIGURE 3.13.   *Titration of a (A) peat and (B) soil humic acid. The small wavy lines on the curves indicate endpoints for dissociation of weak acid groups having different but overlapping dissociation constants. From F. J. Stevenson (1982), "Humus Chemistry." Copyright 1982 John Wiley & Sons, Inc. Reprinted by permission of John Wiley & Sons, Inc.*

The weak acid character of SOM is ascribed to complexation with free metals, such as Al^{3+}, Fe^{3+}, and Cu^{2+}, and hydroxy-Al and hydroxy-Fe materials. Thus, important functional groups such as carboxyls are not always found as free groups in soils but are complexed with metals (Martin and Reeve, 1958).

Humic Substance–Metal Interactions

The complexation of metal ions by SOM is extremely important in affecting the retention and mobility of metal contaminants in soils and waters. Several different types of SOM–metal reactions can occur (Fig. 3.14). These include reaction between dissolved organic carbon (DOC), which is the organic carbon passing through a 0.45-μm silver or glass fiber filter (Thurman, 1985) and metal ions, and complexation reactions between suspended organic matter (SOC, organic carbon retained on a 0.45-μm silver filter) and metal ions and bottom sediment and metal ions. The functional groups of SOM have different affinities for metal ions (Charberek and Martell, 1959) as follows:

$$—O^- \quad > \quad —NH_2 \quad > \quad —N{=}N \quad > \quad \backslash N /$$

(enolate)   (amine)   (azo compounds)   ring N

$$> \quad COO^- \quad > \quad —O^- \quad > \quad —C{=}O.$$

(carboxylate)   (ether)   (carbonyl)

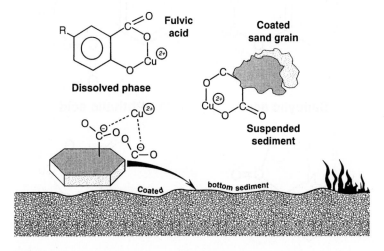

FIGURE 3.14. *Complexation of metal ions by organic matter in suspended sediment, bottom sediment, colloidal, and dissolved phases. From Thurman (1985). Reprinted by permission of Kluwer Academic Publishers.*

If two or more organic functional groups (e.g., carboxylate) coordinate the metal ion, forming an internal ring structure, chelation, a form of complexation, occurs (Fig. 3.15). The total binding capacity of HA for metal ions is about 200–600 $\mu mol\ g^{-1}$. About 33% of this total is due to retention on cation complexing sites. The major complexing sites are carboxyl and phenolic groups.

Factors Affecting Metal–Complexant (Ligand) Interactions

The types of interactions between metal ions and complexants such as inorganic (anions) and organic (e.g., carboxyl and phenolic groups of SOM) ligands (a ligand is an atom, functional group, or molecule attached to the central atom of a coordination compound) can be predicted based on: (1) the hydrolysis properties of elements, and (2) the concept of "hardness" and "softness" of metals and electron-donor atoms of complexing sites (Buffle and Stumm, 1994).

HYDROLYSIS PROPERTIES

Inorganic elements in the periodic table can be divided into three groups based on their reactions with OH^- or O^{2-} (Fig. 3.16). Group 1 elements form nondissociated oxocomplexes (e.g., SO_4^{2-}, PO_4^{3-}) and oxyacids (e.g.,

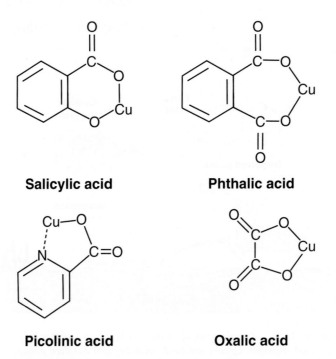

Salicylic acid   **Phthalic acid**

Picolinic acid   **Oxalic acid**

FIGURE 3.15.   *Chelation of metal ions by organic compounds. From Thurman (1985). Reprinted by permission of Kluwer Academic Publishers.*

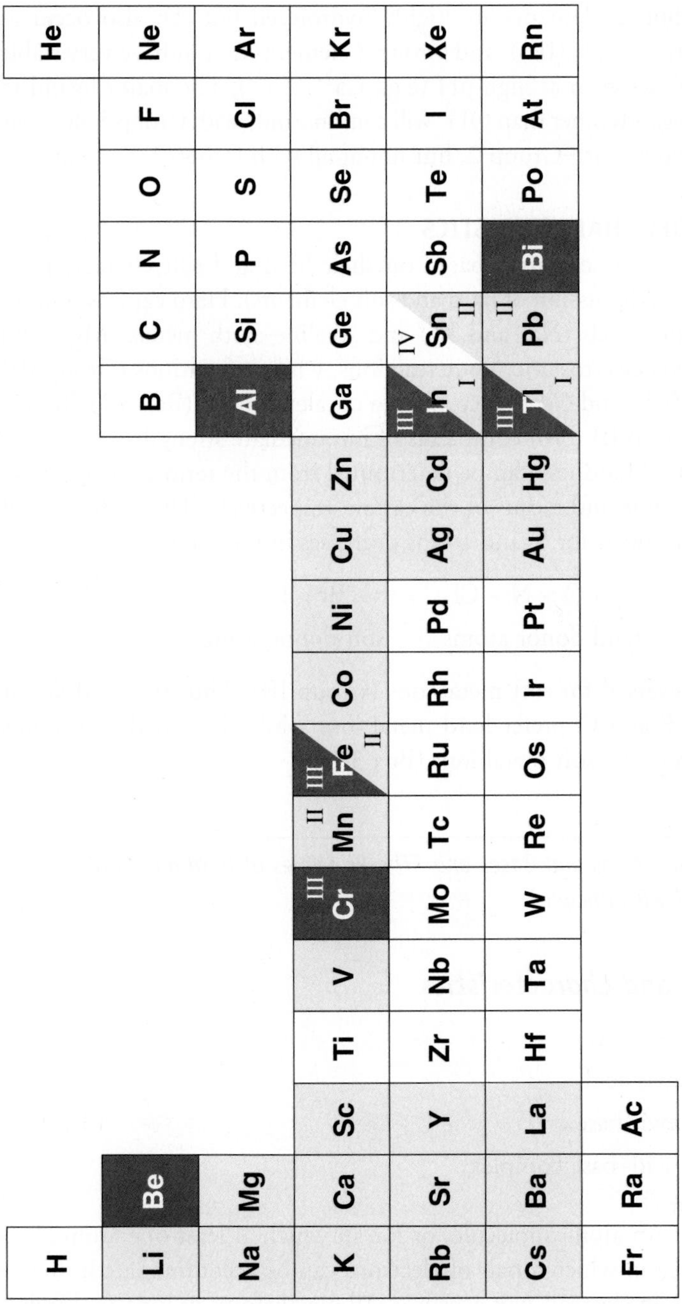

FIGURE 3.16. Group 1 elements (white squares), Group 2 elements (dark shaded squares), and Group 3 elements (light shaded squares). Group elements 1, 2, and 3 are defined in text. Reprinted with permission from Buffle and Stumm (1994). Copyright Lewis Publishers, an imprint of CRC Press, Boca Raton, Florida.

As(OH)$_3$). Group 2 elements are highly hydrolyzed but can also occur as hydrated cations (e.g., Fe(III)), and Group 3 elements do not have very stable hydroxo complexes even at high pH (e.g., Ca^{2+}, Zn^{2+}). The major ligand in water is OH$^-$; ligands other than OH$^-$ will combine only with Group 3 elements and to some extent with Group 2, but not at all with Group 1 elements.

HARD AND SOFT CHARACTERISTICS

Elements can also be classified based on their hard and soft characteristics (see Box 3.1 for a discussion of hard and soft elements). Hard cations (Group I) such as alkali metals (Na$^+$ and K$^+$) and alkaline earth metals (Mg^{2+} and Ca^{2+}) interact via electrostatic, ionic reactions, while soft cations (Group III) such as Cu^{2+}, Zn^{2+}, and Cd^{2+} react to form covalent bonds (Box 3.1). Transition metals (Group II) form complexes of intermediate strength.

The degree of hardness can be determined from the term Z^2/r, where Z and r are the charge and radius of the cation, respectively. The preference of hard metals (Group I) for ligand atoms decreases in the order

$$F > O > N \sim Cl \quad > \quad Br > I > S$$

hard donor atoms   soft donor atoms

This order is reversed for soft metal ions (Group III). Thus the hard donor atoms such as F and O prefer hard metal ions while the soft donor atoms such as I and S prefer soft metal ions (Box 3.1).

BOX 3.1   *Lewis Acids and Bases and "The Principles of Hard and Soft Acids and Bases" (HSAB Principle)*

Definitions and Characteristics

A + :B → A:B

A = Lewis acid

:B = Lewis base   (3.1a)

A:B = acid–base complex

Lewis Acid.   An atom, molecule, or ion in which at least one atom has a vacant orbital in which a pair of electrons can be accommodated; thus, a Lewis acid is an electron-pair acceptor. All metal atoms or ions are Lewis acids. Most cations are Lewis acids. In Eq. (3.1a) above, when A is a metal ion, B is referred to as a ligand. Lewis acids are coodinated to Lewis bases or ligands.

Lewis Base.   An atom, molecule, or ion which has at least one pair of valence electrons not being shared in a covalent bond; thus, a Lewis base is an electron-pair donor. Most anions are Lewis bases.

Hard Acid.   The acceptor atom is of high positive charge and small size, and does not have easily excited outer electrons. A hard acid is not polarizable and associates with hard bases through ionic bonds.

Soft Acid.   The acceptor atom is of low positive charge and large size, and has several easily excited outer electrons. A soft acid is polarizable and prefers soft bases through covalent bonds.

Hard Base.   The donor atom is of low polarizability and high electronegativity, is hard to reduce, and is associated with empty orbitals of high energy and thus is accessible.

Soft Base.   The donor atom is of high polarizability and low electronegativity, is easily oxidized, and is associated with empty, low-lying orbitals.

Classification of Lewis Acids and Bases

Lewis acids	Lewis bases
Hard acids: Group I metals[b]	**Hard bases**
H^+, Li^+, Na^+, K^+, Be^{2+}, Mg^{2+}, Ca^{2+}, Sr^{2+}, Fe^{2+}, Al^{3+}, Se^{3+}	H_2O, OH^-, F^-, PO_4^{3-}, SO_4^{2-}, Cl^-, CO_3^{2-}, ClO_4^-, NO_3^-
Transition acids: Group II metals	**Transition bases**
Cr^{2+}, Mn^{2+}, Fe^{2+}, Co^{2+}, Ni^{2+}, Cu^{2+}	Br^-, NO_2^-, N_2
Soft acids: Group III metals	**Soft bases**
Ag^+, Au^+, Tl^+, Cu^+, Zn^{2+}, Cd^{2+}, Hg^{2+}, Pb^{2+}, Sn^{2+}	I^-, CN^-, CO

[a] Adapted from Pearson (1963, 1968) and Buffle and Stumm (1994).
[b] Refers to metal groups in text discussion of hard and soft characteristics of elements.

Complexants or ligands can be classified as (Fig. 3.17) (Buffle, 1984; Buffle and Stumm, 1994): (1) simple inorganic ligands, X, which are major anions — their donor atom is oxygen and they prefer hard metals; (2) "hard" sites of natural organic matter (NOM) referred to as L_H — they are mainly carboxyl and phenolic sites; and (3) "soft" sites of SOM, denoted as L_s, which are N- and S-containing sites. Based on the concentration of ligands and metals in aqueous systems one can make predictions about metal–complexant interactions.

Group I metals (mainly alkali metal and alkaline earth metal cations) prefer hard ligands (Fig. 3.17), but form weak complexes with them. Thus, complexation would occur when the concentrations of metal or ligand are high and the predominant complexant would be inorganic ligands (anions), X.

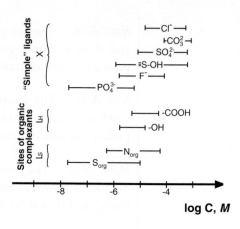

FIGURE 3.17. *Ranges of concentrations of ligands or complexing sites in natural freshwaters. ≡ S – OH refers to inorganic solid surface sites. – COOH and – OH refer to total concentrations of carboxyl and phenolic sites in natural organic matter. N_{org}, S_{org} refer to total concentrations of organic nitrogen and sulfur. Reprinted from Buffle (1984), by courtesy of Marcel Dekker, Inc.*

Group II metals, especially divalent transition metals, have affinity for both hard and soft sites and can react with all three groups of ligands (Fig. 3.17). These metals will compete for L_H sites with metals from Group I, which are less strongly bound but at higher concentrations, and for L_s sites with Group III metals, which are at lower concentrations but more strongly bound.

Group III metals have greater affinity for soft sites (L_s) than for hard sites (L_H) or X ligands.

Determination of Stability Constants of Metal–HS Complexes

The determination of stability constants for HS–metal complexes provides information on the affinity of the metal for the organic ligand, and they also provide valuable insights into the fate of heavy metals in the environment. Box 3.2 outlines how stability constants are calculated. Stability constants can be determined using potentiometric titration, chromatography, ultrafiltration, equilibrium dialysis, ion specific electrodes (ISE), differential pulse anodic stripping voltammetry (DPASV), fluorescence spectrometry, and modeling (Scatchard method). For more details on these methods the reader should consult Buffle (1984) and Thurman (1985).

Stability constants are affected by the source of the HS and extraction or isolation procedure employed, concentration of HS, ionic strength (see Chapter 4) of the solution, temperature, pH, method of analysis of the complex, and method of data manipulation and stability constant calculation (Thurman, 1985).

Schnitzer and Hansen (1970) calculated conditional stability constants (K_i^{cond}) for metal–FA complexes, based on continuous variations and ion exchange equilibrium methods, and found that the order of stability was $Fe^{3+} > Al^{3+} > Cu^{2+} > Ni^{2+} > Co^{2+} > Pb^{2+} > Ca^{2+} > Zn^{2+} > Mn^{2+} > Mg^{2+}$. The stability constants were slightly higher at pH 5.0 than at pH 3.5, which is due to the higher dissociation of functional groups, particularly carboxyl groups, at pH 5.0. Also, H^+ and the metal ions compete for binding sites on the ligand, and less metal is bound at the lower pH.

BOX 3.2 *Determination of Stability Constants for Metal–HS Complexes*

The complexation reaction between a metal ion M and the ith deprotonated ligand (L_i) or binding site in a multiligand mixture at constant pH can be written as (Perdue and Lytle, 1983)

$$M + L_i \rightleftharpoons ML_i. \tag{3.2a}$$

The stability constant for this reaction can be expressed as

$$K_i^{\text{cond}} = \frac{[ML_i]}{[M][L_i]}, \tag{3.2b}$$

where brackets indicate the concentration.

It is usually not possible to analytically determine L_i because of H$^+$ competition and the possibility that the ith metal binding ligand may bind several protons. A more realistic expression of Eq. (3.2b) is

$$K_i^{\prime\text{cond}} = \frac{[ML_i]}{[M][H_{x_i} L_i]}, \tag{3.2c}$$

where $[H_{x_i} L_i]$ is the concentration of all forms of the ith ligand not bound to M. Both K_i^{cond} and $K_i^{\prime\text{cond}}$ are referred to as conditional stability constants. Conditional stability constants are valid only for the conditions (i.e., pH, neutral salt concentration) stated for a particular system. By definition, the value of K_i^{cond} and $K_i^{\prime\text{cond}}$ will change with changes in pH or neutral salt concentration. Note also that $K_i^{\prime\text{cond}}$ is related to K_i^{cond} through the expression

$$K_i^{\prime\text{cond}} = \frac{[ML_i]}{[M][L_i]} \frac{[L_i]}{[H_{x_i} L_i]} = K_i^{\text{cond}} \frac{[L_i]}{[H_{x_i} L_i]}. \tag{3.2d}$$

The conditional stability constant $(K_i^{\prime\text{cond}}) = K_i^{\text{cond}}$ times the fraction of $[H_{x_i} L_i]$ not protonated. This fraction is constant at constant pH and neutral salt concentration.

It is not possible to determine individual values of K_i^{cond} or $K_i^{\prime\text{cond}}$ in complex mixtures of ligands, such as often occurs with natural systems. Rather, average K_i^{cond} ($\bar{K}^{\text{cond}}$) or $K_i^{\prime\text{cond}}$ ($\bar{K}^{\prime\text{cond}}$) values are determined from experimental data,

$$\bar{K}^{\text{cond}} = \frac{\sum_i [ML_i]}{[M] \sum_i [L_i]} = \frac{\sum_i K_i^{\text{cond}} [L_i]}{\sum_i [L_i]} \tag{3.2e}$$

$$\bar{K}^{\prime\text{cond}} = \frac{\sum_i [ML_i]}{[M] \sum_i [H_{x_i} L_i]} = \frac{\sum_i K_i^{\prime\text{cond}} [H_{x_i} L_i]}{\sum_i [H_{x_i} L_i]}$$

$$= \frac{(C_M - [M])}{[M]((C_L - C_M) + [M])}, \tag{3.2f}$$

where C_M and C_L are stoichiometric concentrations of metal and total ligand, respectively. Note that as before $\bar{K}'^{cond}$ is related to $\bar{K}^{cond}$ through an expression similar to Eq. (3.2d):

$$\bar{K}'^{cond} = \bar{K}^{cond} \frac{\sum_i [L_i]}{\sum_i [H_{x_i} L_i]}. \qquad (3.2g)$$

The values of $\bar{K}^{cond}$ stability constants are the easiest to determine experimentally, and are the most common values published in the scientific literature. Their dependence on the total concentration of metal and ligand present in a given system (Eq. 3.2f), however, reinforces the fact that these are conditional stability constants and are only valid for the environmental conditions under which they were determined. Application of the concept of stability constants for metal–HS complexes requires using average stability constants determined under similar environmental conditions.

Effect of HS–Metal Complexation on Metal Transport

The complexation of HS with metals can beneficially as well as deleteriously affect the fate of metals in soils and waters. The speciation (see Chapter 4) of metals can be affected by these complexes as well as oxidation–reduction reactions (see Chapter 8). For example, HA can act as a reducing agent and reduce Cr(VI), the more toxic form of Cr, to Cr(III). As a Lewis hard acid, Cr(III) will form a stable complex with the carboxyl groups on the HA, further limiting its availability for plant or animal uptake.

Humic substances can serve as carriers of toxic metals, forming complexes that are stable and enhance transport of toxic metals in waters. In water treatment scenarios, these stable, soluble complexes can diminish the removal effectiveness of heavy metals from leachate waters of hazardous waste sites when precipitation techniques are employed as a remediation technique (Manahan, 1991). Additionally, the efficiency of membrane processes (reverse osmosis) and resin processes (ion exchange) used to treat the leachate waters is diminished. On the other hand, the binding of heavy metals to HS can enhance biological treatment of inorganic contaminants by reducing the toxicity of the heavy metals to the microbes. Additionally, inorganic anions such as phosphate and cyanide can be removed from water as well by mixed ligand complexation.

Effect of HS–Al³⁺ Complexes on Plant Growth

The importance of HS–Al^{3+} complexes on plant growth is illustrated in studies that have shown that crops often grow well on soils with a pH of < 4 in the surface horizon (Evans and Kamprath, 1970; Thomas, 1975). Organic matter contents of the surface horizons, which were enhanced due to no-tillage practices, were fairly high (4–5%). Exchangeable Al^{3+} (Al^{3+} bound electrostatically on soil colloids such as clay minerals and SOM) and the activity of Al^{3+} (see Chapter 4 for a discussion of ion activity) in the soil solution were low.

Thomas (1975) studied $HS–Al^{3+}$ complexation effects on plant growth with a Maury silt loam soil from Kentucky with an average pH of 6.1 and an organic matter content ranging from 5.11% in the 0- to 7.5-cm layer to 0.80% in the 37.5- to 45-cm layer. Various amounts of 0.1 M HNO_3 were added such that the pH of the soils was 3.10–3.30. No exchangeable H^+ was found in the soils. The relationship between organic matter percentage and exchangeable Al^{3+} at several pH values is shown in Fig. 3.18. One sees that the influence of SOM on exchangeable Al^{3+} at any given pH is greatest up to 2.5% with a smaller effect at higher SOM contents. The effect of organic matter on exchangeable Al^{3+} was greater at the lower pH values. At pH 3.5, an increase from 1 to 2% organic matter lowered exchangeable Al from 6 to 4.2 cmol kg^{-1}. These results show that even small increases in SOM result in a significant reduction in exchangeable Al^{3+} and the activity of Al^{3+} in the soil solution. Therefore, the lack of any deleterious effect on plant growth even at low pH may be due to the complexation of Al^{3+} with SOM and the removal of Al^{3+} and perhaps other toxic metals from the soil solution.

Effects of HS on Mineral Dissolution

Humic substances can also cause mineral dissolution. Lichens, bacteria, and fungi, which often grow on mineral surfaces, enhance this breakdown by producing organic complexing materials. At pH > 6.5 HA and FA can attack and degrade minerals and form water-soluble and water-insoluble metal complexes (Schnitzer, 1986). Humic acids can extract metals from galena (PbS), pyrolusite (MnO_2), calcite ($CaCO_3$), and malachite ($Cu_2(OH)_2CO_3$). Humic acid can also extract metals from silicates (Baker, 1973), but to a lesser degree than from previously mentioned minerals. Various metal sulfides and carbonates can be solubilized by HA, including Pb(II), Zn(II), Cu(II), Ni, Fe(III), and Mn(IV). Solubilization ranges from 2100 µg for PbS to 95 µg for ZnS.

SOM–Clay Complexes

Soil organic matter complexes can also occur with clay minerals coated with metal oxides such as Al and Fe oxides. Useful reviews on this topic can be found in a number of sources (Mortland, 1970; Greenland, 1971; Mortland *et al.*, 1986; Schnitzer, 1986). Clays tend to stabilize SOM, and a correlation is often observed between the clay content and SOM. Some organic substances bridge soil particles together, resulting in stable aggregates. In some soils, particularly those high in humic substances, all of the clay may be coated with HS. It has been estimated that from 52 to 98% of all C in soils occurs as clay–SOM complexes (Stevenson, 1982). Even though such a complex exists, the organic surface is still reactive and can retain ions and other materials such as pesticides.

Other organic acids, organic amine cations, and amino acids can also interact with clays. Organic acids are negatively charged in the pH range of most soils but they are pH-dependently charged; some adsorption can occur through H bonding and van der Waals forces (see definitions in next section) when the pH is below the pK_a of the acidic group, and when the organic acid is in the undissociated form. Organic materials such as proteins and charged organic cations can also be adsorbed in the interlayers of expansible layer silicates like montmorillonite.

Mechanisms of Interactions

The types of interactions involved in SOM–clay complexes include physical adsorption or interactions via van der Waals forces, electrostatic interactions (cation and anion exchange processes), cation and anion bridges whereby the polyvalent metal forms a bridge between the organic molecule and the inorganic surface to which it is bound (clay mineral–metal–HS), often referred to as coordination complexes, chemical adsorption, and H bonding. Two or more of these mechanisms may occur simultaneously, depending on the type of organic material, the nature of the exchangeable ion on the clay surface, the surface acidity, and the moisture content of the system (Schnitzer, 1986).

Physical adsorption or retention via van der Waals forces is weak and results from changes in the electric charge density of individual atoms. An electrically positive fluctuation in one atom causes an electrically negative fluctuation in a neighboring atom, resulting in a net attractive force. Adsorption due to physical forces occurs with neutral polar and nonpolar molecules, particularly those with high molecular weight (Schnitzer, 1986).

Electrostatic bonding can occur via cation or anion exchange or protonation. The cationic property of a weakly basic organic molecule is strongly pH-dependent. Thus, this mechanism is dependent on the basic character of the organic molecule, the pH of the system, the properties of the organic cation or chain length, and the type of cation on the clay surface. Sorption

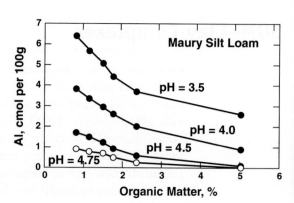

FIGURE 3.18. *The relation between percentage organic matter and exchangeable Al at several pH levels. From Thomas (1975), with permission.*

of HS on negatively charged clays occurs only when polyvalent metal cations such as Al^{3+}, Fe^{3+}, and Ca^{2+} are present on the clay exchanger since they can neutralize the negative charge on the clay and the charge on the deprotonated organic functional group, e.g., COO^-. Calcium is weakly held as a cation bridge and can be easily displaced, whereas Al^{3+} and Fe^{3+} are usually bound nonelectrostatically and are difficult to remove (Schnitzer, 1986). It is probably through this latter mechanism and through coordination that FA and HA are primarily retained in soils (Stevenson, 1982).

Hydrogen bonding results from linkage between two electronegative atoms through bonding with a single H^+ ion. The H^+ ion is a bare nucleus with a 1^+ charge and a propensity to share electrons with those atoms that contain an unshared electron pair such as O. The hydrogen bond is weaker than ionic or covalent bonds, but stronger than van der Waals attractive forces (Stevenson, 1982).

Retention of Pesticides and Other Organic Substances by Humic Substances

Pesticides have a strong affinity for SOM. Soil organic matter has an important effect on the bioactivity, persistence, biodegradability, leachability, and volatility of pesticides. In fact, perhaps SOM is the soil component most important in pesticide retention. Thus, the amount of pesticide that must be added to soils is strongly affected by the quantity of SOM.

Factors that affect the retention of pesticides by SOM are number, type, and accessibility of HS functional groups, nature of the pesticides, properties of the soil including types and quantities of clay minerals and the other soil components, pH, exchangeable cations, moisture, and temperature (Stevenson, 1982). Bailey and White (1970) noted several properties of pesticides that affect retention and type of bonding on soil components: (1) type of the functional group, i.e., carboxyl, alcoholic hydroxyl, amine; (2) nature of the substituting group which alters the functional group, and the position of the substituting groups in relation to the functional group which may affect bonding; and (3) the presence and magnitude of unsaturation in the molecule, which affects the lyophilic and lyophobic balance. These factors determine important chemical properties such as the acidity or basicity of the compound as suggested by the pK_a of the compound.

The mechanisms by which pesticides are retained by SOM are not clearly understood. However, the pesticide may be strongly retained in the internal voids of humus molecules that are sieve-like (Khan, 1973). Adsorption of pesticides on HS occurs via ion exchange and protonation, H bonding, van der Waals forces, ligand exchange (an organic functional group such as a carboxyl or hydroxyl displaces an inorganic hydroxyl or water molecule of a metal ion at the surface of a soil mineral such as a metal oxide, resulting in an inner-sphere complex; see discussions on ligand exchange and inner-sphere complexes in Chapter 5), and cation and water bridging. These mechanisms are discussed in detail in Koskinen and Harper (1990).

Humic substances can also combine with nonpolar organic compounds via a partitioning mechanism (see discussion in Chapter 5). This partitioning on hydrophobic SOM surfaces results from a weak solute–solvent interaction, or the low solubility or hydrophobic nature of the solute. Important hydrophobic sites on HS include fats, waxes, and resins and aliphatic side chains. Since humus has an aromatic framework and contains polar groups it may have both hydrophobic and hydrophilic sites. More details on pesticide and organic chemical reaction mechanisms with SOM are given in Chapter 5.

Soluble humic substances can enhance the transport of pesticides in soils and into groundwaters. Fulvic acids, which have low molecular weight and high acidities and are more soluble than HA, can transport pesticides and other organic materials quite effectively. For example, the downwater movement of the insecticide DDT in the organic layers of some forest soils has been ascribed to water-soluble, humic substances (Ballard, 1971).

Humic substances can also serve as reducing agents and chemically alter pesticides. The alteration is enhanced by the presence of reactive groups such as phenolic, carboxyl, enolic, heterocyclic, aliphatic-OH, and semiquinone like those contained in FA and HA. The presence of stable free radicals in HS would also indicate that they can effect chemical alterations of pesticides. The hydroxylation of the chloro-*s*-triazines is an example of nonbiological transformation of a pesticide by HS (Stevenson, 1982).

Suggested Reading

Aiken, G. R., McKnight, D. M., and Wershaw, R. L., Eds. (1985a). "Humic Substances in Soil, Sediments, and Water." Wiley–Interscience, New York.

Buffle, J. (1984). Natural organic matter and metal–organic interactions in aquatic systems. *In* "Metal Ions in Biological Systems" (H. Sigel, Ed.), pp. 165–221. Dekker, New York.

Buffle, J., and Stumm, W. (1994). General chemistry of aquatic systems. *In* "Chemical and Biological Regulation of Aquatic Systems" (J. Buffle and R. R. DeVitre, Eds.), pp. 1–42. CRC Press, Boca Raton, FL.

Cheng, H. H., Ed. (1990). "Pesticides in the Soil Environment: Processes, Impacts, and Modeling," SSSA Book Ser. 2. Soil Sci. Soc. Am., Madison, WI.

Christman, R. F., and Gjessing, E. T., Eds. (1983). "Aquatic and Terrestrial Humic Materials." Ann Arbor Sci., Ann Arbor, MI.

Greenland, D. L. (1971). Interactions between humic and fulvic acids and clays. *Soil Sci.* **111**, 34–41.

Hayes, M. H. B., MacCarthy, P., Malcolm, R. L., and Swift, R. S., Eds. (1989). "Humic Substances. II: In Search of Structure." Wiley, New York.

Huang, P. M., and Schnitzer, M., Eds. (1986). "Interactions of Soil Minerals with Natural Organics and Microbes." SSSA Spec. Publ. 17. Soil Sci. Soc. Am., Madison, WI.

Lal, R. (2001). World cropland soils as a source or sink for atmospheric carbon. *Adv. Agron.* **71**, 145–191.

Mortland, M. M. (1970). Clay–organic complexes and interactions. *Adv. Agron.* **22**, 75–117.

Schnitzer, M. (2000). A lifetime perspective on the chemistry of soil organic matter. *Adv. Agron.* **68**, 1–58.

Schnitzer, M., and Khan, S. U., Eds. (1972). "Humic Substances in the Environment." Dekker, New York.

Stevenson, F. J. (1982). "Humus Chemistry." Wiley, New York.

Suffet, I. H., and MacCarthy, P., Eds. (1989). "Aquatic Humic Substances. Influence of Fate and Treatment of Pollutants," Adv. Chem. Ser. Vol. 219. Am. Chem. Soc., Washington, DC.

Thurman, E. M. (1985). "Organic Geochemistry of Natural Waters." Kluwer Academic, Hingham, MA.

4 Soil Solution–Solid Phase Equilibria

Introduction

There are a number of complex chemical reactions that can occur in soils. These various reactions interact with each other through the soil solution (Fig. 4.1). The soil solution is defined as the aqueous liquid phase of the soil and its solutes (Glossary of Soil Science Terms, 1997). The majority of solutes in the soil solution is ions, which occur either as free hydrated ions (e.g., Al^{3+} which is expressed as $Al(H_2O)_6^{3+}$) or as various complexes with organic or inorganic ligands. When metal ions and ligands directly interact in solution (no water molecules are present between the metal ion and the ligand) they form what are known as inner-sphere complexes. An outer-sphere complex is formed when a water molecule is positioned between the metal ion and the ligand. Outer-sphere complexes are not as tightly bound as inner-sphere complexes. An uncharged outer-sphere complex is often referred to as an ion pair (e.g., Ca^{2+} plus SO_4^{2-} ions are known to form ion pairs, $CaSO_4^0$). Soil solutions, therefore, are composed of a variety of ion species, either complexed or noncomplexed. The speciation of the soil solution refers to determining the distribution of ions in their various chemical forms.

115

The soil solution is the medium from which plants take up ions (1 in Fig. 4.1) and in which plant exudates reside (2). Ions in the soil solution can be sorbed on organic and inorganic components of the soil (3) and sorbed ions can be desorbed (released) into the soil solution (4). If the soil solution is supersaturated with any mineral in the soil, the mineral can precipitate (5) until equilibrium is reached. If the soil solution becomes undersaturated with any mineral in the soil, the mineral can dissolve (6) until equilibrium is reached. Ions in the soil solution can be transported through the soil (7) into groundwater or removed through surface runoff processes. Through evaporation and drying upward movement of ions can also occur (8). Microorganisms can remove ions from the soil solution (9) and when the organisms die and organic matter is decomposed, ions are released to the soil solution (10). Gases may be released to the soil atmosphere (11) or dissolved in the soil solution (12).

Measurement of the Soil Solution

The concentration of a particular ion in the soil solution (intensity factor) and the ability of solid components in soils to resupply an ion that is depleted from the soil solution (capacity factor) are both important properties of a

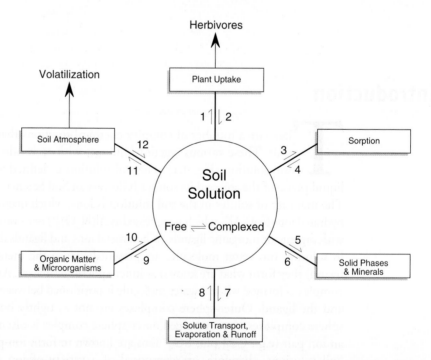

FIGURE 4.1. *Dynamic equilibria reactions in soils. From W. L. Lindsay, "Chemical Equilibria in Soils." Copyright © 1979 John Wiley & Sons. Reprinted by permission of John Wiley & Sons, Inc.*

given soil. The measurement of the intensity factor is critical in predicting and understanding reactions in the soil environment such as weathering and the bioavailability, mobility, and geochemical cycling of nutrients and inorganic and organic chemicals in soils (Wolt, 1994). However, the measurement of soil solution components *in situ* is very difficult, and, technically, not possible for most ion species. In addition, the actual concentration of the ion species in the soil solution changes with changes in soil moisture content (Wolt, 1994).

A number of laboratory methods are used to obtain samples of the soil solution. These are described in detail by Adams (1971) and Wolt (1994) and will be discussed only briefly here. The methods can be classified as displacement techniques including column displacement (pressure or tension displacement, with or without a displacing solution), centrifugation (perhaps with an immiscible liquid), and saturation extracts (e.g., saturation pastes).

In the column displacement method, a field moist soil is packed into a glass column and a displacing solution is added to the soil. The displaced soil solution is collected at the bottom of the column and analyzed for its ion content. The column displacement method is probably the most reliable method for obtaining a sample of the soil solution. However, the procedure is tedious, requiring an analyst who is experienced in packing the column with soil. Furthermore, the procedure is very time-consuming.

The centrifugation method uses an immiscible liquid, such as carbon tetrachloride or ethyl benzoylacetate, to physically remove the soil solution (which is essentially water) from the solid matrix of the moist soil. During centrifugation (>1 MPa for from 30 min to 2 hr), the immiscible liquid passes through the soil and soil solution is displaced upward. The displaced soil solution is then decanted and filtered through a phase-separating filter paper (to remove traces of the immiscible liquid) and then through a 0.2-μm membrane filter to remove suspended clay particles (Wolt, 1994). Centrifugation methods are the most widely used because they are relatively simple and equipment to carry out the procedure is usually available.

In the saturation extract method a saturated soil paste is prepared by adding distilled water to a soil sample in a beaker, stirring with a spatula, and tapping the beaker occasionally to consolidate the soil–water mixture. At saturation, the soil paste should glisten. Prior to extraction the saturated soil paste should be allowed to stand for 4–16 hr. Then, the soil paste is transferred to a funnel containing filter paper. The funnel is connected to a flask, vacuum is applied, and the filtrate is collected.

Table 4.1 shows the average ion composition of soil solutions for soils from around the world. One will note that Ca is the most prevalent metal cation in the soil solution, which is typical for most soils. Nitrate, chloride, and sulfate are the common anions (ligands) in more acidic soils while carbonate can be important in some basic soils.

TABLE 4.1. *Total Ion Composition of Soil Solutions from World Soils*[a]

Location of soils	pH	Total ion composition of soil solution (mmol liter^{-1})										
		Metals							Ligands			
		Ca	Mg	K	Na	NH_4	Al	Si	HCO_3	SO_4	Cl	NO_3
California[b]	7.43	12.86	2.48	2.29	3.84	0.86	—	—	3.26	5.01	3.07	21.36
Georgia[c]	6.15	0.89	0.29	0.07	0.16	—	0.0004	0.13	—	1.18	0.21	0.10
United Kingdom[d]	7.09	1.46	0.12	0.49	0.31	—	0.01	0.34	—	0.32	0.75	0.69
Australia[e]	5.75	0.27	0.40	0.38	0.41	2.60	—	—	0.85	0.87	1.67	0.32

[a] Adapted from J. Wolt, "Soil Solution Chemistry." Copyright © 1994 John Wiley & Sons. Adapted by permission of John Wiley & Sons, Inc.
[b] From Eaton *et al.* (1960); soil solution obtained by pressure plate extraction at 10 kPa moisture. The pH and total composition for each ion represent the mean for seven soils.
[c] From Gillman and Sumner (1987); soil solution obtained by centrifugal displacement at 7.5 kPa moisture. The pH and total composition for each ion represent the mean for four Ap (surface) horizons.
[d] From Kinniburgh and Miles (1983); soil solution obtained by centrifugal displacement at field moisture contents. The pH and total composition for each ion represent the mean for 10 surface soils.
[e] From Gillman and Bell (1978); soil solution obtained by centrifugal displacement at 10 kPa moisture. The pH and total composition for each ion represent the mean value for six soils.

Speciation of the Soil Solution

While total ion concentration (sum of free and complexed ions) of the soil solution, as measured by analytical techniques such as spectrometry, chromatography, and colorimetry, provides important information on the quantities of ions available for plant uptake and movement through the soil profile, it is important that the speciation or chemical forms of the free and complexed ions be known. Ions in the soil solution can form a number of species due to hydrolysis, complexation, and redox (see Chapter 8) reactions. Hydrolysis reactions involve the splitting of a H^+ ion from a water molecule which forms an inner-sphere complex with a metal ion (Baes and Mesmer, 1976). The general hydrolysis reaction for a hydrated metal ($[M(H_2O)_x]^{n+}$) in solution is

$$[M(H_2O)_x]^{n+} \rightleftharpoons [M(OH)_y\,(H_2O)_{x-y}]^{(n-y)+} + \gamma H^+, \tag{4.1}$$

where n^+ refers to the charge on the metal ion, x the coordination number, and y the number of H^+ ions released to solution. The degree of hydrolysis for a solvated metal ion [extent of the reaction in Eq. (4.1)] is a function of pH. A general complexation reaction between a hydrated metal $[(M(H_2O)_x]^{n+}$ and a ligand, L^{l-}, to form an outer-sphere complex is

$$a[M(H_2O)_x]^{n+} + bL^{l-} \rightleftharpoons [M(H_2O)_x]_a\,L_b, \tag{4.2}$$

where *a* and *b* are stoichiometric coefficients.

Analytically, it is not possible to determine all the individual ion species that can occur in soil solutions. To speciate the soil solution, one must apply ion association or speciation models using the total concentration data for each metal and ligand in the soil solution. A mass balance equation for the total concentration of a metal, M_T^{n+} in the solution phase can be written as

$$M_T^{n+} = M^{n+} + M_{ML},\qquad(4.3)$$

where M^{n+} represents the concentration of free hydrated-metal ion species, and M_{ML} is the concentration of metal ion associated with the remaining metal–ligand complexes. The ligands in the metal–ligand complexes could be inorganic anions or humic substances such as humic and fulvic acids. Similar mass balance equations can be written for the total concentration of each metal and ligand in the soil solution.

The formation of each metal–ligand complex in the soil solution can be described using an expression similar to Eq. (4.2) and a conditional stability or conditional equilibrium constant, K^{cond}. As mentioned in the previous chapter, conditional stability constants vary with the composition and total electrolyte concentration of the soil solution. Conditional stability constants for inorganic complexes, however, can be related to thermodynamic stability or thermodynamic equilibrium constants (K^o) which are independent of chemical composition of the soil solution at a particular temperature and pressure. Box 4.1 illustrates the relationship between thermodynamic and conditional equilibrium constants.

BOX 4.1   *Thermodynamic and Conditional Equilibrium Constants*

For any reaction (Lindsay, 1979)

$$aA + bB \rightleftharpoons cC + dD,\qquad(4.1a)$$

a thermodynamic equilibrium constant, K^o, can be written as

$$K^o = \frac{(C)^c\,(D)^d}{(A)^a\,(B)^b},\qquad(4.1b)$$

where the small superscript letters refer to stoichiometric coefficients and A, B, C, and D are chemical species. The parentheses in Eq. (4.1b) refer to the chemical activity of each of the chemical species.

Thermodynamic equilibrium constants are independent of changes in the electrolyte composition of soil solutions because they are expressed in terms of activities, not concentrations. Unfortunately, the activities of ions in solutions generally cannot be measured directly as opposed to their concentration. The activity and concentration of an ion in solution can be related, however, through the expression

$$(B) = [B]\,\gamma_B,\qquad(4.1c)$$

where $[B]$ is the concentration of species B in mol liter^{-1} and γ_B is a single ion activity coefficient for species B.

Using the relationship in Eq. (4.1c), one can rewrite Eq. (4.1b) in terms of concentrations rather than activities,

$$K^\circ = \frac{[C]^c \, \gamma_C^c \, [D]^d \, \gamma_D^d}{[A]^a \, \gamma_A^a \, [B]^b \, \gamma_B^b} = \frac{[C]^c \, [D]^d}{[A]^a \, [B]^b} \left(\frac{\gamma_C^c \, \gamma_D^d}{\gamma_A^a \, \gamma_B^b} \right) = K^{cond} \, \frac{\gamma_C^c \, \gamma_D^d}{\gamma_A^a \, \gamma_B^b} , \quad (4.1d)$$

where K^{cond} is the conditional equilibrium constant. The ratio of the activity coefficients is the correct term that relates K^{cond} to K°. The value of the activity coefficients, and therefore the value of the ratio of the activity coefficients, changes with changes in the electrolyte composition of the soil solution. In very dilute solutions, the value of the activity coefficients approaches 1, and the conditional equilibrium constant equals the thermodynamic equilibrium constant.

To illustrate how mass balance equations like Eq. (4.3) and conditional equilibrium constants can be used to speciate a metal ion in a soil solution, let us calculate the concentration of the different Ca species present in a typical soil solution from a basic soil. Assume that the pH of a soil solution is 7.9 and the total concentration of Ca (Ca_T) is 3.715×10^{-3} mol liter^{-1}. For simplicity's sake, and realizing that other complexes occur, let us assume that Ca exists only as free hydrated Ca^{2+} ions and as complexes with the inorganic ligands CO_3, SO_4, and Cl (i.e., $[CaCO_3^0]$, $[CaHCO_3^+]$, $[CaSO_4^0]$, and $[CaCl^+]$). The Ca_T can be expressed as the sum of free and complexed forms as

$$Ca_T = [Ca^{2+}] + [CaCO_3^0] + [CaHCO_3^+] + [CaSO_4^0] + [CaCl^+], \quad (4.4)$$

where brackets indicate species concentrations in mol liter^{-1}. Each of the above complexes can be described by a conditional stability constant,

$$K_1^{cond} = \frac{[CaCO_3^0]}{[Ca^{2+}][CO_3^{2-}]} \quad (4.5)$$

$$K_2^{cond} = \frac{[CaHCO_3^+]}{[Ca^{2+}][HCO_3^-]} \quad (4.6)$$

$$K_3^{cond} = \frac{[CaSO_4^0]}{[Ca^{2+}][SO_4^{2-}]} \quad (4.7)$$

$$K_4^{cond} = \frac{[CaCl^+]}{[Ca^{2+}][Cl^-]}. \quad (4.8)$$

Since Eqs. (4.5)–(4.8) have $[Ca^{2+}]$ as a common term, Eq. (4.4) can be rewritten in terms of the concentrations of $[Ca^{2+}]$ and each of the inorganic ligands (Sposito, 1989):

$$Ca_T = [Ca^{2+}] \left\{ 1 + \frac{[CaCO_3^0]}{[Ca^{2+}]} + \frac{[CaHCO_3^+]}{[Ca^{2+}]} \right.$$

$$\left. + \frac{[CaSO_4^0]}{[Ca^{2+}]} + \frac{[CaCl^+]}{[Ca^{2+}]} \right\} \quad (4.9)$$

$$= [Ca^{2+}] \{1 + K_1^{cond} [CO_3^{2-}] + K_2^{cond} [HCO_3^-]$$

$$+ K_3^{cond} [SO_4^{2-}] + K_4^{cond} [Cl^-]\}.$$

As noted earlier, mass balance equations like Eq. (4.3) can be expressed for total metal and ligand concentration, M_T^{n+} and L_T^{l-}, respectively, for each metal and ligand species. The mass balance expressions are transformed into coupled algebraic equations like Eq. (4.9) with the free ion concentrations as unknowns by substitution for the complex concentrations. The algebraic equations can be solved iteratively using successive approximation to obtain the free-ion concentrations by using a number of possible computer equilibrium models (Table 4.2). A common iterative approach is the Newton-Raphson algorithm. The computer equilibrium models contain thermodynamic databases (e.g., Martell and Smith, 1976) and computational algorithms. Thorough discussions of these models can be found in Jenne (1979), Melchior and Bassett (1990), and Mattigod and Zachara (1996). The thermodynamic databases contain equilibrium constants for aqueous complex species as well as other equilibrium parameters (e.g., solubility products and redox potential, both of which are discussed later).

The free ion concentrations obtained from the iterative approach can then be used to calculate the concentrations of the metal–ligand complexes. For example, the complex $CaHCO_3^+$ forms based on the reaction

$$Ca^{2+} + H^+ + CO_3^{2-} \rightarrow CaHCO_3^+. \tag{4.10}$$

The K_2^{cond} for $CaHCO_3^+$ is 9.33×10^{10} at 298K. Substituting this value into Eq. (4.6),

TABLE 4.2. *Computer Equilibrium Models Used in Geochemical Research[a]*

Chemical Speciation Models
  REDEQL series models[b]
    REDEQL (Morel and Morgan, 1972)
    REDEQL2 (McDuff *et al.*, 1973)
    MINEQL (Westall *et al.*, 1976)
    GEOCHEM (Sposito and Mattigod, 1980)
    GEOCHEM-PC (Parker *et al.*, 1995)
    REDEQL.EPA (Ingle *et al.*, 1978)
    MICROQL (Westall, 1979)
    REDEQL.UMD. (Harris *et al.*, 1981)
    MINTEQ (Felmy *et al.*, 1984)
    SOILCHEM (Sposito and Coves, 1988)

  WATCHEM series models[c]
    WATCHEM (Barnes and Clarke, 1969)
    SOLMNEQ (Kharaka and Barnes, 1973)
    WATEQ (Truesdall and Jones, 1974)
    WATEQ2 (Ball *et al.*, 1979)
    WATEQ3 (Ball *et al.*, 1981)

[a] Adapted from Baham (1984) and S. V. Mattigod and J. M. Zachara (1996), with permission.
[b] These models use an equilibrium constant approach for describing ion speciation and computing ion activities.
[c] These models use the Gibbs free energy minimization approach.

$$9.33 \times 10^{10} = \frac{[CaHCO_3^+]}{[Ca^{2+}][H^+][CO_3^{2-}]} , \qquad (4.11)$$

and solving for $[CaHCO_3^+]$ yields

$$[CaHCO_3^+] = 9.33 \times 10^{10} [Ca^{2+}][H^+][CO_3^{2-}]. \qquad (4.12)$$

Once the free concentrations of Ca^{2+}, H^+, and CO_3^{2-} are determined, the concentration of the complex, $[CaHCO_3^+]$, can be calculated using Eq. (4.12).

The calculated concentrations for the various species can be checked by inserting them into the mass balance equation [Eq. (4.3)] and determining if the sum is equal to the known total concentration. This can be illustrated for Ca using the data in Table 4.3. Table 4.3 shows the calculated concentrations of seven metals, six ligands, and metal–ligand complexes (99 total complexes were assumed to be possible) of a basic soil solution. To compare the total measured Ca concentration, Ca_T, to the sum of the calculated Ca species concentrations, one would use the following mass balance equation,

$$Ca_T = [Ca^{2+}] + [Ca–CO_3 \text{ complexes}] + [Ca–SO_4 \text{ complexes}]$$

$$+ [Ca–Cl \text{ complexes}] + [Ca–PO_4 \text{ complexes}]$$

$$+ [Ca–NO_3 \text{ complexes}] + [Ca–OH \text{ complexes}]. \qquad (4.13)$$

Using the data in Table 4.3,

$$Ca_T = 3.715 \times 10^{-3} = [3.395 \times 10^{-3}] + [8.396 \times 10^{-5}]$$

$$+ [1.871 \times 10^{-4}] + [1.607 \times 10^{-8}] + [9.050 \times 10^{-6}]$$

$$+ [2.539 \times 10^{-5}] + [3.600 \times 10^{-7}],$$

where all concentrations are expressed in mol liter^{-1},

$$Ca_T = 3.715 \times 10^{-3} \simeq 3.717 \times 10^{-3}. \qquad (4.14)$$

One sees Ca_T compares well to the sum of free and complexed Ca species calculated from the GEOCHEM-PC (Parker et al., 1995) chemical speciation model (Table 4.2).

Table 4.4 shows the primary distribution of free metals and ligands and metal–ligand complexes for the soil solution data in Table 4.3. One sees that Ca, Mg, K, and Na exist predominantly as free metals while Cu mainly occurs complexed with CO_3 and Zn is primarily speciated as the free metal (42.38%) and as complexes with CO_3 (49.25%). The ligands Cl and NO_3 (>98%) occur as free ions, while about 20% of SO_4 and 10% of PO_4 are complexed with Ca. These findings are consistent with the discussion in Chapter 3 (see Box 3.1) on hard and soft acids and bases. Hard acids such as Ca, Mg, K, and Na, and hard bases such as Cl and NO_3, would not be expected to form significant complexes except at high concentrations of the metal or ligand (Cl or NO_3). However, soft acids such as Cu and Zn can form complexes with hard bases such as CO_3.

TABLE 4.3. Concentrations of Free Metals, Free Ligands, Total Metal Complexed by Each Ligand, and Total Concentrations of Metals and Ligands for a Soil Solution[a,b]

	CO_3	SO_4	Cl	PO_4	NO_3	OH	Total metals
Free ligands	1.297×10^{-5}	7.607×10^{-4}	1.971×10^{-3}	6.375×10^{-10}	2.473×10^{-3}	8.910×10^{-7}	
Free metals			Total metal complexed by each ligand[c]				
Ca 3.395×10^{-3}	8.396×10^{-5}	1.871×10^{-4}	1.607×10^{-8}	9.050×10^{-6}	2.539×10^{-5}	3.600×10^{-7}	3.715×10^{-3}
Mg 5.541×10^{-4}	9.772×10^{-6}	2.399×10^{-5}	2.090×10^{-6}	1.995×10^{-6}	1.047×10^{-5}	9.500×10^{-8}	6.026×10^{-4}
K 9.929×10^{-4}	1.412×10^{-6}	3.630×10^{-6}	7.580×10^{-7}	8.300×10^{-8}	1.202×10^{-6}	2.188×10^{-10}	1.000×10^{-3}
Na 4.126×10^{-3}	1.148×10^{-5}	2.398×10^{-5}	5.011×10^{-6}	4.360×10^{-7}	1.995×10^{-6}	1.000×10^{-9}	4.169×10^{-3}
Cu 3.224×10^{-8}	1.659×10^{-6}	2.000×10^{-9}	1.047×10^{-10}	2.900×10^{-8}	1.202×10^{-10}	1.800×10^{-8}	1.738×10^{-6}
Zn 5.335×10^{-6}	6.165×10^{-6}	3.710×10^{-7}	7.000×10^{-9}	1.090×10^{-8}	1.900×10^{-8}	5.490×10^{-7}	1.259×10^{-5}
H^+ 1.400×10^{-7}	2.512×10^{-2}	6.610×10^{-10}	8.710×10^{-20}	1.230×10^{-5}	1.097×10^{-12}		
Total ligands:	2.570×10^{-3}	1.000×10^{-3}	1.995×10^{-3}	2.188×10^{-5}	2.512×10^{-3}		

[a] Adapted, with permission, from S. V. Martigod and J. M. Zachara (1996) using the GEOCHEM-PC (Parker et al., 1995) chemical speciation model.
[b] All concentrations are expressed as mol liter^{-1}.
[c] The concentrations of the metals complexed with the ligands are those data within the area marked by lines and designated as total metal complexed by each ligand. For example, 8.396×10^{-5} mol liter^{-1} Ca is complexed with CO_3.

TABLE 4.4.   *Primary Distribution of Free Metals and Ligands and Metal–Ligand Complexes for a Soil Solution*[a]

Metal	Free Metal	Metal–ligand complexes (%)					
		CO$_3$	SO$_4$	Cl	PO$_4$	NO$_3$	OH
Ca	91.36[b]	2.26	5.03	0.43	0.23	0.68	—
Mg	91.95	1.60	4.02	0.35	0.33	1.72	0.03
K	99.29	0.14	0.36	0.08	—	0.12	—
Na	98.97	0.27	0.58	0.12	0.01	0.05	—
Cu	1.86	95.26	0.13	—	1.68	—	1.06
Zn	42.38	49.25	2.94	0.06	0.87	0.16	4.34

Ligand	Free ligand	Metal–ligand complexes (%)						
		Ca	Mg	K	Na	Cu	Zn	H
CO$_3$	0.50[b]	3.26	0.38	0.06	0.44	0.06	0.24	95.05
SO$_4$	76.07	18.71	2.43	0.36	2.40	—	0.04	—
Cl	98.80	0.80	0.10	0.04	0.25	—	—	—
PO$_4$	38.22	9.19	0.38	0.38	1.98	0.13	0.50	49.59
NO$_3$	98.45	1.01	0.41	0.05	0.08	—	—	—

[a] Distribution for soil solution data in Table 4.3, from S. V. Mattigod and J. M. Zachara (1996), with permission.
[b] Percentages represent the amount of metal or ligand as a free or complexed species. For example, 91.36% of Ca occurs as the free metal ion while 2.26% occurs as a Ca–CO$_3$ metal–ligand complex, and 5.03, 0.43, 0.23, and 0.68% occur as Ca–SO$_4$, Ca–Cl, Ca–PO$_4$, and Ca–NO$_3$ metal–ligand complexes.

Ion Activity and Activity Coefficients

As illustrated in Box 4.1, the activity of an ionic species in solution is related to its concentration in solution through its single-ion activity coefficients [Eq. (4.1c)]. The single-ion activity coefficients are typically calculated using the extended (to account for the effective size of hydrated ions) Debye-Hückel equation (Stumm and Morgan, 1996)

$$\log \gamma_i = -A\, Z_i^2 \left(\frac{I^{1/2}}{1 + Ba_i\, I^{1/2}} \right). \tag{4.15}$$

The A term typically has a value of 0.5 at 298K and is related to the dielectric constant of water. The B term is also related to the dielectric constant of water and has a value of ≈ 0.33 at 298K. The A and B parameters are also dependent on temperature. The a_i term is an adjustable parameter corresponding to the size (Å) of the ion. The Z_i term refers to the valence (charge) of the ion in solution, and I represents the ionic strength. Ionic strength is related to the total electrolyte concentration in solution, and is a measure of the degree of interaction between ions in solution, which, in turn, influences the activity of ions in solution. The ionic strength is calculated using the expression

$$I = \frac{1}{2} \Sigma_i C_i Z_i^2, \tag{4.16}$$

where C_i is the concentration in mol liter^{-1} of the ith species and Z_i is its charge or valence. The Σ extends over all the ions in solution. Values for the a_i parameter (in nm) in Eq. (4.15) and single-activity coefficients at various I values for an array of inorganic ions are listed in Table 4.5. Sample calculations for I and γ_i are given in Box 4.2.

The extended Debye-Hückel equation [Eq. (4.15)] is one of several equations that have been derived to account for ion–ion interactions in aqueous solutions (Stumm and Morgan, 1996). Single activity coefficients calculated using the extended Debye-Hückel equation begin to deviate from actual measurements, however, when the I of the soil solution is >0.2 mol liter^{-1}.

TABLE 4.5. *Values for Single-Ion Activity Coefficients at Various Ionic Strengths at 298K Calculated Using Eq. (4.15)*[a]

Ion	a_i (nm)[b]	Ionic strength, I (mol liter^{-1})				
		0.001	0.005	0.01	0.05	0.1
Inorganic ions of charge 1$^+$						
H^+	0.90	0.967	0.933	0.914	0.860	0.830
Li^+	0.60	0.965	0.929	0.907	0.835	0.800
Na^+	0.45	0.964	0.928	0.902	0.820	0.775
HCO_3^-, $H_2PO_4^-$, $H_2AsO_4^-$	0.40	0.964	0.927	0.901	0.815	0.770
OH^-, F^-, ClO_4^-, MnO_4	0.35	0.964	0.926	0.900	0.810	0.760
K^+, Cl^-, Br^-, I^-, CN^-, NO_2^-, NO_3^-	0.30	0.964	0.925	0.899	0.805	0.755
Rb^+, Cs^+, NH_4^+, Tl^+, Ag^+	0.25	0.964	0.924	0.898	0.800	0.750
Inorganic ions of charge 2$^+$						
Mg^{2+}, Be^{2+}	0.80	0.872	0.755	0.690	0.520	0.450
Ca^{2+}, Cu^{2+}, Zn^{2+}, Sn^{2+}. Fe^{2+}, Ni^{2+}, Co^{2+}	0.60	0.870	0.749	0.765	0.485	0.405
Sr^{2+}, Ba^{2+}, Ra^{2+}, Cd^{2+}, Hg^{2+}, S^{2-}	0.50	0.868	0.744	0.670	0.465	0.380
Pb^{2+}, CO_3^{2-}, MoO_4^{2-}	0.45	0.867	0.742	0.665	0.455	0.370
Hg_2^{2+}, SO_4^{2-}, SeO_4^{2-}, CrO_4^{2-}, HPO_4^{2-}	0.40	0.867	0.740	0.660	0.455	0.355
Inorganic ions of charge 3$^+$						
Al^{3+}, Fe^{3+}, Cr^{3+}, Se^{3+}, Y^{3+}, La^{3+}, In^{3+}, Ce^{3+}, Pr^{3+}, Nd^{3+}, Sm^{3+}	0.90	0.730	0.540	0.445	0.245	0.180
PO_4^{3-}	0.40	0.725	0.505	0.395	0.160	0.095

[a] Reprinted with permission from Kielland (1937). Copyright © 1937 American Chemical Society.
[b] Adjustable parameter corresponding to size of ion.

BOX 4.2   *Calculation of Ionic Strength and Single-Ion Activity Coefficients*

1. Calculate the I of 0.10 mol liter^{-1} KNO$_3$ and 0.10 mol liter^{-1} K$_2$SO$_4$ solutions.

   I for 0.10 mol liter^{-1} KNO$_3$:

   $$I = {}^1\!/_2\, \Sigma_i\, C_i\, Z_i^2$$

   $$I = {}^1\!/_2\, \{(0.10)(+1)^2 + (0.10)(-1)^2\} = 0.10 \text{ mol liter}^{-1}.$$

   I for 0.10 mol liter^{-1} K$_2$SO$_4$:

   $$I = {}^1\!/_2\, \Sigma_i\, C_i\, Z_i^2$$

   $$I = {}^1\!/_2\, \{2(0.10)(+1)^2 + (0.10)(-2)^2\} = 0.30 \text{ mol liter}^{-1}.$$

   KNO$_3$ is a 1:1 electrolyte since the cation and anion both have a valence of 1. With these simple electrolytes, I equals the molarity of the solution. For other stoichiometries (e.g., K$_2$SO$_4$, a 2:1 electrolyte), I is greater than the molarity of the solution (Harris, 1987).

2. Calculate γ_i at 298K for Cd^{2+} when I = 0.01 mol liter^{-1}. Using Table 4.5, one sees that γ_{Cd2+} is 0.67. One can also calculate $\gamma_{Cd^{2+}}$ using Eq. (4.15) as

   $$\log \gamma_{Cd^{2+}} = \frac{(-0.5)(2)^2(0.01)^{1/2}}{1 + (.33)(0.5 \times 10)*(0.01)^{1/2}}$$

   $$\log \gamma_{Cd^{2+}} = -0.1717$$

   $$\gamma_{Cd^{2+}} = 0.67.$$

* The value of a$_i$ in Eq. (4.15) is reported in Å. To convert from nm (unit used for a$_i$ in Table 4.5) to Å, multiple by 10.

For more concentrated solutions (I <0.5 mol liter^{-1}), the Davies (1962) equation is often used:

$$\log \gamma_i = -A\, Z_i^2 \left(\frac{I^{1/2}}{1 + I^{1/2}} - 0.3I \right). \tag{4.17}$$

Special equations for calculating single-ion activity coefficients in more concentrated solutions (I > 0.5 mol liter^{-1}) also exist, but are beyond the scope of this chapter. Such concentrated solutions are seldom, if ever, encountered in soil solutions.

Dissolution and Solubility Processes

Dissolution (the dissolving of a soil solid) and precipitation (the process of depositing a substance from the soil solution) are the chemical reactions that determine the fate of inorganic mineral components in soils. Soil formation, weathering processes, and contaminant mobility are all affected by the dissolution–precipitation equilibria of the solid phase. Precipitation of minerals occurs only when supersaturation conditions exist in the soil solution (the soil solution contains more solute than should be present at equilibrium), while dissolution occurs only when the soil solution is undersaturated with respect to the inorganic mineral components in the soil, which means that these soils are in a continuous state of dissolution.

Calcareous minerals such as gypsum dissolve relatively rapidly and congruently, i.e., after dissolution the stoichiometric proportions in the solution are the same as those in the dissolving mineral. Many soil weathering reactions are incongruent, meaning dissolution is nonstoichiometric. This incongruency is often caused by precipitation of a new solid phase (Stumm, 1992), either separately or on the surface of the original mineral that is undergoing weathering.

Silicate minerals dissolve to form various secondary minerals that are important in soils. Overall weathering reactions that can result in the formation of secondary minerals are shown below (Stumm and Wollast, 1990):

$$CaAl_2Si_2O_8 + 2CO_2 + 3H_2O = Al_2Si_2O_5(OH)_4 + Ca^{2+} + 2HCO_3^-,$$
$$\text{(anorthite)} \qquad\qquad\qquad \text{(kaolinite)} \qquad\qquad\qquad (4.18)$$

$$2NaAlSi_3O_8 + 2CO_2 + 6H_2O$$
$$\text{(albite)}$$

$$= Al_2Si_4O_{10}(OH)_2 + 2Na^+ + 2HCO_3^- + 2H_4SiO_4. \qquad (4.19)$$
$$\text{(smectite)}$$

$$KMgFe_2AlSi_3O_{10}(OH)_2 + {}^1\!/_2\,O_2 + 3CO_2 + 11H_2O$$
$$\text{(biotite)}$$

$$= Al(OH)_3 + 2Fe(OH)_3 + K^+ + Mg^{2+} + 3HCO_3^-$$
$$\text{(gibbsite)} \quad \text{(goethite)}$$

$$+ 3H_4SiO_4. \qquad\qquad\qquad (4.20)$$

These reactions illustrate that water and carbonic acid are the major reactants and through dissolution Ca^{2+}, Mg^{2+}, K^+, Na^+, and HCO_3^- are released to the soil solution. The extent to which these reactions occur in actual soils depends upon the soil forming factors.

Thermodynamic dissolution constants (K_{dis}^o) can be calculated for mineral reactions. For example, the dissolution of gibbsite can be expressed as

$$Al(OH)_3 + 3H^+ \rightleftharpoons Al^{3+} + 3H_2O. \qquad (4.21)$$
$$\text{(gibbsite)}$$

The K_{dis}^o for this reaction can be expressed as

$$K_{dis}^o = (Al^{3+})\,(H_2O)^3/(Al(OH)_3)(H^+)^3. \qquad (4.22)$$

The activity of $Al(OH)_3$ or any other solid phase is defined as 1 if it exists in a pure form (has no structural imperfections) and is at standard temperature (T = 298.15 K) and pressure (0.101 MPa). (If the solid phase is not pure, as is often the case in soils, one cannot equate activity to 1.) The activity of water is also assumed to be 1 since soil solutions are essentially infinitely dilute compared to the concentration of water molecules. Assuming an activity of 1 for both the solid phase and water, Eq. (4.22) reduces to

$$K_{dis}^o = (Al^{3+})/(H^+)^3 = K_{so}^o. \qquad (4.23)$$

The K_{so}^o term is the thermodynamic solubility product constant (Sposito, 1989) and is numerically equal to K_{dis}^o when the solid phase is pure (there are no structural imperfections) and the aqueous solution phase is infinitely dilute (activity of water = 1). The K_{so}^o values for a number of inorganic mineral components found in soils are known and have been tabulated (Lindsay, 1979).

The right side of Eq. (4.23) is often referred to as the ion activity product (IAP) and, together with the K_{so}^o, is an index of whether the soil solution is in equilibrium with a given inorganic mineral component. If the ratio of the $IAP/K_{so}^o = 1$, the soil solution is said to be in equilibrium with a given soil solid phase. If the ratio of $IAP/K_{so}^o > 1$, the solution is supersaturated with respect to the solid phase. A ratio of $IAP/K_{so}^o < 1$ indicates that the soil solution is undersaturated.

Stability Diagrams

A more useful and informative way of determining whether a solid phase controls the concentration of an element in the soil solution, and if it does, what the solid is, is to use a stability diagram. Stability diagrams are constructed by converting solubility relationships such as Eq. (4.23) into log terms and rearranging terms to form a straight line relationship.

An example of this process for gibbsite dissolution is shown below. The log K_{dis}^o value can be calculated from the standard free energy accompanying the reaction (ΔG_r^o), which is equal to (Lindsay, 1979)

$$\Delta G_r^o = \Sigma\, \Delta G_f^o\,\text{products} - \Sigma\, \Delta G_f^o\,\text{reactants},$$

where $\Sigma\, \Delta G_f^o$ is the standard free energy of formation.

For Reaction 7, in Table 4.6,

$$\gamma\text{-Al(OH)}_3(\text{gibbsite}) + 3H^+ \rightleftharpoons Al^{3+} + 3H_2O, \qquad (4.24)$$

using published ΔG_f^o values (Lindsay, 1979) one finds that

$$\Delta G_r^o = [(-491.61 \text{ kJ mol}^{-1}) + 3(-237.52 \text{ kJ mol}^{-1})]$$

$$- [(-1158.24 \text{ kJ mol}^{-1}) + 3(0 \text{ kJ mol}^{-1})] \qquad (4.25)$$

TABLE 4.6. *Equilibrium Reactions of Aluminum Oxides and Hydroxides at 298K with Corresponding log K_{dis}^o* [a]

Reaction number	Equilibrium reaction	log K_{dis}^o
1	$0.5\ \gamma - Al_2O_3(c) + 3H^+ \rightleftharpoons Al^{3+} + 1.5\ H_2O$	11.49
2	$0.5\ \alpha - Al_2O_3(corundum) + 3H^+ \rightleftharpoons Al^{3+} + 1.5\ H_2O$	9.73
3	$Al(OH)_3(amorphous) + 3H^+ \rightleftharpoons Al^{3+} + 3H_2O$	9.66
4	$\alpha - Al(OH)_3(bayerite) + 3H^+ \rightleftharpoons Al^{3+} + 3H_2O$	8.51
5	$\gamma - AlOOH(boehmite) + 3H^+ \rightleftharpoons Al^{3+} + 2H_2O$	8.13
6	$Al(OH)_3(norstrandite) + 3H^+ \rightleftharpoons Al^{3+} + 3H_2O$	8.13
7	$\gamma - Al(OH)_3(gibbsite) + 3H^+ \rightleftharpoons Al^{3+} + 3H_2O$	8.04
8	$\alpha - AlOOH(diaspore) + 3H^+ \rightleftharpoons Al^{3+} + 2H_2O$	7.92

[a] From W. L. Lindsay, "Chemical Equilibria in Soils." Copyright © 1979 John Wiley & Sons. Reprinted by permission of John Wiley & Sons, Inc.

$$= [-1204.16\ kJ\ mol^{-1}] - [1158.24\ kJ\ mol^{-1}]$$

$$= -45.92\ kJ\ mol^{-1}.$$

One can relate ΔG_r^o to K_{dis}^o using Eq. (4.26),

$$\Delta G_r^o = -RT \ln K_{dis}^o, \tag{4.26}$$

where R is the universal gas constant and T is absolute temperature. At 298K,

$$\Delta G_r^o = -(0.008314\ kJ\ K^{-1}\ mol^{-1})(298.15K)(2.303) \log K_{dis}^o. \tag{4.27}$$

Then

$$\log K_{dis}^o = \frac{-\Delta G_r^o}{5.71}. \tag{4.28}$$

Thus, for Reaction 7 in Table 4.6,

$$\log K_{dis}^o = \frac{-(-45.92)}{5.71} = 8.04. \tag{4.29}$$

The solubility line in Fig. 4.2 for Reaction 7 can be determined as

$$\frac{(Al^{3+})}{(H^+)^3} = 10^{8.04} \tag{4.30}$$

or

$$\log Al^{3+} = 8.04 - 3pH, \tag{4.31}$$

where $\log Al^{3+}$ is plotted on the y axis, pH is plotted on the x axis, and -3 and 8.04 are the slope and y intercept, respectively.

The advantage of a stability diagram is that the solubility of several different solid phases can be compared at one time. Figure 4.2 is an example of a stability diagram constructed for a range of aluminum oxides and hydroxides listed in Table 4.6. The positions of the different lines correspond to the solubility of each solid, with the line nearest the axes being the most insoluble.

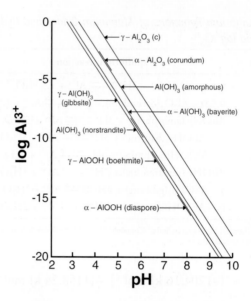

FIGURE 4.2. *Solubility diagram for various aluminum oxides and hydroxides. Reprinted from W. L. Lindsay, "Chemical Equilibria in Soils." Copyright © 1979 John Wiley & Sons. Reprinted by permission of John Wiley & Sons, Inc.*

The other solubility lines in Fig. 4.2 can be similarly developed. The anhydrous aluminum oxides γ-$Al_2O_3(c)$ and α-Al_2O_3 (corundum) depicted in Table 4.6 and Fig. 4.2 are high-temperature minerals that are not usually formed in soils (Lindsay, 1979). The order of decreasing solubility among the remaining minerals is α-$Al(OH)_3$ (bayerite), γ-$AlOOH$ (boehmite), $Al(OH)_3$ (norstrandite), γ-$Al(OH)_3$ (gibbsite), and α-$AlOOH$ (diaspore). However, the differences in solubility between the last four minerals are very slight. It is also evident from Fig. 4.2 that the activity of Al^{3+} in equilibrium with any of the minerals is dependent on pH. The activity of Al^{3+} decreases 1000-fold for each unit increase in pH (slope = 3).

The solubility diagram for several phyllosilicate and oxide minerals found in soils is shown in Fig. 4.3. Note that the x axis is plotted as a function of log $H_4SiO_4^0$ as opposed to pH in Fig. 4.2, and that the y axis is expressed as the log Al^{3+} + 3pH as opposed to log Al^{3+}. This illustrates that solubility diagrams can be constructed in a number of different ways as long as the terms used for each axis are consistent between the different mineral phases of interest.

The stability line for kaolinite in Fig. 4.3 can be obtained as follows, assuming the following dissolution reaction (Lindsay, 1979)

$$Al_2Si_2O_5(OH)_4 + 6H^+ \rightleftharpoons 2Al^{3+} + 2H_4SiO_4^0 + H_2O \qquad (4.32)$$
$$\text{(kaolinite)}$$

$$\log K_{dis}^o = 5.45.$$

Rearranging terms and assuming an activity of 1 for kaolinite and water yields

$$\frac{(Al^{3+})^2(H_4SiO_4^0)^2}{(H^+)^6} = 10^{5.45}. \tag{4.33}$$

Taking the log of both sides of Eq. (4.33) and solving for the terms on the axes in Fig. 4.3 yields the desired linear relationship:

$$2 \log Al^{3+} + 6 \text{ pH} = 5.45 - 2 \log H_4SiO_4^0 \tag{4.34}$$

$$\log Al^{3+} + 3 \text{ pH} = 2.73 - \log H_4SiO_4^0. \tag{4.35}$$

Stability diagrams such as Fig. 4.3 are very useful in predicting the presence of stable solid phases in different soil systems. Remembering that the portion of straight line relationships closest to the x and y axes represents the most stable mineral phase, it is evident that 2:1 clay minerals such as Mg-montmorillonite are only stable when the concentration of $H_4SiO_4^0$ in solution is controlled by solid phases more soluble than quartz. (The vertical lines for quartz and the other sources of SiO_2 are straight lines in Fig. 4.3 because there is no Al^{3+} in the crystalline structure, and the dissolution reaction is essentially independent of pH.) At lower concentrations in the soil solution ($\log H_4SiO_4^0 < -4$), the 1:1 clay mineral kaolinite becomes the most stable solid phase. This is consistent with the observation that kaolinite is the dominant phyllosilicate mineral in acid soils that receive substantial amounts of rainfall (extensive soil leaching). Under more extreme leaching conditions ($\log H_4SiO_4^0 < -5.3$), gibbsite becomes the most stable solid phase.

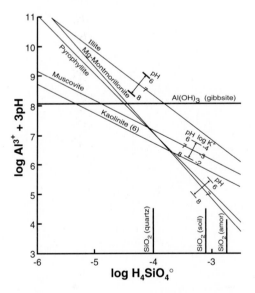

FIGURE 4.3. *Solubility diagram of several primary and secondary minerals in equilibrium with 10^{-3} M K^+, 10^{-3} M Mg^{2+}, and soil-Fe with indicated changes for pH and K^+. Reprinted from W. L. Lindsay, "Chemical Equilibria in Soils." Copyright © 1979 John Wiley & Sons. Reprinted by permission of John Wiley & Sons, Inc.*

Suggested Reading

Adams, F. (1971). Ionic concentrations and activities in the soil solution. *Soil Sci. Soc. Am. Proc.* **35**, 420–426.

Drever, J. I., Ed. (1985). "The Chemistry of Weathering," NATO ASI Ser. C, Vol. 149. Reidel, Dordrecht, The Netherlands.

Garrels, R. M., and Christ, C. L. (1965). "Solutions, Minerals, and Equilibria." Freeman, Cooper, San Francisco.

Lindsay, W. L. (1979). "Chemical Equilibria in Soils." Wiley, New York.

Lindsay, W. L. (1981). Solid-phase solution equilibria in soils. *ASA Spec. Publ.* **40**, 183–202.

Mattigod, S. V., and Zachara, J. M. (1996). Equilibrium modeling in soil chemistry. *In* "Methods of Soil Analysis: Chemical Methods," Part 3 (D. L. Sparks, Ed.), Soil Sci. Soc. Am. Book Ser. 5, pp. 1309–1358. Soil Sci. Soc. Am., Madison, WI.

Stumm, W., and Wollast, R. (1990). Coordination chemistry of weathering. Kinetics of the surface-controlled dissolution of oxide minerals. *Rev. Geophys.* **28**, 53–69.

Wolt, J. (1994). "Soil Solution Chemistry." Wiley, New York.

5 Sorption Phenomena on Soils

Introduction and Terminology

Adsorption can be defined as the accumulation of a substance or material at an interface between the solid surface and the bathing solution. Adsorption can include the removal of solute (a substance dissolved in a solvent) molecules from the solution and of solvent (continuous phase of a solution, in which the solute is dissolved) from the solid surface, and the attachment of the solute molecule to the surface (Stumm, 1992). Adsorption does not include surface precipitation (which will be discussed later in this chapter) or polymerization (formation of small multinuclear inorganic species such as dimers or trimers) processes. Adsorption, surface precipitation, and polymerization are all examples of sorption, a general term used when the retention mechanism at a surface is unknown. There are various sorption mechanisms involving both physical and chemical processes that could occur at soil mineral surfaces (Fig. 5.1). These will be discussed in detail later in this chapter and in other chapters.

It would be useful before proceeding any further to define a number of terms pertaining to retention (adsorption/sorption) of ions and molecules in soils. The adsorbate is the material that accumulates at an interface, the solid surface on which the adsorbate accumulates is referred to as the adsorbent, and the molecule or ion in solution that has the potential of being adsorbed is the adsorptive. If the general term sorption is used, the material that accumulates at the surface, the solid surface, and the molecule or ion in solution that can be sorbed are referred to as sorbate, sorbent, and sorptive, respectively (Stumm, 1992).

Adsorption is one of the most important chemical processes in soils. It determines the quantity of plant nutrients, metals, pesticides, and other organic chemicals retained on soil surfaces and therefore is one of the primary processes that affects transport of nutrients and contaminants in soils. Adsorption also affects the electrostatic properties, e.g., coagulation and settling, of suspended particles and colloids (Stumm, 1992).

Both physical and chemical forces are involved in adsorption of solutes from solution. Physical forces include van der Waals forces (e.g., partitioning) and electrostatic outer-sphere complexes (e.g., ion exchange). Chemical forces resulting from short-range interactions include

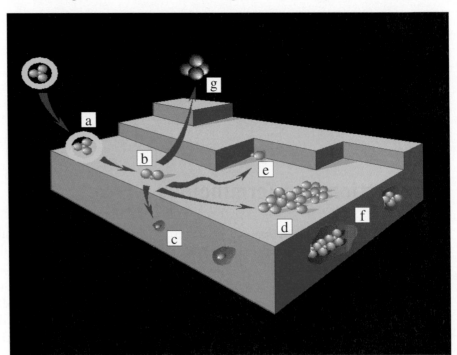

FIGURE 5.1. *Various mechanisms of sorption of an ion at the mineral/water interface: (1) adsorption of an ion via formation of an outer-sphere complex (a); (2) loss of hydration water and formation of an inner-sphere complex (b); (3) lattice diffusion and isomorphic substitution within the mineral lattice (c); (4) and (5) rapid lateral diffusion and formation either of a surface polymer (d), or adsorption on a ledge (which maximizes the number of bonds to the atom) (e). Upon particle growth, surface polymers end up embedded in the lattice structure (f); finally, the adsorbed ion can diffuse back in solution, either as a result of dynamic equilibrium or as a product of surface redox reactions (g). From Charlet and Manceau (1993), with permission. Copyright CRC Press, Boca Raton, FL.*

TABLE 5.1. *Sorption Mechanisms for Metals and Oxyanions on Soil Minerals*

Metal	pH	Sorbent	Sorption mechanism	Molecular probe	Reference
Cd(II)	7.4–9.8	Manganite	Inner-sphere	XAFS	Bochatay and Persson (2000a)
Co(II)	8.1	Al_2O_3	Multinuclear complexes (low loading)	XAFS	Towle et al. (1997)
			Co–Al hydroxide surface precipitates (high loading)		
	6.8–9	Silica	Co-hydroxide precipitates	XAFS	O'Day et al. (1996)
	5.3–7.9	Rutile	Small multinuclear complexes (low loading)	XAFS	O'Day et al. (1996)
			Large multinuclear complexes (high loading)		
	7.8	Kaolinite	Co–Al hydroxide surface precipitates	XAFS	Thompson et al. (1999a)
	4.0	Humic substances	Inner-sphere	XAFS	Xia et al. (1997b)
Cr(III)	4	Goethite, hydrous ferric oxide	Inner-sphere and Cr-hydroxide surface precipitates	XAFS	Charlet and Manceau (1992)
	6	Silica	Inner-sphere monodentate (low loading)	XAFS	Fendorf et al. (1994a)
			Cr hydroxide surface precipitates (high loading)		
Cu(II)	6.5	Bohemite	Inner-sphere (low loading)	EPR, XAFS	Weesner and Bleam (1997)
			Outer-sphere (high loading)		
	4.3–4.5	γ-Al_2O_3	Inner-sphere bidentate	XAFS	Cheah et al. (1998)
	5	Ferrihydrite	Inner-sphere bidentate	XAFS	Scheinost et al. (2001)
	5.5	Silica	Cu-hydroxide clusters	XAFS, EPR	Xia et al. (1997c)
	4.4–4.6	Amorphous silica	Inner-sphere monodentate	XAFS	Cheah et al. (1998)
	4–6	Soil humic substance	Inner-sphere	XAFS	Xia et al. (1997a)
Ni	7.5	Pyrophyllite, kaolinite, gibbsite, and montmorillonite	Mixed Ni–Al hydroxide (LDH) surface precipitates	XAFS	Scheidegger et al. (1997)
	7.5	Pyrophyllite	Mixed Ni–Al hydroxide (LDH) surface precipitates	XAFS	Scheidegger et al. (1996)

TABLE 5.1. *Sorption Mechanisms for Metals and Oxyanions on Soil Minerals (contd)*

Metal	pH	Sorbent	Sorption mechanism	Molecular probe	Reference
	7.5	Pyrophyllite–montmorillonite mixture (1:1)	Mixed Ni–Al hydroxide (LDH) surface precipitates	XAFS	Elzinga and Sparks (1999)
	6–7.5	Illite	Mixed Ni–Al hydroxide (LDH) surface precipitates at pH >6.25	XAFS	Elzinga and Sparks (2000)
	7.5	Pyrophyllite (in presence of citrate and salicylate)	Ni–Al hydroxide (LDH) surface precipitates	DRS	Yamaguchi *et al.* (2001)
	7.5	Gibbsite/amorphous silica mixture	γ-Ni(OH)$_2$ surface precipitate transforming with time to Ni-phyllosilicate	XAFS–DRS	Scheckel and Sparks (2000)
	7.5	Gibbsite (in presence of citrate and salicylate)	α-Ni hydroxide surface precipitate	DRS	Yamaguchi *et al.* (2001)
	7.5	Soil clay fraction	α-Ni–Al hydroxide surface precipitate	XAFS	Roberts *et al.* (1999)
Pb(II)	6	γ-Al$_2$O$_3$	Inner-sphere monodentate mononuclear	XAFS	Chisholm-Brause *et al.* (1990a)
	6.5	γ-Al$_2$O$_3$	Inner-sphere bidentate (low loading) Surface polymers (high loading)	XAFS	Strawn *et al.* (1998)
	7	α-Alumina (0001 single crystal)	Outer-sphere	Grazing incidence XAFS (GI-XAFS)	Bargar *et al.* (1996)
		α-Alumina (1T02 single crystal)	Inner-sphere	Grazing incidence XAFS (GI-XAFS)	
	6 and 7	Al$_2$O$_3$ powders	Inner-sphere bidentate mononuclear (low loading) Dimeric surface complexes (high loading)	XAFS	Bargar *et al.* (1997a)
	6–8	Goethite and hematite powders	Inner-sphere bidentate binuclear	XAFS	Bargar *et al.* (1997b)
	Variable	Goethite	Inner-sphere (low loading)	XAFS	Roe *et al.* (1991)

TABLE 5.1. Sorption Mechanisms for Metals and Oxyanions on Soil Minerals (contd)

Metal	pH	Sorbent	Sorption mechanism	Molecular probe	Reference
	3–7	Goethite (in presence of SO_4^{2-})	Inner-sphere bidentate due to ternary complex formation	XAFS, ATR-FTIR	Ostergren et al. (2000a)
	5 and 6	Goethite (in absence and presence of SO_4^{2-})	Inner-sphere bidentate mononuclear (pH 6) (in absence of SO_4^{2-})	XAFS, ATR-FTIR	Elzinga et al. (2001)
			Inner-sphere bidentate mononuclear and binuclear (pH 5) (in absence of SO_4^{2-})		
			Inner-sphere bidentate binuclear due to ternary complex formation (in the presence of SO_4^{2-})		
	5.7	Goethite (in presence of CO_3^{2-})	Inner-sphere bidentate	XAFS, ATR-FTIR	Ostergren et al. (2000b)
	5	Ferrihydrite	Inner-sphere bidentate	XAFS	Scheinost et al. (2001)
	3.5	Birnessite	Inner-sphere mononuclear	XAFS	Matocha et al. (2001)
	6.7	Manganite	Inner-sphere mononuclear	XAFS	
	6.77	Montmorillonite	Inner-sphere	XAFS	Strawn and Sparks (1999)
	6.31–6.76	Montmorillonite	Inner-sphere and outer-sphere	XAFS	
	4.48–6.40	Montmorillonite	Outer-sphere		
Sr(II)	7	Ferrihydrite	Outer-sphere	XAFS	Axe et al. (1997)
		Kaolinite, amorphous silica, goethite	Outer-sphere	XAFS	Sahai et al. (2000)
Zn(II)	7–8.2	Alumina powders	Inner-sphere bidentate (low loading)	XAFS	Trainor et al. (2000)
			Mixed metal–Al hydroxide surface precipitates (high loading)		
	6.17–9.87	Manganite	Multinuclear hydroxo-complexes or Zn–hydroxide phases	XAFS	Bochatay and Persson (2000b)
	7.5	Pyrophyllite	Mixed Zn–Al hydroxide surface precipitates	XAFS	Ford and Sparks (2001)

TABLE 5.1. *Sorption Mechanisms for Metals and Oxyanions on Soil Minerals (contd)*

Metal	pH	Sorbent	Sorption mechanism	Molecular probe	Reference
Oxyanion					
Arsenite (As(III))	5.5, 8	γ-Al$_2$O$_3$	Inner-sphere bidentate binuclear and outer-sphere	XAFS	Arai et al. (2001)
	5.8	Fe(OH)$_3$	Inner-sphere	ATR-FTIR, DRIFT	Suarez (1998)
	5.5	Goethite	Inner-sphere bidentate binuclear	ATR-FTIR (dry)	Sun and Doner (1996)
	7.2–7.4	Goethite	Inner-sphere bidentate binuclear	XAFS	Manning et al. (1998)
	5, 10.5	Amorphous Fe oxides	Inner-sphere and outer-sphere	ATR-FTIR and Raman	Goldberg and Johnston (2001)
		Amorphous Al oxides	Outer-sphere	ATR-FTIR and Raman	Goldberg and Johnston (2001)
Arsenate (As(V))	5, 9	Amorphous Al and Fe oxides	Inner-sphere	ATR-FTIR and Raman	Goldberg and Johnston (2001)
	5.5	Gibbsite	Inner-sphere bidentate binuclear	XAFS	Ladeira et al. (2001)
	4, 8, 10	γ-Al$_2$O$_3$	Inner-sphere bidentate binuclear	XAFS	Arai et al. (2001)
	5.5	Goethite	Inner-sphere bidentate binuclear	ATR-FTIR	Sun and Doner (1996)
	6	Goethite	Inner-sphere bidentate binuclear	XAFS	O'Reilly et al. (2001)
	5, 8	Fe(OH)$_3$	Inner-sphere	ATR-FTIR DRIFT-FTIR	Suarez (1998)
	8	Goethite	Inner-sphere bidentate binuclear, inner-sphere monodentate	XAFS	Waychunas et al. (1993)
	6, 8, 9	Goethite	Inner-sphere monodentate (low loading) Inner-sphere bidentate binuclear (high loading)	XAFS	Fendorf et al. (1997)
Boron (trigonal (B(OH)$_3$) and tetrahedral (B(OH)$_4^-$)	7	Green rust lepidocrocite	Inner-sphere bidentate	XAFS	Randall et al. (2001)
	7, 11	Amorphous Fe(OH)$_3$	Inner-sphere	ATR-FTIR DRIFT-FTIR	Su and Suarez (1995)
	7, 10	Amorphous Al(OH)$_3$	Inner-sphere	ATR-FTIR DRIFT-FTIR	Su and Suarez (1995)

TABLE 5.1. *Sorption Mechanisms for Metals and Oxyanions on Soil Minerals (contd)*

Metal	pH	Sorbent	Sorption mechanism	Molecular probe	Reference
Carbonate	4.1–7.8	Amorphous Al and Fe oxides gibbsite goethite	Inner-sphere monodentate	ATR-FTIR	Su and Suarez (1997)
	5.2–7.2	γ-Al_2O_3	Inner-sphere monodentate	ATR-FTIR and DRIFT-FTIR	Winja and Schulthess (1999)
	4–9.2	Goethite	Inner-sphere monodentate	ATR-FTIR	Villalobos and Leckie (2000)
	4.8–7	Goethite	Inner-sphere monodentate	ATR-FTIR	Winja and Schulthess (2001)
Chromate (Cr(VI))	5, 6	Goethite	Inner-sphere bidentate mononuclear (pH 5, 5 mM Cr(VI)) Inner-sphere bidentate binuclear (pH 6, 3 mM Cr(VI)) Inner-sphere monodentate (pH 6, 2 mM Cr(VI))	XAFS	Fendorf et al. (1997)
Phosphate	4–11	Boehmite	Inner-sphere	MAS-NMR	Bleam et al. (1991)
	3–12.8	Goethite	Inner-sphere monodentate	DRIFT-FTIR	Persson et al. (1996)
	4–8	Goethite	Inner-sphere bidentate and monodentate	ATR-FTIR	Tejedor-Tejedor and Anderson (1990)
	4–9	Ferrihydrite	Inner-sphere nonprotonated bidentate binuclear (pH >7.5) Inner-sphere protonated (pH 4–6)	ATR-FTIR	Arai and Sparks (2001)
Selenate (Se(VI))	4	Goethite	Outer-sphere	XAFS	Hayes et al. (1987)
	Variable	Goethite	Inner-sphere monodentate (pH <6) Outer-sphere (pH >6)	ATR-FTIR and Raman	Winja and Schulthess (2000)
		Al oxide	Outer-sphere		
	3.5–6.7	Goethite $Fe(OH)_3$	Inner-sphere binuclear	XAFS	Manceau and Charlet (1994)

TABLE 5.1.   *Sorption Mechanisms for Metals and Oxyanions on Soil Minerals (contd)*

Metal	pH	Sorbent	Sorption mechanism	Molecular probe	Reference
Selenite (Se(IV))	4	Goethite	Inner-sphere bidentate	XAFS	Hayes *et al.* (1987)
	3	Goethite Fe(OH)$_3$	Inner-sphere bidentate	XAFS	Manceau and Charlet (1994)
Sulfate	3.5–9	Goethite	Outer-sphere and inner-sphere monodentate (pH <6)	ATR-FTIR	Peak *et al.* (1999)
			Outer-sphere (pH >6)		
	Variable	Goethite	Inner-sphere monodentate (pH <6)	ATR-FTIR and Raman	Winja and Schulthess (2000)
			Outer-sphere (pH >6)		
		Al oxide	Outer-sphere		
	3–6	Hematite	Inner-sphere monodentate	ATR-FTIR	Hug (1997)

inner-sphere complexation that involves a ligand exchange mechanism, covalent bonding, and hydrogen bonding (Stumm and Morgan, 1981). The physical and chemical forces involved in adsorption are discussed in sections that follow.

Surface Functional Groups

Surface functional groups in soils play a significant role in adsorption processes. A surface functional group is "a chemically reactive molecular unit bound into the structure of a solid at its periphery such that the reactive components of the unit can be bathed by a fluid" (Sposito, 1989). Surface functional groups can be organic (e.g., carboxyl, carbonyl, phenolic) or inorganic molecular units. The major inorganic surface functional groups in soils are the siloxane surface groups associated with the plane of oxygen atoms bound to the silica tetrahedral layer of a phyllosilicate and hydroxyl groups associated with the edges of inorganic minerals such as kaolinite, amorphous materials, and metal oxides, oxyhydroxides, and hydroxides. A cross section of the surface layer of a metal oxide is shown in Fig. 5.2. In Fig. 5.2a the surface is unhydrated and has metal ions that are Lewis acids and that have a reduced coordination number. The oxide anions are Lewis bases. In Fig. 5.2b, the surface metal ions coordinate to H_2O molecules forming a Lewis acid site, and then a dissociative chemisorption (chemical bonding to the surface) leads to a hydroxylated surface (Fig. 5.2c) with surface OH groups (Stumm, 1987, 1992).

The surface functional groups can be protonated or deprotonated by adsorption of H^+ and OH^-, respectively, as shown below:

$$S - OH + H^+ \rightleftharpoons S - OH_2^+ \tag{5.1}$$

$$S - OH \rightleftharpoons S - O^- + H^+. \tag{5.2}$$

Here the Lewis acids are denoted by S and the deprotonated surface hydroxyls are Lewis bases. The water molecule is unstable and can be exchanged for an inorganic or organic anion (Lewis base or ligand) in the solution, which then bonds to the metal cation. This process is called ligand exchange (Stumm, 1987, 1992).

The Lewis acid sites are present not only on metal oxides such as on the edges of gibbsite or goethite, but also on the edges of clay minerals such as kaolinite. There are also singly coordinated OH groups on the edges of clay minerals. At the edge of the octahedral sheet, OH groups are singly coordinated to Al^{3+}, and at the edge of the tetrahedral sheet they are singly coordinated to Si^{4+}. The OH groups coordinated to Si^{4+} dissociate only protons; however, the OH coordinated to Al^{3+} dissociate and bind protons. These edge OH groups are called silanol (SiOH) and aluminol (AlOH), respectively (Sposito, 1989; Stumm, 1992).

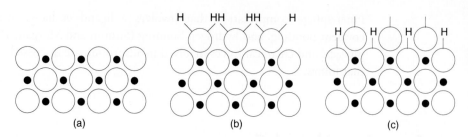

FIGURE 5.2. *Cross section of the surface layer of a metal oxide. (•) Metal ions, (0) oxide ions. (a) The metal ions in the surface layer have a reduced coordination number and exhibit Lewis acidity. (b) In the presence of water, the surface metal ions may coordinate H_2O molecules. (c) Dissociative chemisorption leads to a hydroxylated surface. From Schindler (1981), with permission.*

Spectroscopic analyses of the crystal structures of oxides and clay minerals show that different types of hydroxyl groups have different reactivities. Goethite (α-FeOOH) has four types of surface hydroxyls whose reactivities are a function of the coordination environment of the O in the FeOH group (Fig. 5.3). The FeOH groups are A-, B-, or C-type sites, depending on whether the O is coordinated with 1, 3, or 2 adjacent Fe(III) ions. The fourth type of site is a Lewis acid-type site, which results from chemisorption of a water molecule on a bare Fe(III) ion. Sposito (1984) has noted that only A-type sites are basic; i.e., they can form a complex with H^+, and A-type and Lewis acid sites can release a proton. The B- and C-type sites are considered unreactive. Thus, A-type sites can be either a proton acceptor or a proton donor (i.e., they are amphoteric). The water coordinated with Lewis acid sites may be a proton donor site, i.e., an acidic site.

Clay minerals have both aluminol and silanol groups. Kaolinite has three types of surface hydroxyl groups: aluminol, silanol, and Lewis acid sites (Fig. 5.4).

Surface Complexes

When the interaction of a surface functional group with an ion or molecule present in the soil solution creates a stable molecular entity, it is called a surface complex. The overall reaction is referred to as surface complexation. There are two types of surface complexes that can form, outer-sphere and inner-sphere. Figure 5.5 shows surface complexes between metal cations and siloxane ditrigonal cavities on 2:1 clay minerals. Such complexes can also occur on the edges of clay minerals. If a water molecule is present between the surface functional group and the bound ion or molecule, the surface complex is termed outer-sphere (Sposito, 1989).

Surface Hydroxyls

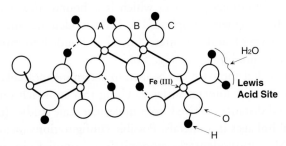

FIGURE 5.3. *Types of surface hydroxyl groups on goethite: singly (A-type), triply (B-type), and doubly (C-type) hydroxyls coordinated to Fe(III) ions (one Fe–O bond not represented for type B and C groups); and a Lewis acid site (Fe(III) coordinated to an H_2O molecule). The dashed lines indicate hydrogen bonds. From Sposito (1984), with permission.*

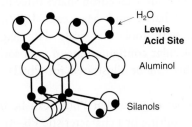

FIGURE 5.4. *Surface hydroxyl groups on kaolinite. Besides the OH groups on the basal plane, there are aluminol groups, Lewis acid sites (at which H_2O is adsorbed), and silanol groups, all associated with ruptured bonds along the edges of the kaolinite. From Sposito (1984), with permission.*

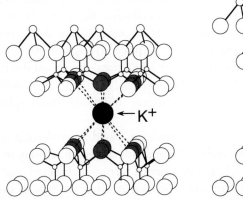
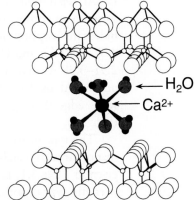

INNER-SPHERE SURFACE COMPLEX:
K^+ ON VERMICULITE

OUTER-SPHERE SURFACE COMPLEX:
$Ca(H_2O)_6^{2+}$ ON MONTMORILLONITE

FIGURE 5.5. *Examples of inner- and outer-sphere complexes formed between metal cations and siloxane ditrigonal cavities on 2:1 clay minerals. From Sposito (1984), with permission.*

If there is not a water molecule present between the ion or molecule and the surface functional group to which it is bound, this is an inner-sphere complex. Inner-sphere complexes can be monodentate (metal is bonded to only one oxygen) and bidentate (metal is bonded to two oxygens) and mononuclear and binuclear (Fig. 5.6).

A polyhedral approach can be used to determine molecular configurations of ions sorbed on mineral surfaces. Using this approach one can interpret metal–metal distances derived from molecular scale studies (e.g., XAFS) and octahedral linkages in minerals. Possible configurations include: (1) a single corner (SC) monodentate mononuclear complex in which a given octahedron shares one oxygen with another octahedron; (2) a double corner (DC) bidentate binuclear complex in which a given octahedron shares two nearest oxygens with two different octahedra; (3) an edge (E) bidentate mononuclear complex in which an octahedron shares two nearest oxygens with another octahedron; and (4) a face (F) tridentate mononuclear complex in which an octahedron shares three nearest neighbors with another octahedron (Charlet and Manceau, 1992). A polyhedral approach can be applied, with molecular scale data (e.g., EXAFS), to determine the possible molecular configurations of ions sorbed on mineral surfaces. An example of this can be illustrated for Pb(II) sorption on Al oxides (Bargar et al., 1997).

There are a finite number of ways that Pb(II) can be linked to Al_2O_3 surfaces, with each linkage resulting in a characteristic Pb–Al distance. These configurations are shown in Fig. 5.7. Pb(II) ions could adsorb in monodentate, bidentate, or tridentate fashion. Using the average EXAFS derived Pb–O bond distance of 2.25 Å and using known Al–O bond distances for AlO_6 octahedra of 1.85 to 1.97 Å and AlO_6 octahedron edge lengths (i.e., O–O separations) of 2.52 to 2.86 Å, the range of Pb–Al separations for Pb(II) sorbed to AlO_6 octahedra is monodentate sorption to corners of AlO_6 octahedra ($R_{Pb–Al} \approx 4.10$ to 4.22 Å); bridging bidentate sorption to corners of neighboring AlO_6 octahedra ($R_{Pb–Al} \approx$ 3.87–3.99 Å); and bidentate sorption to edges of AlO_6 octahedra ($R_{Pb–Al} \approx$ 2.91–3.38 Å). Based on the EXAFS data, the Pb–Al distances for Pb sorbed on the Al oxides were between 3.20 and 3.32 Å, which are consistent with edge-sharing mononuclear bidentate inner-sphere complexation (Fig. 5.7).

The type of surface complexes, based on molecular scale investigations, that occur with metals and metalloids sorbed on an array of mineral surfaces is given in Table 5.1. Environmental factors such as pH, surface loading, ionic strength, type of sorbent, and time all affect the type of sorption complex or product. An example of this is shown for Pb sorption on montmorillonite over an I range of 0.006–0.1 and a pH range of 4.48–6.77 (Table 5.2). Employing XAFS analysis, at pH 4.48 and $I = 0.006$, outer-sphere complexation on basal planes in the interlayer regions of the montmorillonite predominated. At pH 6.77 and $I = 0.1$, inner-sphere complexation on edge sites of montmorillonite was most prominent, and at pH 6.76, $I = 0.006$ and

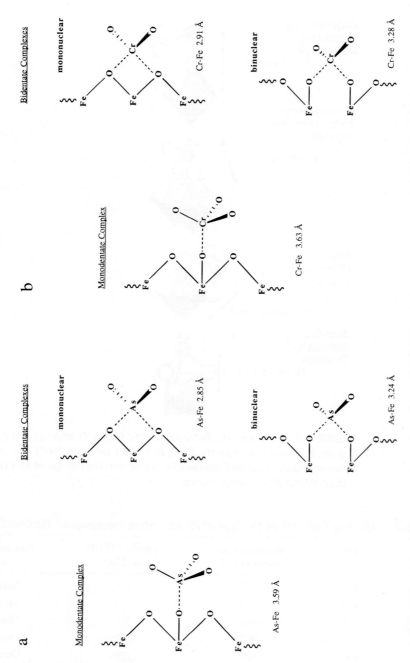

FIGURE 5.6. *Schematic illustration of the surface structure of (a) As(V) and (b) Cr(VI) on goethite based on the local coordination environment determined with EXAFS spectroscopy. From Fendorf et al. (1997), with permission. Copyright 1997 American Chemical Society.*

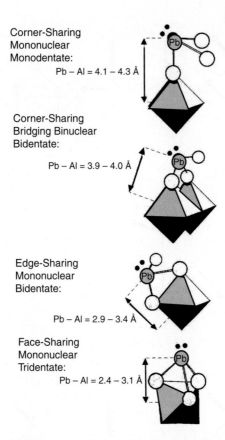

Corner-Sharing
Mononuclear
Monodentate:
    Pb – Al = 4.1 – 4.3 Å

Corner-Sharing
Bridging Binuclear
Bidentate:
    Pb – Al = 3.9 – 4.0 Å

Edge-Sharing
Mononuclear
Bidentate:

    Pb – Al = 2.9 – 3.4 Å

Face-Sharing
Mononuclear
Tridentate:
    Pb – Al = 2.4 – 3.1 Å

FIGURE 5.7. *Characteristic Pb–Al separations for Pb(II) adsorbed to AlO_6 octahedra. In order to be consistent with the EXAFS and XANES data, the Pb(II) ions are depicted as having trigonal pyramidal coordination geometries. From Bargar et al. (1997a), with permission from Elsevier Science.*

TABLE 5.2.   *Effect of I and pH on the Type of Pb Adsorption Complexes on Montmorillonite[a]*

I (M)	pH	Removal from solution (%)	Adsorbed Pb(II) (mmol kg^{-1})	Primary adsorption complex[b]
0.1	6.77	86.7	171	Inner-sphere
0.1	6.31	71.2	140	Mixed
0.006	6.76	99.0	201	Mixed
0.006	6.40	98.5	200	Outer-sphere
0.006	5.83	98.0	199	Outer-sphere
0.006	4.48	96.8	197	Outer-sphere

[a] From Strawn and Sparks (1999), with permission from Academic Press, Orlando, FL.
[b] Based on results from XAFS data analysis.

pH 6.31, I = 0.1, both inner- and outer-sphere complexation occurred. These data are consistent with other findings that inner-sphere complexation is favored at higher pH and ionic strength (Elzinga and Sparks, 1999). Clearly, there is a continuum of adsorption complexes that can exist in soils.

Outer-sphere complexes involve electrostatic coulombic interactions and are thus weak compared to inner-sphere complexes in which the binding is covalent or ionic. Outer-sphere complexation is usually a rapid process that is reversible, and adsorption occurs only on surfaces of opposite charge to the adsorbate.

Inner-sphere complexation is usually slower than outer-sphere complexation, it is often not reversible, and it can increase, reduce, neutralize, or reverse the charge on the sorptive regardless of the original charge. Adsorption of ions via inner-sphere complexation can occur on a surface regardless of the surface charge. It is important to remember that outer- and inner-sphere complexations can, and often do, occur simultaneously.

Ionic strength effects on sorption are often used as indirect evidence for whether an outer-sphere or inner-sphere complex forms (Hayes and Leckie, 1986). For example, strontium [Sr(II)] sorption on γ-Al$_2$O$_3$ is highly dependent on the I of the background electrolyte, NaNO$_3$, while Co(II) sorption is unaffected by changes in I (Fig. 5.8). The lack of I effect on Co(II) sorption would suggest formation of an inner-sphere complex, which is consistent with findings from molecular scale spectroscopic analyses (Hayes and Katz, 1996; Towle et al., 1997). The strong dependence of Sr(II) sorption on I, suggesting outer-sphere complexation, is also consistent with spectroscopic findings (Katz and Boyle-Wight, 2001).

Adsorption Isotherms

One can conduct an adsorption experiment as explained in Box 5.1. The quantity of adsorbate can then be used to determine an adsorption isotherm.

An adsorption isotherm, which describes the relation between the activity or equilibrium concentration of the adsorptive and the quantity of adsorbate on the surface at constant temperature, is usually employed to describe adsorption. One of the first solute adsorption isotherms was described by van Bemmelen (1888), and he later described experimental data using an adsorption isotherm.

Adsorption can be described by four general types of isotherms (S, L, H, and C), which are shown in Fig. 5.9. With an S-type isotherm the slope initially increases with adsorptive concentration, but eventually decreases and becomes zero as vacant adsorbent sites are filled. This type of isotherm indicates that at low concentrations the surface has a low affinity for the adsorptive, which increases at higher concentrations. The L-shaped (Langmuir) isotherm is characterized by a decreasing slope as concentration increases since vacant adsorption sites decrease as the adsorbent becomes covered. Such adsorption behavior could be explained by the high affinity of the adsorbent for the adsorptive at low concentrations, which then decreases as concentration increases. The H-type (high-affinity) isotherm is indicative of strong adsorbate–adsorptive interactions such as inner-sphere complexes.

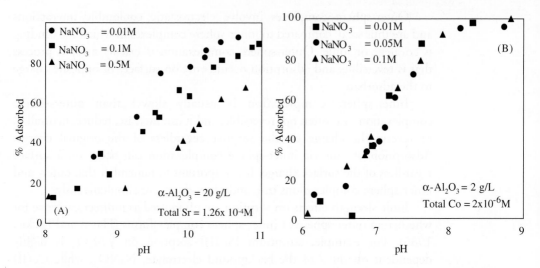

FIGURE 5.8.   *Effect of increasing ionic strength on pH adsorption edges for (A) a weakly sorbing divalent metal, Sr(II), and (B) a strongly sorbing divalent metal ion, Co(II). From Katz and Boyle-Wight (2001), with permission.*

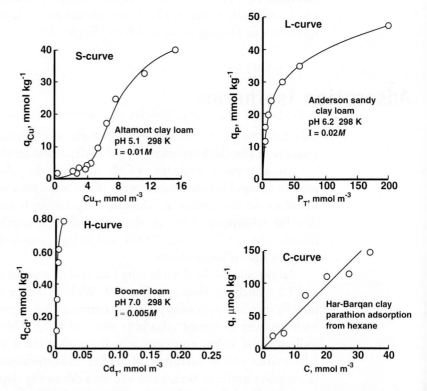

FIGURE 5.9.   *The four general categories of adsorption isotherms. From Sposito (1984), with permission.*

The C-type isotherms are indicative of a partitioning mechanism whereby adsorptive ions or molecules are distributed or partitioned between the interfacial phase and the bulk solution phase without any specific bonding between the adsorbent and adsorbate (see Box 5.2 for discussion of partition coefficients).

One should realize that adsorption isotherms are purely descriptions of macroscopic data and do not definitively prove a reaction mechanism. Mechanisms must be gleaned from molecular investigations, e.g., the use of spectroscopic techniques. Thus, the conformity of experimental adsorption data to a particular isotherm does not indicate that this is a unique description of the experimental data, and that only adsorption is operational. Thus, one cannot differentiate between adsorption and precipitation using an adsorption isotherm even though this has been done in the soil chemistry literature. For example, some researchers have described data using the Langmuir adsorption isotherm and have suggested that one slope at lower adsorptive concentrations represents adsorption and a second slope observed at higher solution concentrations represents precipitation. This is an incorrect use of an adsorption isotherm since molecular conclusions are being made and, moreover, depending on experimental conditions, precipitation and adsorption can occur simultaneously.

BOX 5.1   *Conducting an Adsorption Experiment*

Adsorption experiments are carried out by equilibrating (shaking, stirring) an adsorptive solution of a known composition and volume with a known amount of adsorbent at a constant temperature and pressure for a period of time such that an equilibrium (adsorption reaches a steady state or no longer changes after a period of time) is attained. The pH and ionic strength are also controlled in most adsorption experiments.

After equilibrium is reached (it must be realized that true equilibrium is seldom reached, especially with soils), the adsorptive solution is separated from the adsorbent by centrifugation, settling, or filtering, and then analyzed.

It is very important to equilibrate the adsorbent and adsorptive long enough to ensure that steady state has been reached. However, one should be careful that the equilibration process is not so lengthy that precipitation or dissolution reactions occur (Sposito, 1984). Additionally, the degree of agitation used in the equilibration process should be forceful enough to effect good mixing but not so vigorous that adsorbent modification occurs (Sparks, 1989). The method that one uses for the adsorption experiment, e.g., batch or flow, is also important. While batch techniques are simpler, one should be aware of their pitfalls, including the possibility of secondary precipitation and alterations in equilibrium states. More details on these techniques are given in Chapter 7.

One can determine the degree of adsorption by using the following mass balance equation,

$$\frac{(C_0 V_0) - (C_f V_f)}{m} = q, \tag{5.1a}$$

where q is the amount of adsorption (adsorbate per unit mass of adsorbent) in mol kg^{-1}, C_0 and C_f are the initial and final adsorptive concentrations, respectively, in mol liter^{-1}, V_0 and V_f are the initial and final adsorptive volumes, respectively, in liters, and m is the mass of the adsorbent in kilograms. Adsorption could then be described graphically by plotting C_f or C (where C is referred to as the equilibrium or final adsorptive concentration) on the x axis versus q on the y axis.

BOX 5.2   *Partitioning Coefficients*

A partitioning mechanism is usually suggested from linear adsorption isotherms (C-type isotherm, Fig. 5.9), reversible adsorption/desorption, a small temperature effect on adsorption, and the absence of competition when other adsorptives are added; i.e., adsorption of one of the adsorptives is not affected by the inclusion of a second adsorptive.

A partition coefficient, K_p, can be obtained from the slope of a linear adsorption isotherm using the equation

$$q = K_p C, \tag{5.2a}$$

where q was defined earlier and C is the equilibrium concentration of the adsorptive. The K_p provides a measure of the ratio of the amount of a material adsorbed to the amount in solution.

Partition mechanisms have been invoked for a number of organic compounds, particularly for NOC and some pesticides (Chiou *et al.*, 1977, 1979, 1983).

A convenient relationship between K_p and the fraction of organic carbon (f_{oc}) in the soil is the organic carbon–water partition coefficient, K_{oc}, which can be expressed as

$$K_{oc} = K_p / f_{oc}. \tag{5.2b}$$

Equilibrium-based Adsorption Models

There is an array of equilibrium-based models that have been used to describe adsorption on soil surfaces. These include the widely used Freundlich equation, a purely empirical model, the Langmuir equation, and double-layer models including the diffuse double-layer, Stern, and surface complexation models, which are discussed in the following sections.

Freundlich Equation

The Freundlich equation, which was first used to describe gas phase adsorption and solute adsorption, is an empirical adsorption model that has been widely used in environmental soil chemistry. It can be expressed as

$$q = K_d C^{1/n}, \tag{5.3}$$

where q and C were defined earlier, K_d is the distribution coefficient, and n is a correction factor. By plotting the linear form of Eq. (5.3), log q = $1/n$ log C + log K_d, the slope is the value of $1/n$ and the intercept is equal to log K_d. If $1/n$ = 1, Eq. (5.3) becomes equal to Eq. (5.2a) (Box 5.2), and K_d is a partition coefficient, K_p. One of the major disadvantages of the Freundlich equation is that it does not predict an adsorption maximum. The single K_d term in the Freundlich equation implies that the energy of adsorption on a homogeneous surface is independent of surface coverage. While researchers have often used the K_d and $1/n$ parameters to make conclusions concerning mechanisms of adsorption, and have interpreted multiple slopes from Freundlich isotherms (Fig. 5.10) as evidence of different binding sites, such interpretations are speculative. Plots such as those of Fig. 5.10 cannot be used for delineating adsorption mechanisms at soil surfaces.

Langmuir Equation

Another widely used sorption model is the Langmuir equation. It was developed by Irving Langmuir (1918) to describe the adsorption of gas molecules on a planar surface. It was first applied to soils by Fried and Shapiro (1956) and Olsen and Watanabe (1957) to describe phosphate sorption on soils. Since that time, it has been heavily employed in many fields to describe sorption on colloidal surfaces. As with the Freundlich equation, it best describes sorption at low sorptive concentrations. However, even here, failure occurs. Beginning in the late 1970s researchers began to question the validity of its original assumptions and consequently its use in describing sorption on heterogeneous surfaces such as soils and even soil components (see references in Harter and Smith, 1981).

To understand why concerns have been raised about the use of the Langmuir equation, it would be instructive to review the original assumptions that Langmuir (1918) made in the development of the equation. They are (Harter and Smith, 1981): (1) Adsorption occurs on planar surfaces that have a fixed number of sites that are identical and the sites can hold only one molecule. Thus, only monolayer coverage is permitted, which represents maximum adsorption. (2) Adsorption is reversible. (3) There is no lateral movement of molecules on the surface. (4) The adsorption energy is the same for all sites and independent of surface coverage (i.e., the surface is homogeneous), and there is no interaction between adsorbate molecules (i.e., the adsorbate behaves ideally).

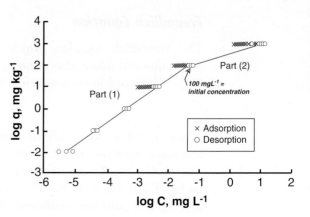

FIGURE 5.10. *Use of the Freundlich equation to describe zinc adsorption (x)/desorption (0) on soils. Part 1 refers to the linear portion of the isotherm (initial Zn concentration <100 mg liter⁻¹) while Part 2 refers to the nonlinear portion of the isotherm. From Elrashidi and O'Connor (1982), with permission.*

Most of these assumptions are not valid for the heterogeneous surfaces found in soils. As a result, the Langmuir equation should only be used for purely qualitative and descriptive purposes.

The Langmuir adsorption equation can be expressed as

$$q = kCb/(1 + kC), \tag{5.4}$$

where q and C were defined previously, k is a constant related to the binding strength, and b is the maximum amount of adsorptive that can be adsorbed (monolayer coverage). In some of the literature x/m, the weight of the adsorbate/unit weight of adsorbent, is plotted in lieu of q. Rearranging to a linear form, Eq. (5.4) becomes

$$C/q = 1/kb + C/b. \tag{5.5}$$

Plotting C/q vs C, the slope is $1/b$ and the intercept is $1/kb$. An application of the Langmuir equation to sorption of zinc on a soil is shown in Fig. 5.11. One will note that the data were described well by the Langmuir equation when the plots were resolved into two linear portions.

A number of other investigators have also shown that sorption data applied to the Langmuir equation can be described by multiple, linear portions. Some researchers have ascribed these to sorption on different binding sites. Some investigators have also concluded that if sorption data conform to the Langmuir equation, this indicates an adsorption mechanism, while deviations would suggest precipitation or some other mechanism. However, it has been clearly shown that the Langmuir equation can equally well describe both adsorption and precipitation (Veith and Sposito, 1977). Thus, mechanistic information cannot be derived from a purely macroscopic model like the Langmuir equation. While it is admissible to calculate maximum sorption (b) values for different soils and to compare them in a qualitative sense, the calculation of binding strength (k) values seems questionable. A better approach for calculating these parameters is to determine energies of activation from kinetic studies (see Chapter 7).

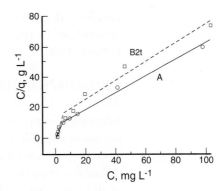

FIGURE 5.11. *Zinc adsorption on the A and B2t horizons of a Cecil soil as described by the Langmuir equation. The plots were resolved into two linear portions. From Shuman (1975), with permission.*

Some investigators have also employed a two-site or two-surface Langmuir equation to describe sorption data for an adsorbent with two sites of different affinities. This equation can be expressed as

$$q = \frac{b_1 k_1 C}{1 + k_1 C} + \frac{b_2 k_2 C}{1 + k_2 C} , \qquad (5.6)$$

where the subscripts refer to sites 1 and 2, e.g., adsorption on high- and low-energy sites. Equation (5.6) has been successfully used to describe sorption on soils of different physicochemical and mineralogical properties. However the conformity of data to Eq. (5.6) does not prove that multiple sites with different binding affinities exist.

Double-Layer Theory and Models

Some of the most widely used models for describing sorption behavior are based on the electric double-layer theory developed in the early part of the 20th century. Gouy (1910) and Chapman (1913) derived an equation describing the ionic distribution in the diffuse layer formed adjacent to a charged surface. The countercharge (charge of opposite sign to the surface charge) can be a diffuse atmosphere of charge, or a compact layer of bound charge together with a diffuse atmosphere of charge. The surface charge and the sublayers of compact and diffuse counterions (ions of opposite charge to the surface charge) constitute what is commonly called the double layer. In 1924, Stern made corrections to the theory accounting for the layer of counterions nearest the surface. When quantitative colloid chemistry came into existence, the "Kruyt" school (Verwey and Overbeek, 1948) routinely employed the Gouy–Chapman and Stern theories to describe the diffuse layer of counterions adjacent to charged particles. Schofield (1947) was among the first persons in soil science to apply the diffuse double-layer (DDL) theory to study the thickness of water films on mica surfaces. He used the theory to calculate negative adsorption of anions (exclusion of anions from the area adjacent to a negatively charged surface) in a bentonite (montmorillonite) suspension.

The historical development of the electrical double-layer theory can be found in several sources (Verwey, 1935; Grahame, 1947; Overbeek, 1952).

Excellent discussions of DDL theory and applications to soil colloidal systems can be found in van Olphen (1977), Bolt (1979), and Singh and Uehara (1986).

GOUY–CHAPMAN MODEL

The Gouy–Chapman model (Gouy, 1910; Chapman, 1913) makes the following assumptions: the distance between the charges on the colloid and the counterions in the liquid exceeds molecular dimensions; the counterions, since they are mobile, do not exist as a dense homoionic layer next to the colloidal surface but as a diffuse cloud, with this cloud containing both ions of the same sign as the surface, or coions, and counterions; the colloid is negatively charged; the ions in solution have no size, i.e., they behave as point charges; the solvent adjacent to the charged surface is continuous (same dielectric constant[1]) and has properties like the bulk solution; the electrical potential is a maximum at the charged surface and drops to zero in the bulk solution; the change in ion concentration from the charged surface to the bulk solution is nonlinear; and only electrostatic interactions with the surface are assumed (Singh and Uehara, 1986).

Figure 5.12 shows the Gouy–Chapman model of the DDL, illustrating the charged surface and distribution of cations and anions with distance from the colloidal surface to the bulk solution. Assuming the surface is negatively charged, the counterions are most concentrated near the surface and decrease (exponentially) with distance from the surface until the distribution of coions is equal to that of the counterions (in the bulk solution). The excess positive ions near the surface should equal the negative charge in the fixed layer; i.e., an electrically neutral system should exist. Coions are repelled by the negative surface, forcing them to move in the opposite direction so there is a deficit of anions close to the surface (van Olphen, 1977; Stumm, 1992).

A complete and easy-to-follow derivation of the Gouy–Chapman theory is found in Singh and Uehara (1986) and will not be given here. There are a number of important relationships and parameters that can be derived from the Gouy–Chapman theory to describe the distribution of ions near the charged surface and to predict the stability of the charged particles in soils. These include:

1. The relationship between potential (ψ) and distance (x) from the surface,

$$\tanh [Ze\psi/4kT] = \tanh [Ze\psi_0/4kT]e^{-\kappa x}, \tag{5.7}$$

where Z is the valence of the counterion, e is the electronic charge (1.602×10^{-19} C, where C refers to Coulombs), ψ is the electric potential in V, k is Boltzmann's constant (1.38×10^{-23} J K^{-1}), T is absolute temperature in

---

[1] The dielectric constant of a solvent is an index of how well the solvent can separate oppositely charged ions. The higher the dielectric constant, the smaller the attraction between ions. It is a dimensionless quantity (Harris, 1987).

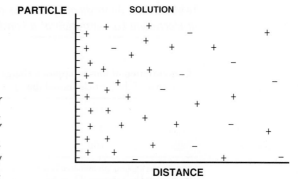

FIGURE 5.12. *Diffuse electric double-layer model according to Gouy. From H. van Olphen, "An Introduction to Clay Colloid Chemistry," 2nd ed. Copyright 1977 © John Wiley and Sons, Inc. Reprinted by permission of John Wiley and Sons, Inc.*

degrees Kelvin, tanh is the hyperbolic tangent, ψ_0 is the potential at the surface in V, κ is the reciprocal of the double-layer thickness in m^{-1}, and x is the distance from the surface in m.

2. The relationship between number of ions (n_i) and distance from the charged surface (x),

$$n_i = n_{i_o} \left[\frac{1 - \tanh(-Ze\psi_o/4kT)e^{-\kappa x}}{1 + \tanh(-Ze\psi_o/4kT)e^{-\kappa x}} \right]^2 , \qquad (5.8)$$

where n_i is the concentration of the ith ions (ions m^{-3}) at a point where the potential is ψ, and n_{i_o} is the concentration of ions (ions m^{-3}) in the bulk solution.

3. The thickness of the double layer is the reciprocal of κ ($1/\kappa$) where

$$\kappa = \left(\frac{1000 \ dm^3 m^{-3} e^2 N_A \Sigma_i Z_i^2 M_i}{\varepsilon kT} \right)^{1/2} , \qquad (5.9)$$

where N_A is Avogadro's number, Z_i is the valence of ion i, M_i is the molar concentration of ion i, and ε is the dielectric constant. It should be noted that when SI units are used, $\varepsilon = \varepsilon_r \ \varepsilon_o$, where $\varepsilon_o = 8.85 \times 10^{-12} \ C^2 \ J^{-1} \ m^{-1}$ and ε_r is the dielectric constant of the medium. For water at 298 K, $\varepsilon_r = 78.54$. Thus, in Eq. (5.9) $\varepsilon = (78.54) \ (8.85 \times 10^{-12} \ C^2 \ J^{-1} \ m^{-1})$.

The Gouy–Chapman theory predicts that double-layer thickness ($1/\kappa$) is inversely proportional to the square root of the sum of the product of ion concentration and the square of the valency of the electrolyte in the external solution and directly proportional to the square root of the dielectric constant. This is illustrated in Table 5.3. The actual thickness of the electrical double layer cannot be measured, but it is defined mathematically as the distance of a point from the surface where $d\psi/dx = 0$.

Box 5.3 provides solutions to problems illustrating the relationship between potential and distance from the surface and the effect of concentration and electrolyte valence on double-layer thickness.

TABLE 5.3. *Approximate Thickness of the Electric Double Layer as a Function of Electrolyte Concentration at a Constant Surface Potential*[a]

Concentration of ions of opposite charge to that of the particle (mmol dm^{-3})	Thickness of the double layer (nm)	
	Monovalent ions	Divalent ions
0.01	100	50
1.0	10	5
100	1	0.5

[a] From H. van Olphen, "An Introduction to Clay Colloid Chemistry," 2nd ed. Copyright © 1977 John Wiley & Sons, Inc. Reprinted by permission of John Wiley & Sons, Inc.

The type of colloid (i.e., variable charge or constant charge) affects various double-layer parameters including surface charge, surface potential, and double-layer thickness (Fig. 5.13). With a variable charge surface (Fig. 5.13a) the overall diffuse layer charge is increased at higher electrolyte concentration (n'). That is, the diffuse charge is concentrated in a region closer to the surface when electrolyte is added and the total net diffuse charge, $C'A'D$, which is the new surface charge, is greater than the surface charge at the lower electrolyte concentration, CAD. The surface potential remains the same (Fig. 5.13a) but since $1/\kappa$ is less, ψ decays more rapidly with increasing distance from the surface.

In variable charge systems the surface potential is dependent on the activity of PDI (potential determining ions, e.g., H$^+$ and OH$^-$) in the solution phase. The ψ_0 is not affected by the addition of an indifferent electrolyte solution (e.g., NaCl; the electrolyte ions do not react nonelectrostatically with the surface) if the electrolyte solution does not contain PDI and if the activity or concentration of PDI is not affected by the indifferent electrolyte.

BOX 5.3   *Electrical Double-Layer Calculations*

Problem 1. Plot the relationship between electrical potential (ψ) and distance from the surface (x) for the following values of x: $x = 0$, 5×10^{-9}, 1×10^{-8}, and 2×10^{-8} m according to the Gouy–Chapman theory. Given $\psi_0 = 1 \times 10^{-1}$ J C^{-1}, $M_i = 0.001$ mol dm^{-3} NaCl, $e = 1.602 \times 10^{-19}$ C, $\varepsilon = \varepsilon_r\varepsilon_0$, $\varepsilon_r = 78.54$, $\varepsilon_0 = 8.85 \times 10^{-12}$ C^2 J^{-1} m^{-1}, N_A (Avogadro's constant) $= 6.02 \times 10^{23}$ ions mol^{-1}, $k = 1.381 \times 10^{-23}$ J K^{-1}, $R = 8.314$ J K^{-1} mol^{-1}, $T = 298$ K.

First calculate κ, using Eq. (5.9):

$$\kappa = \left(\frac{1000e^2 N_A \Sigma_i Z_i^2 M_i}{\varepsilon kT}\right)^{1/2}, \tag{5.3a}$$

Substituting values,

$$\kappa = \left(\frac{(1000 \text{ dm}^3 \text{ m}^{-3})(1.602 \times 10^{19} \text{ C})^2(6.02 \times 10^{23} \text{ ions mol}^{-1}) \times [(1)^2(0.001 \text{ mol dm}^{-3}) + (-1)^2 (0.001 \text{ mol dm}^{-3})]}{(78.54)(8.85 \times 10^{-12} \text{ C}^2 \text{ J}^{-1} \text{ m}^{-1})(1.38 \times 10^{-23} \text{ J K}^{-1})(298 \text{ K})}\right)^{1/2} \tag{5.3b}$$

$$\kappa = (1.08 \times 10^{16} \text{ m}^{-2})^{1/2} = 1.04 \times 10^8 \text{ m}^{-1}. \tag{5.3c}$$

Therefore, $1/\kappa$, or the double-layer thickness, would equal 9.62×10^{-9} m. To solve for ψ as a function of x, one can use Eq. (5.7). For $x = 0$

$$\tanh\left(\frac{Ze\psi}{4kT}\right) = \tanh\left(\frac{(1)(1.602 \times 10^{-19}\ \text{C})(0.1\ \text{J C}^{-1})}{4(1.381 \times 10^{-23}\ \text{J K}^{-1})(298\ \text{K})}\right) \tag{5.3d}$$

$$\times\left(e^{-(1.04\times10^8\ \text{m}^{-1})(0\ \text{m})}\right)$$

$$\tanh\left(\frac{Ze\psi}{4kT}\right) = \tanh\left(\frac{1.60 \times 10^{-20}}{1.64 \times 10^{-20}}\right) e^0 \tag{5.3e}$$

$$\tanh\left(\frac{Ze\psi}{4kT}\right) = \tanh\ (9.76 \times 10^{-1})\ (1) \tag{5.3f}$$

$$\tanh\left(\frac{Ze\psi}{4kT}\right) = 0.75. \tag{5.3g}$$

The inverse tanh ($\tanh^{-1}$) of 0.75 is 0.97. Therefore,

$$\left(\frac{Ze\psi}{4kT}\right) = 0.97. \tag{5.3h}$$

Substituting in Eq. (5.3h),

$$\frac{(1)(1.602 \times 10^{-19}\text{C})(\psi)}{4\ (1.38 \times 10^{-23}\ \text{J K}^{-1})\ (298\ \text{K})} = 0.97. \tag{5.3i}$$

Rearranging, and solving for ψ,

$$\psi = 9.96 \times 10^{-2}\ \text{J C}^{-1}. \tag{5.3j}$$

One can solve for ψ at the other distances, using the approach above. The ψ values for the other x values are $\psi = 4.58 \times 10^{-2}$ J C^{-1} for $x = 5 \times 10^{-9}$ m, $\psi = 2.72 \times 10^{-2}$ J C^{-1} for $x = 1 \times 10^{-8}$ m, and $\psi = 9.62 \times 10^{-3}$ J C^{-1} for $x = 2 \times 10^{-8}$ m. One can then plot the relationship between ψ and x as shown in Fig. 5.B1.

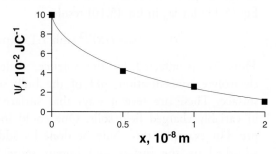

FIGURE 5.B1.

Problem 2. Compare the "thickness" of the double layer ($1/\kappa$) for 0.001 M (mol dm^{-3}) NaCl, 0.01 M NaCl, and 0.001 M CaCl$_2$.

The $1/\kappa$ for 0.001 mol dm^{-3} NaCl was earlier found to be 9.62×10^{-9} m. For 0.01 M NaCl,

$$\kappa = \left(\frac{(1000 \text{ dm}^3 \text{ m}^{-3})(1.602 \times 10^{-19} \text{ C})^2(6.02 \times 10^{23} \text{ ions mol}^{-1})}{\times [(1)^2(0.01 \text{ mol dm}^{-3}) + (-1)^2 (0.01 \text{ mol dm}^{-3})]}{(78.54)(8.85 \times 10^{-12} \text{ C}^2 \text{ J}^{-1} \text{ m}^{-1})(1.38 \times 10^{-23} \text{ J K}^{-1})(298 \text{ K})} \right)^{1/2} \quad (5.3k)$$

$$\kappa = (1.08 \times 10^{17} \text{ m}^{-2})^{1/2} = 3.29 \times 10^8 \text{ m}^{-1} \quad (5.3l)$$

$$\frac{1}{\kappa} = 3.04 \times 10^{-9} \text{ m}. \quad (5.3m)$$

The $1/\kappa$ for 0.001 M CaCl$_2$ is

$$\kappa = \left(\frac{(1000 \text{ dm}^3 \text{ m}^{-3})(1.602 \times 10^{-19} \text{ C})^2(6.02 \times 10^{23} \text{ ions mol}^{-1})}{\times [(2)^2(0.001 \text{ mol dm}^{-3}) + (-1)^2 (2) (0.001 \text{ mol dm}^{-3})]}{(78.54)(8.85 \times 10^{-12} \text{ C}^2 \text{ J}^{-1} \text{ m}^{-1})(1.38 \times 10^{-23} \text{ J K}^{-1})(298 \text{ K})} \right)^{1/2} \quad (5.3n)$$

$$\kappa = (3.24 \times 10^{16} \text{ m}^{-2})^{1/2} = 1.80 \times 10^8 \text{ m}^{-1}. \quad (5.3o)$$

In variable charge systems the surface charge (σ_v) is

$$\sigma_v = \left(\frac{2n\varepsilon kT}{\pi} \right)^{1/2} \sinh \left[\left(\frac{Ze}{2kT} \right) \text{constant } \psi_0, \right] \quad (5.10)$$

where sinh is the hyperbolic sin. If the PDI are H$^+$ and OH$^-$, the constant surface potential is related to H$^+$ by the Nernst equation

$$\psi_0 = \left(\frac{kT}{e} \right) \ln \text{ H}^+/\text{H}_0^+ \quad \text{or} \quad \psi_0 = \left(\frac{kT}{e} \right) 2.303 \text{ (pzc} - \text{pH)}, \quad (5.11)$$

where Z is the valence of the PDI, H$^+$ is the activity of the H$^+$ ion, H$_0^+$ is the hydrogen ion activity when ψ_0 is equal to 0, and pzc is the point of zero charge (suspension pH at which a surface has a net charge of 0). Substituting Eq. (5.11) for ψ_0 in Eq. (5.10) results in

$$\sigma_v = (2n\varepsilon kT/\pi)^{1/2} \sinh 1.15Z(\text{pzc} - \text{pH}). \quad (5.12)$$

Thus, σ_v is affected by the valence, dielectric constant, temperature, electrolyte concentration, pH of the bulk solution, and the pzc of the surface. There are several ways that surface charge can be manipulated in variably charged field soils. One could increase CEC by lowering the pzc. For example, this could be done by adding an anion that would be adsorbed on the surface and impart more negative charge (Wann and Uehara, 1978). The latter investigators added phosphate to an Oxisol, which

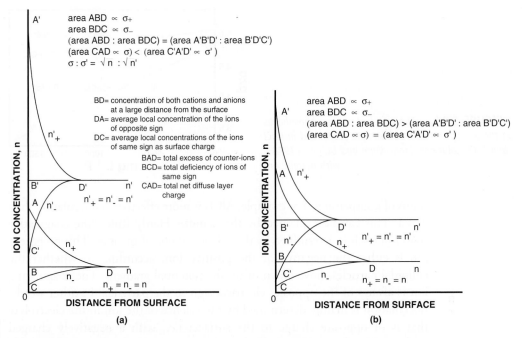

FIGURE 5.13. *Charge distribution in the diffuse double layer of a negatively charged particle surface at two electrolyte concentrations, n (lower) and n' (higher). (a) Variable surface charge mineral. (b) Constant surface charge mineral. From H. van Olphen, "An Introduction to Clay Colloid Chemistry," 2nd ed. Copyright © 1977 John Wiley & Sons, Inc. Reprinted by permission of John Wiley & Sons, Inc.*

resulted in the pzc decreasing with a concomitant increase in net negative charge (Fig. 5.14).

The CEC of a soil containing variably charged components increases as pH increases. Liming the soil would increase the pH and the CEC. However, it is difficult to raise the pH of a variable charge soil above 6.5, particularly if the soil has a high buffering capacity along with a high surface area (Uehara and Gillman, 1981; Singh and Uehara, 1986). This is because the OH^- ion, which is produced from the hydrolysis of the CO_3^{2-} ion when $CaCO_3$ is added, can raise the soil pH. However, in a variably charged system containing hydroxylated surfaces, H^+ ions, which neutralize the OH^-, are released. This results in significant resistance to pH change or buffering.

With a constant surface charge mineral, e.g., vermiculite, the total net surface charge, *CAD*, is not affected by higher electrolyte concentration (n'), but $1/\kappa$ is lower and ψ_0 decreases. In fact, double-layer potential, dielectric constant, temperature, and counterion valence do not influence the sign or magnitude of the surface charge, but any change in these parameters would be offset by a reduction or increase in ψ_0 (Fig. 5.13). This is effected by compression of the double layer, which depends on electrolyte concentration and valence of the counterions.

The Gouy–Chapman model of the electric double layer can be used to predict the effect of electrolyte valence on colloidal stability. The valence of the electrolyte significantly affects the stability (a disperse system) or flocculation

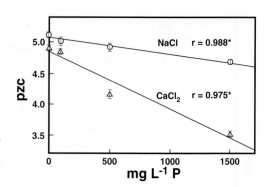

FIGURE 5.14. *The effect of applied phosphorus on the pzc of an Oxisol soil sample suspended in NaCl and CaCl₂ solutions. From Wann and Uehara (1978), with permission.*

status of a suspension. For example, $AlCl_3$ is more effective in flocculation than NaCl. This can be explained by the Schulze–Hardy Rule: "the coagulative power of a salt is determined by the valency of one of the ions. This prepotent ion is either the negative or the positive ion, according to whether the colloidal particles move down or up the potential gradient. The coagulating ion is always of the opposite electrical sign to the particle." In other words, flocculation is mainly determined by the valence of the ion in the electrolyte that is of opposite charge to the surface; i.e., with a negatively charged surface, the valence of the cation in the electrolyte is important. Thus, the higher the valence, the greater the flocculating power of the electrolyte, and hence the lower the electrolyte concentration needed to cause flocculation.

For example, the following concentration ranges are needed to cause flocculation of suspensions: 25–150 mmol dm^{-3} for monovalent ions, 0.5–2.0 mmol dm^{-3} for divalent ions, and 0.01–0.1 mmol dm^{-3} for trivalent ions. This rule is operational only for indifferent electrolytes that do not undergo any specific (nonelectrostatic such as inner-sphere complexation) adsorption with the surface. Thus, the electrolyte cannot contain any PDI or other specifically adsorbed ions; no chemical interactions with the sorptive can occur. The flocculating power of group I cations in the periodic table for negative suspensions decreases slightly in the order $Cs^+ > Rb^+ > NH_4^+ > K^+ > Na^+ > Li^+$.

There are a number of problems with using the Gouy–Chapman double-layer model to describe sorption reactions at soil surfaces. It often fails to describe the distribution of ions adjacent to soil colloidal surfaces. For example, the assumption that the ions behave as point charges results in too high of a concentration of counterions calculated to be adjacent to the charged surface. This in turn results in calculated surface potentials being much too high and unrealistic for soil systems. It is also known that the dielectric constant of water next to a charged surface is about 6 versus a value of 80 in the bulk solution. This means there is much more interaction between ions adjacent to the charged surface than assumed present with simple electrostatic attraction of the ions. There is also no provision in the Gouy–Chapman model for surface complexation, and the charged surface in the original model was assumed to be plate-shaped and very large. Soil particles have a much more complex shape and vary in size.

STERN THEORY

A diagram of the Stern model for describing the electric double layer is shown in Fig. 5.15. Stern (1924) modified the Gouy–Chapman model as follows. The first layer of ions (Stern layer) is not immediately at the surface, but at a distance away from it. The counterion charge is separated from the surface charge by a layer of thickness δ in which no charge exists. As a result, the concentration and potential in the diffuse part of the double layer drops to values low enough to warrant the approximation of ions as point charges. Stern further considered the possibility of specific adsorption of ions and assumed that these ions were located in the Stern layer. The Stern model then assumes that the charge that exists at the surface (σ) is balanced by charge in solution, which is distributed between the Stern layer (σ_1) and a diffuse layer (σ_2). The total surface charge, σ, in C m^{-2}, is

$$\sigma = -(\sigma_1 + \sigma_2), \tag{5.13}$$

where σ_1 is the Stern layer charge in C m^{-2} and σ_2 is the diffuse layer charge in C m^{-2}.

The charge in the Stern layer is

$$\sigma_1 = N_i Ze/\{1 + (N_A\, w/Mn)\, \exp\,[-(Ze\psi_\delta + \varphi)/kT]\}, \tag{5.14}$$

where N_i is the number of available sites m^{-2} for absorption, M is the molecular weight of the solvent in g mol^{-1}, w is the solvent density in kg m^{-3}, n is the electrolyte concentration in ions m^{-3}, ψ_δ is the Stern potential or electrical potential at the boundary between the Stern layer and the diffuse layer in V, φ is the specific adsorption potential in the Stern layer (in J), and the other terms have already been defined.

The charge in the diffuse layer is given by the Gouy–Chapman theory except the reference is now the Stern potential instead of the surface potential,

$$\sigma_2 = (2n\varepsilon kT/\pi)^{1/2}\, \sinh\,(Ze\psi_\delta/2kT). \tag{5.15}$$

In Eqs. (5.14) and (5.15) σ_1 and σ_2 are valid for both constant and variable charge systems. Thus, the Stern model considers that the distance of closest approach of a counterion to the charged surface is limited by the size of the ion. With this model, the addition of electrolyte results in not only a compression of the diffuse part of the double layer but also in a shift of the counterions from the diffuse layer to the Stern layer and hence in a decrease of the Stern potential, ψ_δ.

Few studies to compare the prediction of surface charge behavior on soil colloids using the Gouy–Chapman and Stern theories have been conducted. In a classic study, van Raji and Peech (1972) compared the Gouy–Chapman and Stern models for describing surface charge on variably charged soils from the tropics (Fig. 5.16). The relationship between the net negative surface charge as a function of the surface potential for the two models at several electrolyte strengths is shown. The surface potential, ψ_0, was calculated using the Nernst equation, Eq. (5.11). The net surface charge was determined by

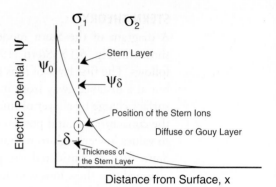

FIGURE 5.15.   *Diagram of the Stern model.*
From H. van Olphen, "An Introduction to Clay Colloid
Chemistry," 2nd ed. Copyright © 1977 John Wiley &
Sons, Inc. Reprinted by permission of John Wiley
and Sons, Inc.

δ = Specific adsorption potential
ψ_0 = Surface potential
ψ_δ = Stern potential
σ_1 = Stern layer charge
σ_2 = Diffuse layer charge

experimentally measuring point of zero salt effect (PZSE) values. (See the section on points of zero charge and Fig. 5.34 for experimental approach for determining PZSE.) The surface charge was calculated using the Gouy–Chapman theory, Eq. (5.10), after introducing the value for ψ_0 from the Nernst equation. Surface charge from the Stern model was calculated using Eq. (5.14) for σ_1, Eq. (5.15) for σ_2, and Eq. (5.13). The agreement between the net surface charge, calculated from theory and experimentally measured values, was better when the Stern theory was used than when the Gouy–Chapman theory was employed, except at the 0.001 M NaCl concentration (Fig. 5.16). The net surface charge values were lower than those predicted by the Gouy–Chapman theory, although the agreement improved with lower electrolyte concentration and with decreasing surface potential.

Surface Complexation Models

Due to the deficiencies inherent in the Gouy–Chapman and Stern models, the retention of ions on soil surfaces has been described using surface complexation models. These models have been widely used to describe an array of chemical reactions including proton dissociation, metal cation and anion adsorption reactions on oxides, clays, and soils, organic ligand adsorption on oxides, and competitive adsorption reactions on oxides. Applications and theoretical aspects of surface complexation models are extensively reviewed in Goldberg (1992). Surface complexation models often are chemical models based on molecular descriptions of the electric double layer using equilibrium-derived adsorption data (Goldberg, 1992). They include the constant capacitance (CCM), triple-layer (TLM), and modified triple-layer, Stern variable surface charge–variable surface potential

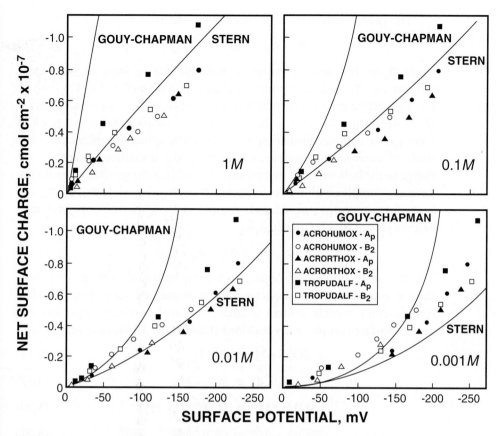

FIGURE 5.16. *Comparison of the net negative surface charge of soils as determined by potentiometric titration (determining PZSE values, see Fig. 5.34 for experimental approach) with that calculated by Gouy–Chapman and Stern theories. The ratio δ/ε', where δ is the thickness of the Stern layer and ε' is the average dielectric constant, used in calculating the net surface charge was 0.015. From van Raji and Peech (1972), with permission.*

(VSC–VSP), generalized two-layer, and the one-pK models. Differences in the surface complexation models lie in the descriptions of the electrical double layer, i.e., in the definition and assignment of ions to the planes or layers of adsorption and in differences in the electrostatic equations and the relations between the surface potential and surface charge. These models provide some information on the physical description of the electric double layer, including the capacitance and location of adsorbed ions, and they can describe data over a broad range of experimental conditions such as varying pH and *I*. Two kinds of data can be derived from these models: material balance data, i.e., the quantity of a material adsorbed, and information that can be used to describe electrokinetic phenomena (Westall and Hohl, 1980). Common characteristics of surface complexation models are consideration of surface charge balance, electrostatic potential terms, equilibrium constants, capacitances, and surface charge density. A summary of characteristics of the different surface complexation models is given in Table 5.4.

The general balance of surface charge model is

$$\sigma_o + \sigma_H + \sigma_{is} + \sigma_{os} + \sigma_d = 0, \tag{5.16}$$

where σ_o is the constant charge in minerals due to ionic or isomorphic substitution, σ_H is the net proton charge, which is equal to $\Gamma_H - \Gamma_{OH}$, where Γ is the surface excess concentration (σ_H is equivalent to the dissociation of H^+ in the diffuse layer), σ_{is} is the inner-sphere complex charge as a result of inner-sphere complex formation, σ_{os} is the outer-sphere complex charge as a result of outer-sphere complexes, and σ_d is the dissociated charge or the charge in the bulk solution that balances the surface charge (these ions do not form any complex with the surface). The σ_o is negative while σ_H, σ_{is}, σ_{os} can be positive, negative, or neutral. Another way to describe the net total particle charge (σ_p) on a colloid is

$$\sigma_p = \sigma_o + \sigma_H + \sigma_{is} + \sigma_{os}. \tag{5.17}$$

The σ_p can be positive or negative but must be balanced by σ_d or ions in the soil solution or in the dissociated form. As pointed out earlier, all surface complexation models contain a balance of surface charge equation and general surface complexation reactions (Hohl et al., 1980; Goldberg, 1992),

$$SOH + H^+ \equiv SOH_2^+ \tag{5.18}$$

$$SOH \equiv SO^- + H^+ \tag{5.19}$$

$$SOH + M^{n+} \equiv SOM^{(n-1)} + H^+ \tag{5.20}$$

$$2SOH + M^{n+} \equiv (SO)_2\, M^{(n-2)} + 2H^+ \tag{5.21}$$

$$SOH + L^{l-} \equiv SL^{(l-1)-} + OH^- \tag{5.22}$$

$$2SOH + L^{l-} \equiv S_2L^{(l-2)-} + 2OH^-, \tag{5.23}$$

where SOH is the surface functional group and S represents the metal bound to the surface functional group, e.g., OH of an oxide surface or of the aluminol or silanol group of a clay mineral, M is the metal ion, n^+ is the charge on the metal ion, L is a ligand, and l^- is the ligand charge. The intrinsic equilibrium constants (see Box 5.4 for a discussion of intrinsic and conditional equilibrium constants) for reactions in Eqs. (5.18)–(5.23) are (Hohl et al., 1980; Goldberg, 1992)

$$K_+^{int} = \frac{[SOH_2^+]}{[SOH][H^+]}\, \exp[F\psi_i\, /\, RT] \tag{5.24}$$

$$K_-^{int} = \frac{[SO^-][H^+]}{[SOH]}\, \exp[-F\psi_i\, /\, RT] \tag{5.25}$$

$$K_{M1}^{int} = \frac{[SOM^{(n-1)}][H^+]}{[SOH][M^{n+}]}\, \exp[(n-1)\, F\psi_i\, /\, RT] \tag{5.26}$$

$$K_{M2}^{int} = \frac{[(SO)_2M^{(n-2)}][H^+]^2}{[SOH]^2[M^{n+}]}\, \exp[(n-2)\, F\psi_i\, /\, RT] \tag{5.27}$$

TABLE 5.4. *Characteristics of Surface Complexation Models[a]*

Model	Complexation	Reference state	Relationship between surface charge and surface potential	Balance of surface charge expression
Constant capacitance model	Inner-sphere	Constant ionic medium reference state determines the activity coefficients of the aqueous species in the conditional equilibrium constants.	$\sigma = (CAa/F)\psi_o^b$ (linear relationship)	$\sigma = \sigma_H + \sigma_{is}$
Original triple-layer model	Outer-sphere (metal ions, ligands, cations, and anions); inner-sphere (H^+ and OH^-)	Infinite dilution reference state determines the activity coefficients of the aqueous species in the conditional equilibrium constants.	$\psi_o - \psi_\beta = \sigma_o/C_1$ $\psi_\beta - \psi_d = -\sigma_d/C_2$ $\sigma_d = -(8RTC\varepsilon_o\varepsilon_r)^{1/2}$ $\sinh(F\psi_d/2RT)$	$\sigma = \sigma_H + \sigma_{os}$
Modified triple-layer model	Inner-sphere in o plane; outer-sphere in β plane	Infinite dilution and zero surface charge.		$\sigma = \sigma_H + \sigma_{is} + \sigma_{os}$
Stern variable surface charge-variable surface potential model (four-layer model)	Inner-sphere (H^+, OH^-, and strongly adsorbed oxyanions and metals); outer-sphere (cations and anions)	Defines no surface reactions, no equilibrium constant expressions and specific surface species.	$\psi_o - \psi_a = \sigma_o/C_{oa}$ $\psi_a - \psi_\beta = (\sigma_o + \sigma_a)/C_{a\beta}$ $\psi_\beta - \psi_d = -\sigma_d/C_{\beta d}$	$\sigma = \sigma_H + \sigma_{is} + \sigma_{os}$
Generalized two-layer model	Inner-sphere	Infinite dilution for the solution and a reference state of zero charge and potential for the surface.	$\sigma = (CAa/F)\psi_o^b$ $\sigma_d = -(8RTC\varepsilon_o\varepsilon_r)^{1/2}$ $\sinh(F\psi_d/2RT)$	$\sigma = \sigma_H + \sigma_{is}$
One-pK Stern model	Inner-sphere (H^+ and OH^-); outer-sphere (metal ions, ligands, cations and anions)		$\sigma_o = C(\psi_o - \psi_d)$ $\sigma_d = -(8RTC\varepsilon_o\varepsilon_r)^{1/2}$ $\sinh(F\psi_d/2RT)$	$\sigma = \sigma_H$

[a] Information to prepare table taken from Goldberg (1992).
[b] C refers to capacitance density.

$$K_{L1}^{int} = \frac{[SL^{(l-1)-}][OH^-]}{[SOH][L^{l-}]} \exp[-(l-1)\,F\psi_i\,/\,RT] \qquad (5.28)$$

$$K_{L2}^{int} = \frac{[S_2L^{(l-2)-}][OH^-]^2}{[SOH]^2[L^{l-}]} \exp[-(l-2)\,F\psi_i\,/\,RT], \qquad (5.29)$$

where brackets indicate concentrations in mol liter^{-1}, ψ_i is the surface potential in V in the ith surface plane, F is the Faraday constant in C mol^{-1}, R is the gas constant in J mol^{-1} K^{-1}, and T is absolute temperature in degrees Kelvin. The log of the intrinsic equilibrium constants can be obtained by plotting the log of the conditional equilibrium constants versus surface charge (σ) and extrapolating to zero surface charge (Stumm $et\ al.$, 1980). The term $e_i^{-F\psi_i/RT}$ considers surface charge effects on surface complexation. Surface complexation models also contain several adjustable parameters including K_i, equilibrium constants; C_i, capacitance density for the ith surface plane; and $[SOH]_T$, the total number of reactive surface hydroxyl groups. Details on the determination of these parameters can be found in Goldberg (1992).

BOX 5.4 *Calculation of Conditional and Intrinsic Equilibrium Constants*

Consider the following ionization reactions illustrating the variable charge behavior of a surface (Dzombak and Morel, 1990),

$$SOH_2^+ \rightleftharpoons SOH + H^+ \qquad (5.4a)$$

$$SOH \rightleftharpoons SO^- + H^+, \qquad (5.4b)$$

where SOH_2^+, SOH, and SO^- are positively charged, neutral, and negatively charged surface hydroxyl groups. One can write conditional acidity constants, K_{a1}^{cond} for Eq. (5.4a) and K_{a2}^{cond} for Eq. (5.4b) as

$$K_{a1}^{cond} = \frac{[SOH][H^+]}{[SOH_2^+]} \qquad (5.4c)$$

$$K_{a2}^{cond} = \frac{[SO^-][H^+]}{[SOH]}. \qquad (5.4d)$$

The activity coefficients of the surface species are assumed to be equal. The K_{a1}^{cond} and K_{a2}^{cond} are conditional equilibrium constants because they include effects of surface charge and are thus dependent on the degree of surface ionization, which is dependent on pH. Conditional equilibrium constants thus hold under given experimental conditions, e.g., a given pH.

The total free energy of sorption (ΔG_{tot}^o) can be separated into component parts,

$$\Delta G_{tot}^o = \Delta G_{int}^o + \Delta G_{coul}^o, \qquad (5.4e)$$

where ΔG^o_{int} is the chemical or "intrinsic" free-energy term and ΔG^o_{coul} is the variable electrostatic or "coulombic" term. An expression for the ΔG^o_{coul}, determined by taking into account the electrostatic work required to move ions through an interfacial potential gradient (Morel, 1983), is substituted for ΔG^o_{coul} such that

$$\Delta G^o_{tot} = \Delta G^o_{int} + \Delta ZF\psi_o, \qquad (5.4f)$$

where ΔZ is the change in the charge of the species resulting from sorption. Since $\Delta G^o_r = -RT \ln K^o$, Eq. (5.4f) can be rewritten as:

$$K^{int} = K^{cond} \exp (\Delta ZF\psi_o / RT) \qquad (5.4g)$$

where K^{int} is referred to as an intrinsic equilibrium constant. The exponential term in Eq. (5.4g) is referred to as an electrostatic or coulombic correction factor. This term allows one to consider the effects of surface charge variations on surface complexation reactions.

If the specific and nonspecific interactions for the surface acidity reactions (Eqs. (5.4a and 5.4b)) are separated, one obtains

$$K^{int}_{a1} = \frac{[SOH][H^+_s]}{[SOH_2]} = K^{cond}_{a1} \exp (-F\psi_0 / RT) \qquad (5.4h)$$

$$K^{int}_{a2} = \frac{[SOH^-][H^+_s]}{[SOH]} = K^{cond}_{a2} \exp (-F\psi_o / RT), \qquad (5.4i)$$

where H^+_s is a proton released at the surface but not moved to the bulk solution. This differentiation is made since K^{int} refers only to specific chemical interactions at the surface.

CONSTANT CAPACITANCE MODEL (CCM)

The CCM model was formulated by Schindler, Stumm, and co-workers (Schindler and Gamsjager, 1972; Hohl and Stumm, 1976). It assumes that (Goldberg, 1992): (1) all surface complexes are inner-sphere and anions are adsorbed by a ligand exchange mechanism, (2) a constant ionic strength reference state determines the activity coefficient of the aqueous species in the conditional equilibrium constant, and (3) there is a linear relationship between surface charge and surface potential (ψ_o), which is expressed as

$$\sigma = (CAa/F)\psi_0, \qquad (5.30)$$

where C is capacitance density in F m^{-2}, A is specific surface area in m^2 g^{-1}, a is suspension density in g liter^{-1}, and σ is expressed in units of mol liter^{-1}. The balance of surface charge equation is

$$\sigma = \sigma_H + \sigma_{is}. \qquad (5.31)$$

An illustration of the CCM is shown in Fig. 5.17, and the application of metal adsorption data to the CCM is shown in Fig. 5.18.

TRIPLE-LAYER MODEL (TLM)

The TLM was developed by Davis and Leckie (1978, 1980). It consists of two capacitance layers and a diffuse layer (Fig. 5.19). The assumption of the TLM is that all metals and ligands are retained as outer-sphere complexes. Only H^+ and OH^- are adsorbed as inner-sphere complexes. The balance of surface charge equation for the TLM is (Goldberg, 1992)

$$\sigma = \sigma_H + \sigma_{os}. \tag{5.32}$$

As illustrated in Fig. 5.19 the PDI at the surface, H^+ and OH^-, are at the o layer next to the surface and all other metals, ligands, major cations (C^+) and anions (A^-) are at the β layer or β plane. The diffuse layer begins at the d plane and extends into the bulk solution. The TLM includes three equations, representing the three layers, that relate potential to charge (Davis *et al.*, 1978),

$$\psi_o - \psi_\beta = \sigma_o/C_1 \tag{5.33}$$

$$\psi_\beta - \psi_d = \sigma_d/C_2 \tag{5.34}$$

$$\sigma_d = -(8RTC\varepsilon_o\varepsilon_r)^{1/2} \sinh(F\psi_d/2RT), \tag{5.35}$$

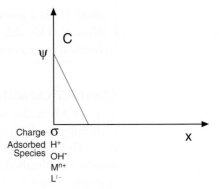

FIGURE 5.17. *Placement of ions, potential, charge, and capacitance for the constant capacitance model. C refers to capacitance density. Reprinted with permission from Westall (1986). Copyright 1986 American Chemical Society.*

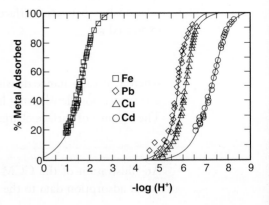

FIGURE 5.18. *Fit of the constant capacitance model to metal adsorption on silica. Model results are represented by solid lines. □, Fe; ◊, Pb; △, Cu; O, Cd. From Schindler et al. (1976), with permission.*

where σ has the units C m^{-2}, C_1 and C_2 are the capacitance densities in the o and β layers, respectively, ε_o is the permitivity of vacuum, ε_r is the dielectric constant of water, and C is the concentration of a 1:1 background electrolyte (e.g., NaClO$_4$; both the cation and anion have a charge of 1). For 1:2, e.g., Na$_2$SO$_4$ (the cation has a charge of 1 and the anion a charge of 2), and for 2:1, e.g., CaCl$_2$ (the cation has a charge of 2 and the anion has a charge of 1), background electrolytes the above equations are more complicated.

The TLM has two characteristics not included in other surface complexation models: the chemical constants are applicable over a wide range of ionic strength and the value of ψ_d can be used as an estimate of the electrokinetic potential (Goldberg, 1992). The electrokinetic (zeta) potential refers to the electrical potential at the surface of shear between immobile liquid attached to a charged particle and mobile liquid further removed from the particle (Glossary of Soil Science Terms, 1987).

The modified TLM was developed by Hayes and Leckie (1986, 1987) and allows metals and ligands to be adsorbed as inner-sphere complexes. Under these conditions the surface balance equation is

$$\sigma = \sigma_H + \sigma_{is} + \sigma_{os}. \tag{5.36}$$

An application of the modified TLM to selenate adsorption on goethite is shown in Fig. 5.20.

STERN VARIABLE SURFACE CHARGE–VARIABLE SURFACE POTENTIAL (VSC–VSP) MODEL

This model was formulated by Bowden et al. (1977, 1980) and Barrow et al. (1980, 1981). It is a hybrid between the CCM and TLM and is commonly known as the four-layer model. The assumptions of the VSC–VSP model are that H$^+$, OH$^-$, and strongly adsorbed oxyanions such as arsenate and metals develop inner-sphere complexes and major cations and anions form outer-sphere complexes. The balance of surface charge equation for this model is the same as for the modified TLM (Eq. 5.36).

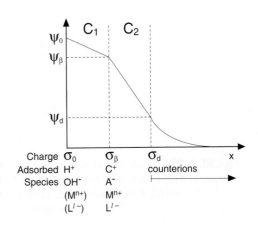

FIGURE 5.19.   Placement of ions, potentials, charges, and capacitances for the triple-layer model. Parentheses represent ion placement allowed only in the modified triple-layer model. Reprinted with permission from Westall (1980). Copyright © 1980 American Chemical Society.

Figure 5.21 illustrates the VSC–VSP model. Hydrogen ions and OH^- are in the o plane, strongly adsorbed oxyanions and metals are in the a plane, major cations and anions are present as outer-sphere complexes and are in the β plane, and the d plane is the start of the diffuse double layer. The surface functional group is $OH–S–OH_2$, or one protonation or dissociation can occur for every two surface OH groups (Barrow *et al.*, 1980; Goldberg, 1992). The VSC–VSP model defines no surface reactions, equilibrium constant expressions, and specific surface species. Consequently, one could consider it not to be a chemical model. There are four charge potential equations and four charge balance equations that are given in Goldberg (1992).

GENERALIZED TWO-LAYER MODEL

This model was promulgated by Dzombak and Morel (1990) and is based on the DDL description of Stumm and co-workers (Huang and Stumm, 1973). It assumes that all surface complexes are inner-sphere. The description of surface charge is identical to that employed in the CCM (Eq. (5.30)). Similarly to the TLM and the VSC–VSP models, the generalized two-layer model includes a diffuse layer and derives its name from two surface planes of charge (Fig. 5.22). Moreover, all surface complexes are found in the surface

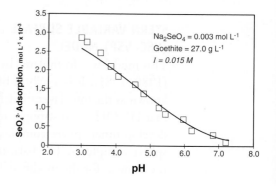

FIGURE 5.20. *Adsorption of SeO_4^{2-} (selenate) on goethite vs pH. The symbols represent the experimental data and the solid line represents TLM conformity, assuming nonspecific adsorption with constant C_1 and $C_2 = 0.13$. Reprinted with permission from Zhang and Sparks (1990b). Copyright © 1990 American Chemical Society.*

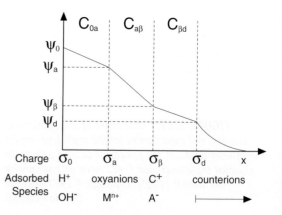

FIGURE 5.21. *Placement of ions, potentials, charges, and capacitances for the Stern VSC–VSP model. From Bowden et al. (1980), with permission.*

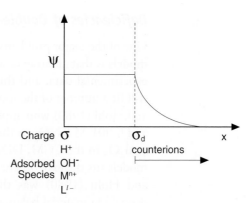

FIGURE 5.22.   *Placement of ions, potential, and charges for the generalized two-layer model. From D. A. Dzombak and F. M. M. Morel, "Surface Complexation Modeling. Hydrous Ferric Oxide." Copyright © 1990 John Wiley & Sons, Inc. Reprinted by permission of John Wiley & Sons, Inc.*

plane and the diffuse layer starts at the d plane and goes to the bulk solution. Metal ions are adsorbed on two types of sites, a small set of high-affinity sites and a large set of low-affinity sites (Goldberg, 1992).

ONE-pK STERN MODEL

This model was proposed by Bolt and van Riemsdijk (1982) and formulated by van Riemsdijk and co-workers (1986, 1987; Hiemstra *et al.*, 1987; van Riemsdijk and van der Zee, 1991). The surface functional group is a singly coordinated oxygen atom that carries one or two H^+, resulting in the surface sites SOH and $(SOH)_2$, respectively. It is similar to the Stern model and assumes that H^+ and OH^- form inner-sphere complexes and are in the o plane and major cations and anions, metals, and ligands form outer-sphere complexes at the Stern plane, or d plane. The Stern plane is at the start of the DDL (Fig. 5.23). The surface charge balance equation for the one pK model is (Goldberg, 1992)

$$\sigma = \sigma_H. \tag{5.37}$$

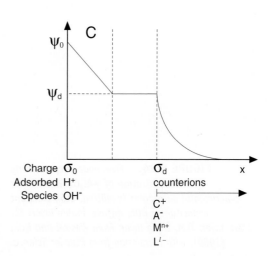

FIGURE 5.23.   *Placement of ions, potentials, charges, and capacitances for the one-pK Stern model. Reprinted with permission from Westall (1986). Copyright 1986 American Chemical Society.*

Deficiencies of Double-Layer and Surface Complexation Models

One of the major problems with many of the DDL and surface complexation models is that an array of adjustable parameters are often employed to fit the experimental data, and thus it may not be surprising that equilibrium data will fit a number of the models equally well. This was pointed out by Westall and Hohl (1980) who applied acid–base titration data for TiO_2 in 0.1, 0.01, and 0.001 M KNO_3 to the CCM and TLM and acid/base titration data for γ-Al_2O_3 to the CCM, DDL, Stern, and TLM. They showed that each of the models fits the experimental data well (Fig. 5.24). The conclusion of Westall and Hohl (1980) was that one model is about as good as another in describing material balance data.

Another disadvantage of many of the equilibrium-based models that have been used to describe sorption is that they do not consider surface precipitation (discussed in the next section). Farley *et al.* (1985) and James and Healy (1972) considered surface precipitation in successfully modeling sorption of hydrolysable metal ions. In fact, Farley *et al.* (1985) incorporated surface precipitation into a thermodynamic sorption model that is a surface complexation model. Major needs in modeling sorption on soils and soil components are to include surface precipitation and other nonadsorption phenomena, molecular scale spectroscopic findings about sorption complexes and surface precipitates, and microbial effects as part of the model description and prediction. A number of recent studies have appeared on incorporating molecular scale findings (Katz and Hayes, 1995a, b; Hayes and Katz, 1996; Katz and Boyle-Wight, 2001) into sorption models.

Sorption of Metal Cations

Sorption of metal cations is pH-dependent and is characterized by a narrow pH range where sorption increases to nearly 100%, traditionally known as an adsorption edge (Fig. 5.25). The pH position of the adsorption edge for a

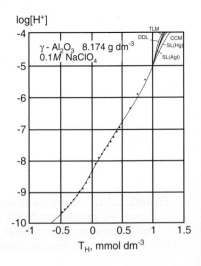

FIGURE 5.24.   *Five models representing acid–base titration of γ-Al_2O_3. Every other experimental data point is plotted. CCM, constant capacitance; DDL, diffuse double layer; SL, Stern layer; TLM, triple layer. From Westall and Hohl (1980), with permission from Elsevier Science.*

particular metal cation is related to its hydrolysis or acid–base characteristics. In addition to pH, sorption of metals is dependent on sorptive concentration, surface coverage, and the type of the sorbent.

One can measure the relative affinity of a cation for a sorbent or the selectivity. The properties of the cation, sorbent, and the solvent all affect the selectivity. With monovalent alkali metal cations, electrostatic interactions predominate and the general order of selectivity is $Li^+ < Na^+ < K^+ < Rb^+ < Cs^+$ (Kinniburgh and Jackson, 1981). This order is related to the size of the hydrated radius. The ion in the above group with the smallest hydrated radius, Cs^+, can approach the surface the closest and be held the most tightly. However, on some hydrous oxides, the reverse order is often observed (Table 5.5). This has been particularly noted for some hydrous metal oxides. The reason for this selectivity is not well understood, but may be related to the effect of the solid on water that is present on the oxide surface (Kinniburgh and Jackson, 1981) or to variation in the solution matrix.

With divalent ions there is little consistency in the selectivity order. Table 5.6 shows the selectivity of alkaline earth cations and divalent transition and heavy metal cations on metal oxide surfaces. The type of surface and the pH both appear to have major effects on the selectivity sequence. Differences in the H^+/M^{n+} could cause reversals in selectivity since the ion with the higher H^+/M^{n+} stoichiometry would be favored at higher pH (Kinniburgh and Jackson, 1981). Divalent transition and heavy metal cations, both of which are often sorbed as inner-sphere complexes (Table 5.1), are more strongly sorbed than alkaline earth cations.

Recently, studies on the sorption of metals on bacterial species have appeared. Figure 5.26 shows data for Cd sorption on various gram-positive and gram-negative bacterial species. The pH-dependent sorption behavior is similar for the different bacterial species with the sorption edge strongly resembling metal cation sorption behavior.

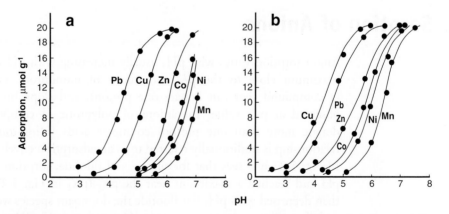

FIGURE 5.25. *Sorption of a range of metals on (a) hematite and (b) goethite when they were added at a rate of 20 μmol g^{-1} of adsorbate. The values for the pK$_1$ for dissociation of the metals to give the monovalent MOH$^+$ ions are Pb, 7.71; Cu, 8; Zn, 8.96; Co, 9.65; Ni, 9.86; and Mn, 10.59. From McKenzie (1980), with permission.*

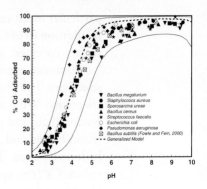

FIGURE 5.26. *Cd sorption onto pure cultures of various gram-positive and gram-negative bacterial species. Each data point represents individual batch experiments with $10^{-4.1}$ M Cd and 1.0 g/liter (dry wt.) bacteria. The dotted curve represents the sorption behavior calculated using the average parameters given in Table 1 of Yee and Fein (2001). The solid lines represent the upper and lower uncertainity limits calculated from the errors in pK_a, site concentration, and log K_{ads} measurements. From Yee and Fein (2001), with permission from Elsevier Science.*

TABLE 5.5. *Selectivity Sequences for the Alkali Metal Cations on Various Hydrous Metal Oxides[a]*

Sequence	Oxide	Reference
Cs > Rb > K > Na > Li[b]	Si gel	Helfferich (1962b)
K > Na > Li	Si gel	Tien (1965)
Cs > K > Li	SiO_2	Altug and Hair (1967)
Cs > K > Na > Li	SiO_2	Abendroth (1970)
Li > Na > K	SiO_2	Bartell and Fu (1929)
K > Na > Li	Al_2O_3	Churms (1966)
Li > K ~ Cs	Fe_2O_3	Breeuwsma and Lyklema (1971)
Li > Na > K ~ Cs	Fe_2O_3	Dumont and Watillon (1971)
Cs ~K > Na > Li	Fe_3O_4	Venkataramani *et al.* (1978)
Li > Na > Cs	TiO_2	Berubé and de Bruyn (1968)
Li > Na > K	$Zr(OH)_4$	Britz and Nancollas (1969)

[a] From Kinniburgh and Jackson (1981), with permission.
[b] All cations are monovalent.

Sorption of Anions

Anion sorption varies with pH, usually increasing with pH and reaching a maximum close to the pK_a for anions of monoprotic conjugate acids (a compound that can donate one proton), and slope breaks have been observed at pK_a values for anions of polyprotic (a compound that can donate more than one proton) conjugate acids (Hingston, 1981). This relationship is traditionally referred to as an adsorption envelope (Fig. 5.27). In Fig. 5.27 one sees that for silicate and fluoride, sorption increased with pH and reached a maximum near the pK_a (pK_1 in Fig. 5.27) of the acid, then decreased with pH. For fluoride the dominant species were HF and F^-, and for silicate the species were H_4SiO_4 and $H_3SiO_4^-$. With selenite and phosphate, sorption decreased with increased pH with the pH decrease more pronounced above pK_2. The ion species for selenite and phosphate were $HSeO_3^-$ and SeO_3^{2-}, and $H_2PO_4^-$ and HPO_4^{2-}, respectively (Barrow,

1985). An example of the correlation between adsorption maxima (pH of inflection) and the pK_a values for conjugate acids is shown in Fig. 5.28.

Generally speaking, the anions NO_3^-, Cl^-, and ClO_4^- are sorbed as outer-sphere complexes and sorbed on surfaces that exhibit a positive charge. Sorption is sensitive to ionic strength. Some researchers have also concluded that SO_4^{2-} (Zhang and Sparks, 1990a) can be sorbed as an outer-sphere complex; however, there is other evidence that SO_4^{2-} can also be sorbed as an inner-sphere complex (Table 5.1). There is direct spectroscopic evidence to show that selenate is sorbed as an outer-sphere complex and selenite is sorbed as an inner-sphere complex (Hayes *et al.*, 1987). XAFS spectroscopy analyses showed that the sorbed selenite was bonded to the surface as a bidentate species with two Fe atoms 3.38 Å from the selenium atom. Selenate had no Fe atom in the second coordination shell of Se, which indicated that its hydration sphere was retained on sorption.

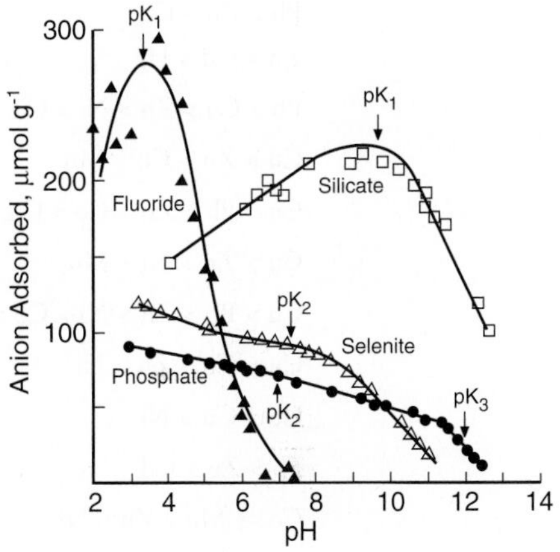

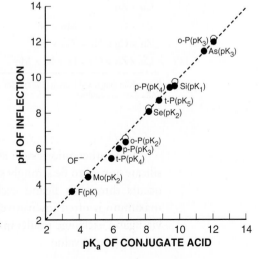

FIGURE 5.27. *Sorption of a range of anions on goethite. Two samples of goethite were used, and the level of addition of sorbing ion differed between the different ions. Redrawn from Hingston et al. (1972), J. Soil Sci. 23, 177–192, with permission of Blackwell Science Ltd.*

FIGURE 5.28. *Relationship between pK_a and pH at the change in slope of sorption envelopes. Sorbents: •, goethite; ○, gibbsite. Sorbates: F, fluoride; Mo, molybdate; t-P, tripolyphosphate; p-P, pyrophosphate; o-P, orthophosphate; Se, selenite; Si, silicate; As, arsenate. From Hingston et al. (1972), J. Soil Sci. 23, 177–192, with permission of Blackwell Science Ltd.*

TABLE 5.6. *Selectivity Sequences for the Adsorption or Coprecipitation of Alkaline Earth Cations and of Divalent Transition and Heavy Metal Cations on Various Hydrous Metal Oxides[a]*

Sequence	Oxide	Reference
Alkaline earth cations		
Ba > Ca[b]	Fe gel	Kurbatov *et al.* (1945)
Ba > Ca > Sr > Mg	Fe gel	Kinniburgh *et al.* (1976)
Mg > Ca > Sr > Ba	α-Fe_2O_3	Breeuwsma and Lyklema (1971)
Mg > Ca > Sr > Ba	Al gel	Kinniburgh *et al.* (1976)
Mg > Ca > Sr > Ba	Al_2O_3	Huang and Stumm (1973)
Ba > Sr > Ca	α-Al_2O_3	Belot *et al.* (1966)
Ba > Sr > Ca > Mg	MnO_2	Gabano *et al.* (1965)
Ba > Sr > Ca > Mg	δ-MnO_2	Murray (1975)
Ba > Ca > Sr > Mg	SiO_2	Tadros and Lyklema (1969)
Ba > Sr > Ca	SiO_2	Malati and Estefan (1966)
Divalent transition and heavy metal cations		
Pb > Zn > Cd	Fe gel	Gadde and Laitinen (1974)
Zn > Cd > Hg	Fe gel	Bruninx (1975)
Pb > Cu > Zn > Ni > Cd > Co	Fe gel	Kinniburgh *et al.* (1976)
Cu > Zn > Co > Mn	α-FeOOH	Grimme (1968)
Cu > Pb > Zn > Co > Cd	α-FeOOH	Forbes *et al.* (1976)
Cu > Zn > Ni > Mn	Fe_3O_4	Venkataramani *et al.* (1978)
Cu > Pb > Zn > Ni > Co > Cd	Al gel	Kinniburgh *et al.* (1976)
Cu > Co > Zn > Ni	MnO_2	Kozawa (1959)
Co > Cu > Ni	MnO_2	Murray *et al.* (1968)
Pb > Zn > Cd	MnO_2	Gadde and Laitinen (1974)
Co $\simeq$ Mn > Zn > Ni	MnO_2	Murray (1975)
Cu > Zn > Co > Ni	δ-MnOOH	McKenzie (1972)
Co > Cu > Zn > Ni	α-MnO_3	McKenzie (1972)
Co > Zn	δ-MnO_2	Loganathan and Burau (1973)
Zn > Cu > Co > Mn > Ni	Si gel	Vydra and Galba (1969)
Zn > Cu > Ni $\simeq$ Co > Mn	Si gel	Taniguechi *et al.* (1970)
Cu > Zn > Co > Fe > Ni > Mn	SnO_2	Donaldson and Fuller (1968)

[a] From Kinniburgh and Jackson (1981), with permission.
[b] All cations are divalent.

Most other anions such as molybdate, arsenate, arsenite, phosphate, and silicate appear to be strongly sorbed as inner-sphere complexes, and sorption occurs through a ligand exchange mechanism (Table 5.1). The sorption maximum is often insensitive to ionic strength changes. Sorption of anions via ligand exchange results in a shift in the pzc (discussed later) of the oxide to a more acid value.

Surface Precipitation

As the amount of metal cation or anion sorbed on a surface (surface coverage or loading, which is affected by the pH at which sorption occurs) increases, sorption can proceed from mononuclear adsorption to surface precipitation (a three-dimensional phase). There are several thermodynamic reasons for surface precipitate formation: (1) the solid surface may lower the energy of nucleation by providing sterically similar sites (McBride, 1991); (2) the activity of the surface precipitate is <1 (Sposito, 1986); and (3) the solubility of the surface precipitate is lowered because the dielectric constant of the solution near the surface is less than that of the bulk solution (O'Day *et al.*, 1994a). There are several types of surface precipitates. They can arise via polymeric metal complexes (dimers, trimers, etc.) that form on mineral surfaces and via the sorption of aqueous polymers (Chisholm-Brause, 1990b). Homogeneous precipitates can form on a surface when the solution becomes saturated and the surface acts as a nucleation site. When adsorption attains monolayer coverage sorption continues on the newly created sites, causing a precipitate on the surface (Farley *et al.*, 1985; McBride, 1991). When the precipitate consists of chemical species derived from both the aqueous solution and dissolution of the mineral, it is referred to as a coprecipitate. The composition of the coprecipitate varies between that of the original solid and a pure precipitate of the sorbing metal. The ionic radius of the sorbing metal and sorbent ions must be similar for coprecipitates to form. Thus Co(II), Mn(II), Ni(II), and Zn(II) form coprecipitates on sorbents containing Al(III) and Si(IV) but not Pb(II), which is considerably larger (0.12 nm). Coprecipitate formation is most limited by the rate of mineral dissolution, rather than the lack of thermodynamic favorability (McBride, 1994; Scheidegger *et al.*, 1998). If the formation of a precipitate occurs under solution conditions that would, in the absence of a sorbent, be undersaturated with respect to any known solid phase, this is referred to as surface-induced precipitation (Towle *et al.*, 1997).

Thus there is often a continuum between surface complexation (adsorption) and surface precipitation. This continuum depends on several factors: (1) ratio of the number of surface sites vs the number of metal ions in solution; (2) the strength of the metal oxide bond; and (3) the degree to which the bulk solution is undersaturated with respect to the metal hydroxide precipitate (McBride, 1991). At low surface coverages surface complexation (e.g., outer- and inner-sphere adsorption) tends to dominate. As surface coverage increases, nucleation occurs and results in the formation of distinct entities or aggregates on the surface. As surface loadings increase further, surface precipitation becomes the dominant mechanism, Fig. 5.29. For example, Fendorf *et al.* (1994a) and Fendorf and Sparks (1994b) used XAFS, FTIR, and HRTEM to study Cr(III) sorption on Si oxide. At low Cr(III) surface coverage (<20%), adsorption was the dominant process with an inner-sphere monodentate complex forming. As Cr(III) surface coverage

adsorption

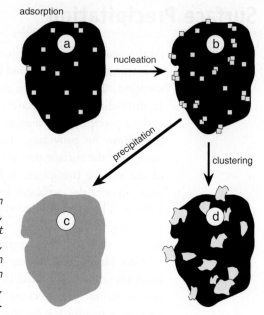

FIGURE 5.29. *An illustration of metal ion sorption reactions on (hydr)oxide. (a) At low surface coverage, isolated site binding (adsorption) is the dominant sorption mechanism; (b) with increased metal loading, M hydroxide nucleation begins. Further increases in metal loadings results in (c) surface precipitation or (d) surface clusters. From Fendorf (1992), with permission.*

increased (>20%), surface precipitation occurred and was the dominant process. Table 5.1 shows that for a number of ions, as surface coverage increases, surface precipitates form.

Using *in situ* XAFS, it has been shown by a number of scientists that multinuclear metal hydroxide complexes and surface precipitates Co(II), Cr(III), Cu(II), Ni(II), and Pb(II) can form on metal oxides, phyllosilicates, soil clays, and soils (Chisholm-Brause *et al.*, 1990a, b; Roe *et al.*, 1991; Charlet and Manceau, 1992; Fendorf *et al.*, 1994a; O'Day *et al.*, 1994a,b; Papelis and Hayes, 1996; Scheidegger *et al.*, 1996, 1997, 1998; Towle *et al.*, 1997; Roberts *et al.*, 1999; Thompson *et al.*, 1999a,b; Elzinga and Sparks, 1999; Ford and Sparks, 2000; Scheckel and Sparks, 2001). These metal hydroxide phases occur at metal loadings below a theoretical monolayer coverage and in a pH range well below the pH where the formation of metal hydroxide precipitates would be expected according to the thermodynamic solubility product (Scheidegger and Sparks, 1996).

Scheidegger *et al.* (1997) were the first to show that sorption of metals, such as Ni, on an array of phyllosilicates and Al oxide, could result in the formation of mixed metal–Al hydroxide surface precipitates which appear to be coprecipitates. The precipitate phase shares structural features common to the hydrotalcite group of minerals and the layered double hydroxides (LDH) observed in catalyst synthesis. The LDH structure is built of stacked sheets of edge-sharing metal octahedra containing divalent and trivalent metal ions separated by anions between the interlayer spaces. The general structural formula can be expressed as $[Me^{2+}_{1-x}\ Me^{3+}_x\ (OH)_2]^{x+} \cdot (x/n)\ A^{-n} \cdot -m\ H_2O$, where, for example, Me^{2+} could be Mg(II), Ni(II), Co(II), Zn(II), Mn(II), and Fe(II) and Me^{3+} is Al(III), Fe(III), and Cr(III). The LDH structure

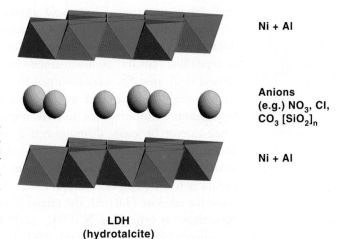

FIGURE 5.30. *Structure of Ni–Al LDH showing brucite-like octahedral layers in which Al^{3+} substitutes for Ni^{2+}, creating a net positive charge balanced by hydrated anions in the interlayer space. From Scheinost et al. (1999), with permission from Elsevier Science.*

exhibits a net positive charge x per formula unit, which is balanced by an equal negative charge from interlayer anions A^{n-} such as Cl^-, Br^-, I^-, NO_3^-, OH^-, ClO_4^-, and CO_3^{2-}; water molecules occupy the remaining interlayer space (Allman, 1970; Taylor, 1984). The minerals takovite, $Ni_6 Al_2(OH)_{16} CO_3 \cdot H_2O$, and hydrotalcite, $Mg_6 Al_2(OH)_{16} CO_3 \cdot H_2O$, are among the most common natural mixed-cation hydroxide compounds containing Al (Taylor, 1984). Figure 5.30 shows a Ni–Al LDH phase.

XAFS data, showing the formation of Ni–Al LDH phases on soil components, are shown in Fig. 5.31 and Table 5.7 (Scheidegger *et al.*, 1997). Radial structure functions (RSFs), collected from XAFS analyses, for Ni sorption on pyrophyllite, kaolinite, gibbsite, and montmorillonite were compared to the spectra of crystalline $Ni(OH)_2$ and takovite. All spectra showed a peak at $R \approx 1.8$ Å, which represents the first coordination shell of Ni. A second peak representing the second Ni shell was observed in the spectra of the Ni sorption samples and takovite (Fig. 5.31). The structural parameters, derived from XAFS analyses, for the various sorption samples and takovite and $Ni(OH)_2$ are shown in Table 5.7. In the first coordination

TABLE 5.7. *Structural Information Derived from XAFS Analysis for Ni Sorption on Various Sorbents and for Known Ni Hydroxides[a,b]*

	Γ ($\mu mol/m^2$)	Ni–O			Ni–Ni			Ni–Si/Al			$N(Ni)/N(Si/Al)$
		R (Å)	N	$2\sigma^2$	R (Å)	N	$2\sigma^2$	R (Å)	N	$2\sigma^2$	
$Ni(OH)_2$		2.06	6.0	0.011	3.09	6.0	0.010				
Takovite		2.03	6.0	0.01	3.01	3.1	0.009	3.03	1.1	0.009	2.8
Pyrophyllite	3.1	2.02	6.1	0.01	3.00	4.8	0.009	3.02	2.7	0.009	1.8
Kaolinite	19.9	2.03	6.1	0.01	3.01	3.8	0.009	3.02	1.8	0.009	2.2
Gibbsite	5.0	2.03	6.5	0.01	3.02	5.0	0.009	3.05	1.8	0.09	2.7
Montmorillonite	0.35	2.03	6.3	0.01	3.03	2.8	0.011	3.07	2.0	0.015	1.4

[a] From Scheidegger *et al.* (1997), with permission from Academic Press, Orlando, FL.
[b] Interatomic distances (R, Å), coordination numbers (N), and Debye–Waller factors ($2\sigma^2$, nm^2). The reported values are accurate to within $R \pm 0.02$ Å, $N_{(Ni-O)} \pm 20\%$, $N_{(Ni-Ni)} \pm 40\%$, and $N_{(Ni-Si/Al)} \pm 40\%$.

shell Ni is surrounded by six O atoms, indicating that Ni(II) is in an octahedral environment. The Ni–O distances for the Ni sorption samples are 2.02–2.03 Å, and similar to those in takovite (2.03 Å). The Ni–O distances in crystalline $Ni(OH)_2(s)$ are distinctly longer (2.06 Å in this study). For the second shell, best fits were obtained by including both Ni and Si or Al as second-neighbor backscatter atoms. Since Si and Al differ in atomic number by 1 (atomic number = 14 and 13, respectively), backscattering is similar. They cannot be easily distinguished from each other as second-neighbor backscatterers. There are 2.8 (for montmorillonite) to 5.8 (gibbsite) Ni second-neighbor (N) atoms, indicative of Ni surface precipitates. The Ni–Ni distances for the sorption samples were 3.00–3.03 Å, which are similar to those for takovite (3.01 Å), the mixed Ni–Al LDH phase, but much shorter than those in crystalline $Ni(OH)_2$ (3.09 Å). There are also 1.8–2.7 Si/Al second-neighbor atoms at 3.02–3.07 Å. The bond distances are in good agreement with the Ni–Al distances observed in takovite (3.03 Å).

Mixed Co–Al and Zn–Al hydroxide surface precipitates also can form on aluminum-bearing metal oxides and phyllosilicates (Towle et al., 1997; Thompson et al., 1999a; Ford and Sparks, 2000). This is not surprising as Co^{2+}, Zn^{2+}, and Ni^{2+} all have radii similar to that of Al^{3+}, enhancing substitution in the mineral structure and formation of a coprecipitate. However, surface precipitates have not been observed with Pb^{2+}, as Pb^{2+} is too large to substitute for Al^{3+} in mineral structures.

Metal hydroxide precipitate phases can also form in the presence of non-Al-bearing minerals (Scheinost et al., 1999). Using diffuse reflectance spectroscopy (DRS), which is quite sensitive for discriminating between Ni–O bond distances, it was shown that α-Ni $(OH)_2$ formed upon Ni^{2+} sorption to talc and silica (Fig. 5.32).

The mechanism for the formation of metal hydroxide surface precipitates is not clearly understood. It is clear that the type of metal ion determines whether metal hydroxide surface precipitates form, and the type of surface precipitate formed, i.e., metal hydroxide or mixed metal hydroxide, is dependent on the sorbent type. The precipitation could be explained by the combination of several processes (Yamaguchi et al., 2001). First, the electric field of the mineral surface attracts metal ions, e.g., Ni, through adsorption, leading to a local supersaturation at the mineral–water interface. Second, the solid phase may act as a nucleation center for polyhydroxy species and catalyze the precipitation process (McBride, 1994). Third, the physical properties of water molecules adsorbed at the mineral surface are different from those of free water (Sposito, 1989), causing a lower solubility of metal hydroxides at the mineral–water interface. With time Al, which is released by weathering of the mineral surface, slowly diffuses into the octahedral layer of the mineral and partially replaces the metal (e.g., Ni) in the octahedral sites. A Ni–Al LDH, which is thermodynamically favored over α-Ni hydroxide, is formed.

The formation of metal hydroxide surface precipitates appears to be an important way to sequester metals. As the surface precipitates age, metal release is greatly reduced. Thus, the metals are less prone to leaching and

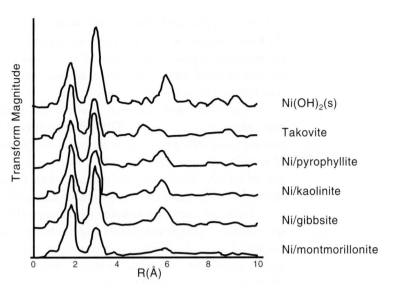

FIGURE 5.31. *Radial structure functions (RSFs) produced by forward Fourier transforms of Ni sorbed on pyrophyllite, kaolinite, gibbsite, and montmorillonite compared to the spectrum of crystalline Ni(OH)$_2$(s) and takovite. The spectra are uncorrected for phase shift. From Scheidegger et al. (1997), with permission.*

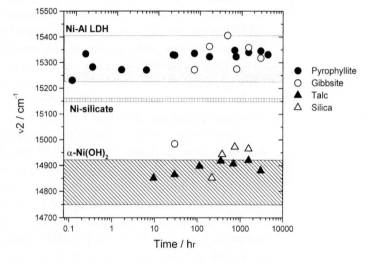

FIGURE 5.32. *Fitted v_2 band positions of the Ni reacted minerals (dots and triangles) over time using diffuse reflectance spectroscopy (DRS). The v_2 band is attributed to crystal field splitting induced within the incompletely filled 3d electronic shell of Ni^{2+} through interaction with the negative charge of nearest-neighbor oxygen ions. For talc and silica, the v_2 band was at ~14,900 cm^{-1}, indicating the formation of an α-Ni(OH)$_2$ phase while for Al-containing pyrophyllite and gibbsite, it appeared at ~15,300 cm^{-1}, indicating a Ni–Al LDH phase. From Scheinost et al. (1999), with permission.*

being taken up by plants. This is due to silication of the interlayer of the LDH phases creating precursor phyllosilicate surface precipitates. More details on the metal release rates from surface precipitates and the mechanisms for the metal sequestration are given in Chapter 7.

Speciation of Metal-Contaminated Soils

The ability to speciate metal-contaminated soils is critical in developing viable and cost effective remediation strategies and in predicting mobility and bioavailability of the metals. Metals in contaminated soils can be found in mineral and sorbed phases. Total metal contents of soils, while indicating the degree of contamination, reveal no information on speciation. Chemical extraction techniques (e.g., sequential extractions) provide useful information on quantities of metals that could be associated with different phases, but transformations in the phases may occur.

X-ray diffraction, thermal gravimetric analysis (TGA), and X-ray photoelectron spectroscopy (XPS) have been used to characterize metals in contaminated soils. Even though they provide a more accurate characterization of metal speciation than extraction methods, they may introduce artifacts due to sample alterations. Additionally, their detection limits are often far above the background concentrations of the metals contained in the soils.

Electron microscopy, coupled with energy-dispersive X-ray fluorescence spectroscopy (EDXS) can enable one to obtain both quantitative (elemental composition) and qualitative information (metal distribution) with good spatial resolution (<1 μm). However, only elemental concentrations are obtained. Thus, one could not, for example, distinguish between S and SO_4^{2-}. The elements also must be in nearly crystalline phases in concentrations greater than 1%. This limits its utility in many contaminated systems. Bulk XAFS has been useful in characterizing metal-bearing soil minerals and soils. However, one can probe an area of only several millimeters, providing only an average speciation of metals in a sample. This can be a problem in heterogeneous soil samples where several metal species can be present over a small area. Moreover, in samples where the metal may be present in numerous phases, the detection limit for minor species is problematic.

To overcome the above difficulties, major advances in the use of spatially resolved micro (μ)-XAFS (both μ-XANES and μ-EXAFS) combined with μ-synchrotron-X-ray fluorescence spectroscopy (μ-SXRF) have occurred. With these techniques, one can study metal speciation in soils on a micron scale (Manceau *et al.*, 2000; Roberts *et al.*, 2002).

An example of using these techniques to speciate Zn in a subsoil is shown in Fig. 5.33. From the μ-SXRF maps one sees that Zn is strongly associated with Mn in the center of the sample, with Fe in other portions of the sample, and with neither Mn or Fe in some portions of the sample (Fig. 5.33a). This microscale heterogeneity in elemental associations suggests Zn could be present in different phases over a small sample area. Bulk XAFS analyses suggested that Zn was bound as an inner-sphere complex to Al, Fe, and Mn oxides but the data interpretation was difficult (Fig. 5.33b).

μ-XAFS data were collected on the three regions of the sample (labeled 1, 2, and 3 on the Zn map, Fig. 5.33a). Spot 1 on the map showed that Zn was octahedrally coordinated and sorbed to an Al oxide phase (Fig. 5.33c). Spots 2 and 3, with Zn in a tetrahedral coordination, revealed that Zn was primarily sorbed to Fe oxide (Spot 2) and to Mn oxide (Spot 3) (Fig. 5.33c). These results suggest that Zn^{2+} could have been released from Zn-bearing mineral phases in the topsoil (primarily sphalerite, ZnS, and franklinite, Zn Fe_2O_4) and transported to the subsoil where the Zn^{2+} was partially readsorbed to Al, Fe, and Mn oxides.

Points of Zero Charge

Definition of Terms

One of the most useful and meaningful chemical parameters that can be determined for a soil or soil component is a point of zero charge, which is abbreviated to pzc to denote a general point of zero charge. The pzc can be defined as the suspension pH at which a surface has a net charge of zero (Parks, 1967). If the measured pH of a colloid is lower than the pzc, the surface is net positively charged; if the pH > pzc, the surface is net negatively charged.

Soil components have a wide range of pzc values (Table 5.8). Oxides (Fe and Al) have high pzc values, while silica and SOM have low pzc values. The pzc values for a whole soil are strongly reflective of the individual pzc values of the soil's components. Generally, surface soil horizons have a pzc lower than that of subsoil horizons since SOM is higher in the surface and it typically has a low pzc. A low pzc would indicate that the soil has a net negative charge over a wide pH range and thus would have the ability to adsorb cations. The pzc of a soil generally increases with depth as clay and oxide contents increase, both of which have pzc values higher than that of

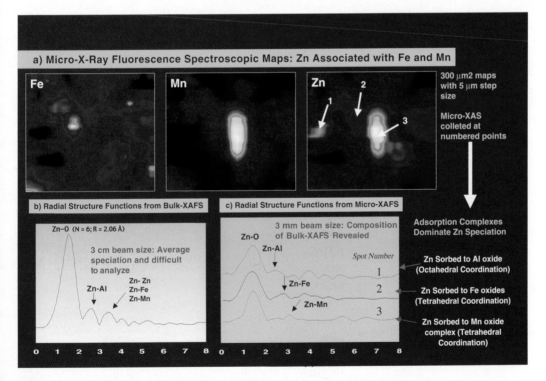

FIGURE 5.33. *Synchrotron-based speciation of Zn in a smelter-contaminated subsurface soil using (a) μ-synchrotron X-ray fluorescence (μ-SXRF) spectroscopy, (b) bulk XAFS, and (c) μ-XAFS techniques. From Roberts (2001), with permission.*

SOM. Thus the soil would exhibit some positive charge and could electrostatically adsorb anions.

There are a number of points of zero charge that can be experimentally measured (Fig. 5.34). The pzc is measured electrophoretically by tracking the movement of the colloidal particles in an applied electric field. The pzc is the pH value at which there is no net movement of the particles or the point at which flocculation occurs. It can be theoretically defined, based on the balance of surface charge, as the pH where the net total particle charge, $\sigma_p = 0$ (Sposito, 1984). The point of zero net charge (PZNC) is the pH at which the difference in CEC and AEC (anion exchange capacity) is zero. One can experimentally measure PZNC by reacting the soil with an electrolyte solution such as NaCl and measuring the quantity of Na^+ and Cl^- adsorbed as a function of pH (Fig. 5.34). The pH where the two curves intersect is the PZNC. If the electrolyte can displace only ions contributing to σ_{os} and σ_d, i.e., nonspecifically adsorbed (outer-sphere complexed) ions, then the PZNC corresponds to the condition $\sigma_{os} + \sigma_d$ (Sposito, 1984). The point of zero net proton charge (PZNPC) is another point of zero charge (Fig. 5.34), which is the pH at which the net proton charge $\sigma_H = 0$ (Sposito, 1984). The point of zero salt effect (PZSE) is the pH at which two or more potentiometric titration curves intersect (Parker *et al.*, 1979). This is referred to as the common intersection point (CIP), or the crossover $(\partial\sigma_H/\partial_I)_T = 0$.

TABLE 5.8. *Point of Zero Charge (pzc) of Minerals[a,b]*

Material	pzc
α-Al_2O_3	9.1
α-$Al(OH)_3$	5.0
γ-$AlOOH$	8.2
CuO	9.5
Fe_3O_4	6.5
α-$FeOOH$	7.8
α-Fe_2O_3	6.7
"$Fe(OH)_3$" (amorphous)	8.5
MgO	12.4
δ-MnO_2	2.8
β-MnO_2	7.2
SiO_2	2.0
$ZrSiO_4$	5.0
Feldspars	2–2.4
Kaolinite	4.6
Montmorillonite	2.5
Albite	2.0
Chrysotile	> 10

[a] The values are from different investigators who have used different methods and thus are not necessarily comparable.
[b] From W. Stumm, and J. J. Morgan, "Aquatic Chemistry." Copyright © 1981 John Wiley & Sons, Inc. Reprinted by permission of John Wiley & Sons, Inc.

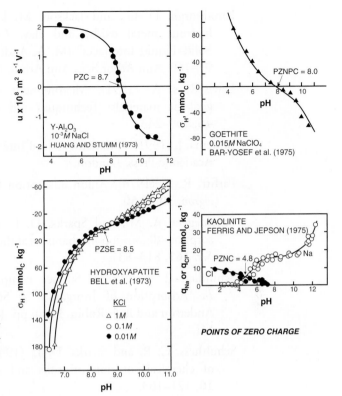

FIGURE 5.34. *Experimental examples of the PZC of γ-Al₂O₃, the PZNPC of goethite, the PZSE of hydroxyapatite, and the PZNC of kaolinite (u is the electrophoretic mobility). From Sposito (1984), with permission.*

This is the most commonly determined point of zero charge, and its experimental determination is shown in Fig. 5.34. Details on methodologies for measuring the different pzcs can be found in Zelazny *et al.* (1996).

Suggested Reading

Anderson, M. A., and Rubin, A. J., Eds. (1981). "Adsorption of Inorganics at Solid–Liquid Interfaces." Ann Arbor Sci., Ann Arbor, MI.

Barrow, N. J. (1985). Reactions of anions and cations with variable-charged soils. *Adv. Agron.* **38**, 183–230.

Bolt, G. H., De Boodt, M. F., Hayes, M. H. B., and McBride, M. B., Eds. (1991). "Interactions at the Soil Colloid–Soil Solution Interface," NATO ASI Ser. E, Vol. 190. Kluwer Academic, Dordrecht, The Netherlands.

Goldberg, S. (1992). Use of surface complexation models in soil chemical systems. *Adv. Agron.* **47**, 233–329.

Hingston, F. J. (1981). A review of anion adsorption. *In* "Adsorption of Inorganics at Solid–Liquid Interfaces" (M. A. Anderson and A. J. Rubin, Eds.), pp. 51–90. Ann Arbor Sci., Ann Arbor, MI.

Kinniburgh, D. G., and Jackson, M. L. (1981). Cation adsorption by hydrous metal oxides and clay. *In* "Adsorption of Inorganics at Solid–Liquid Interfaces" (M. A. Anderson and A. J. Rubin, Eds.), pp. 91–160. Ann Arbor Sci., Ann Arbor, MI.

Lewis-Russ, A. (1991). Measurement of surface charge of inorganic geologic materials: Techniques and their consequences. *Adv. Agron.* **46**, 199–243.

Lyklema, J. (1991). "Fundamentals of Interface and Colloid Science," Vol. 1. Academic Press, London.

Parfitt, R. L. (1978). Anion adsorption by soils and soil materials. *Adv. Agron.* **30**, 1–50.

Scheidegger, A. M., and Sparks, D. L. (1996). A critical assessment of sorption–desorption mechanisms at the soil mineral/water interface. *Soil Sci.* **161**, 813–831.

Schindler, P. W. (1981). Surface complexes at oxide–water interfaces. *In* "Adsorption of Inorganics at Solid–Liquid Interfaces" (M. A. Anderson and A. J. Rubin, Eds.), pp. 1–49. Ann Arbor Sci., Ann Arbor, MI.

Schulthess, C. P., and Sparks, D. L. (1991). Equilibrium-based modeling of chemical sorption on soils and soil constituents. *Adv. Soil Sci.* **16**, 121–163.

Singh, U., and Uehara, G. (1999). Electrochemistry of the double-layer: Principles and applications to soils. *In* "Soil Physical Chemistry" (D. L. Sparks, Ed.), 2nd ed., pp. 1–46. CRC Press, Boca Raton, FL.

Sparks, D. L., and Grundl, T. J., Eds. (1998). "Mineral–Water Interfacial Reactions." ACS Symp. Ser. 715. Am. Chem. Soc., Washington, DC.

Sparks, D. L., Ed. (1999). "Soil Physical Chemistry," 2nd ed. CRC Press, Boca Raton, FL.

Sposito, G. (1984). "The Surface Chemistry of Soils." Oxford Univ. Press, New York.

Stumm, W., Ed. (1987). "Aquatic Surface Chemistry." Wiley, New York.

Stumm, W. (1992). "Chemistry of the Solid–Water Interface." Wiley, New York.

Stumm, W., and Morgan, J. J. (1996). "Aquatic Chemistry," 3rd ed. Wiley, New York.

Theng, B. K. G., Ed. (1980). "Soils with Variable Charge." New Zealand Society of Soil Science, Palmerston.

Uehara, G., and Gillman, G. (1981). "The Mineralogy, Chemistry, and Physics of Tropical Soils with Variable Charge Clays." Westview Press, Boulder, CO.

van Olphen, H. (1977). "An Introduction to Clay Colloid Chemistry," 2nd ed. Wiley, New York.

6 Ion Exchange Processes

Introduction

I on exchange, the interchange between an ion in solution and another ion in the boundary layer between the solution and a charged surface (Glossary of Soil Science Terms, 1997), truly has been one of the hallmarks in soil chemistry. Since the pioneering studies of J. Thomas Way in the middle of the 19th century (Way, 1850), many important studies have occurred on various aspects of both cation and anion exchange in soils. The sources of cation exchange in soils are clay minerals, organic matter, and amorphous minerals. The sources of anion exchange in soils are clay minerals, primarily 1:1 clays such as kaolinite, and metal oxides and amorphous materials.

The ion exchange capacity is the maximum adsorption of readily exchangeable ions (diffuse ion swarm and outer-sphere complexes) on soil particle surfaces (Sposito, 2000). From a practical point of view, the ion exchange capacity (the sum of the CEC (defined earlier; see Box 6.1 for description of CEC measurement) and the AEC (anion exchange capacity, which is the sum of total exchangeable anions that a soil can adsorb, expressed as $cmol_c$ kg^{-1}, where c is the charge; Glossary of Soil Science Terms, 1997)) of a soil is

187

important since it determines the capacity of a soil to retain ions in a form such that they are available for plant uptake and not susceptible to leaching in the soil profile. This feature has important environmental and plant nutrient implications. As an example, NO_3^- is important for plant growth, but if it leaches, as it often does, it can move below the plant root zone and leach into groundwater where it is deleterious to human health (see Chapter 1). If a soil has a significant AEC, nitrate can be held, albeit weakly. Sulfate can be significantly held in soils that have AEC and be available for plant uptake (sulfate accumulations are sometimes observed in subsoils where oxides as discrete particles or as coatings on clays impart positive charge or an AEC to the soil). However, in soils lacking the ability to retain anions, sulfate can leach readily and is no longer available to support plant growth.

BOX 6.1. *Measurement of CEC*

The CEC of a soil is usually measured by saturating a soil or soil component with an index cation such as Ca^{2+}, removing excess salts of the index cation with a dilute electrolyte solution, and then displacing the Ca^{2+} with another cation such as Mg^{2+}. The amount of Ca^{2+} displaced is then measured and the CEC is calculated. For example, let us assume that 200 mg of Ca^{2+} were displaced from 100 g of soil. The CEC would then be calculated as

$$\text{CEC} = \left(\frac{200 \text{ mg Ca}^{2+}}{100 \text{ g}} \right) \left(\frac{20 \text{ mg Ca}^{2+}}{\text{meq}} \right) = 10 \text{ meq/100 g} = 10 \text{ } c\text{mol}_c \text{ kg}^{-1}$$

The CEC values of various soil minerals were provided in Chapter 2. The CEC of a soil generally increases with soil pH due to the greater negative charge that develops on organic matter and clay minerals such as kaolinite due to deprotonation of functional groups as pH increases. Thus, in measuring the CEC of variable charge soils and minerals, if the index cation saturating solution is at a pH greater than the pH of the soil or mineral, the CEC can be overestimated (Sumner and Miller, 1996). The anion exchange capacity increases with decreasing pH as the variable charge surfaces become more positively charged due to protonation of functional groups.

The magnitude of the CEC in soils is usually greater than the AEC. However, in soils that are highly weathered and acidic, e.g., some tropical soils, copious quantities of variable charge surfaces such as oxides and kaolinite may be present and the positive charge on the soil surface may be significant. These soils can exhibit a substantial AEC.

Characteristics of Ion Exchange

Ion exchange involves electrostatic interactions between a counterion in the boundary layer between the solution and a charged particle surface and counterions in a diffuse cloud around the charged particle. It is usually rapid, diffusion-controlled, reversible, and stoichiometric, and in most cases there is some selectivity of one ion over another by the exchanging surface. Exchange reversibility is indicated when the exchange isotherms for the forward and backward exchange reactions coincide (see the later section Experimental Interpretations for discussion of exchange isotherms). Exchange irreversibility or hysteresis is sometimes observed and has been attributed to colloidal aggregation and the formation of quasi-crystals (Van Bladel and Laudelout, 1967). Quasi-crystals are packets of clay platelets with a thickness of a single layer in stacked parallel alignment (Verburg and Baveye, 1994). The quasi-crystals could make exchange sites inaccessible.

Stoichiometry means that any ions that leave the colloidal surface are replaced by an equivalent (in terms of ion charge) amount of other ions. This is due to the electroneutrality requirement. When an ion is displaced from the surface, the exchanger has a deficit in counterion charge that must be balanced by counterions in the diffuse ion cloud around the exchanger. The total counterion content in equivalents remains constant. For example, to maintain stoichiometry, two K^+ ions are necessary to replace one Ca^{2+} ion.

Since electrostatic forces are involved in ion exchange, Coulomb's law can be invoked to explain the selectivity or preference of the ion exchanger for one ion over another. This was discussed in Chapter 5. However, in review, one can say that for a given group of elements from the periodic table with the same valence, ions with the smallest hydrated radius will be preferred, since ions are hydrated in the soil environment. Thus, for the group 1 elements the general order of selectivity would be $Cs^+ > Rb^+ > K^+ > Na^+ > Li^+ > H^+$. If one is dealing with ions of different valence, generally the higher charged ion will be preferred. For example, $Al^{3+} > Ca^{2+} > Mg^{2+} > K^+ = NH_4^+ > Na^+$. In examining the effect of valence on selectivity polarization must be considered. Polarization is the distortion of the electron cloud about an anion by a cation. The smaller the hydrated radius of the cation, the greater the polarization, and the greater its valence, the greater its polarizing power. With anions, the larger they are, the more easily they can be polarized. The counterion with the greater polarization is usually preferred, and it is also least apt to form a complex with its coion. Helfferich (1962b) has given the following selectivity sequence, or lyotropic series, for some of the common cations: $Ba^{2+} > Pb^{2+} > Sr^{2+} > Ca^{2+} > Ni^{2+} > Cd^{2+} > Cu^{2+} > Co^{2+} > Zn^{2+} > Mg^{2+} > Ag^+ > Cs^+ > Rb^+ > K^+ > NH_4^+ > Na^+ > Li^+$.

The rate of ion exchange in soils is dependent on the type and quantity of inorganic and organic components and the charge and radius of the ion being considered (Sparks, 1989). With clay minerals like kaolinite, where only external exchange sites are present, the rate of ion exchange is rapid. With 2:1 clay minerals that contain both external and internal exchange sites, particularly with vermiculite and micas where partially collapsed interlayer space sites exist, the kinetics are slower. In these types of clays, ions such as K^+ slowly diffuse into the partially collapsed interlayer spaces and the exchange can be slow and tortuous. The charge of the ion also affects the kinetics of ion exchange. Generally, the rate of exchange decreases as the charge of the exchanging species increases (Helfferich, 1962a). More details on the kinetics of ion exchange reactions can be found in Chapter 7.

Cation Exchange Equilibrium Constants and Selectivity Coefficients

Many attempts to define an equilibrium exchange constant have been made since such a parameter would be useful for determining the state of ionic equilibrium at different ion concentrations. Some of the better known equations attempting to do this are the Kerr (1928), Vanselow (1932), and Gapon (1933) expressions. In many studies it has been shown that the equilibrium exchange constants derived from these equations are not constant as the composition of the exchanger phase (solid surface) changes. Thus, it is often better to refer to them as *selectivity coefficients* rather than *exchange constants*.

Kerr Equation

In 1928 Kerr proposed an "equilibrium constant," given below, and correctly pointed out that the soil was a solid solution (a macroscopically homogeneous mixture with a variable composition; Lewis and Randall (1961)). For a binary reaction (a reaction involving two ions),

$$vACl_u \text{ (aq)} + uBX_v \text{ (s)} \rightleftharpoons uBCl_v \text{ (aq)} + vAX_u \text{ (s)},\tag{6.1}$$

where A^{u+} and B^{v+} are exchanging cations and X represents the exchanger, (aq) represents the solution or aqueous phase, and (s) represents the solid or exchanger phase.

Kerr (1928) expressed the "equilibrium constant," or more correctly, a selectivity coefficient for the reaction in Eq. (6.1), as

$$K_K = \frac{[BCl_v]^u \{AX_u\}^v}{[ACl_u]^v \{BX_v\}^u},\tag{6.2}$$

where brackets ([ ]) indicate the concentration in the aqueous phase in mol liter^{-1} and braces ({ }) indicate the concentration in the solid or exchanger phase in mol kg^{-1}.

Kerr (1928) studied Ca–Mg exchange and found that the K_K value remained relatively constant as exchanger composition changed. This indicated that the system behaved ideally; i.e., the exchanger phase activity coefficients for the two cations were each equal to 1 (Lewis and Randall, 1961). These results were fortuitous since Ca–Mg exchange is one of the few binary exchange systems where ideality is observed.

Vanselow Equation

Albert Vanselow was a student of Lewis and was the first person to give ion exchange a truly thermodynamical context. Considering the binary cation exchange reaction in Eq. (6.1), Vanselow (1932) described the thermodynamic equilibrium constant as

$$K_{eq} = \frac{(BCl_v)^u \, (AX_u)^v}{(ACl_u)^v \, (BX_v)^u} \, , \tag{6.3}$$

where parentheses indicate the thermodynamic activity. It is not difficult to determine the activity of solution components, since the activity would equal the product of the equilibrium molar concentration of the cation multiplied by the solution activity coefficients of the cation, i.e., $(ACl_u) = (C_A)\,(\gamma_A)$ and $(BCl_v) = (C_B)\,(\gamma_B)$. C_A and C_B are the equilibrium concentrations of cations A and B, respectively, and γ_A and γ_B are the solution activity coefficients of the two cations, respectively.

The activity coefficients of the electrolytes can be determined using Eq. (4.15).

However, calculating the activity of the exchanger phase is not as simple. Vanselow defined the exchanger phase activity in terms of mole fractions, $\bar{N}_A$ and $\bar{N}_B$ for ions A and B, respectively. Thus, according to Vanselow (1932) Eq. (6.3) could be rewritten as

$$K_V = \frac{\gamma_B^u \, C_B^u \, \bar{N}_A^v}{\gamma_A^v \, C_A^v \, \bar{N}_B^u} \, , \tag{6.4}$$

where

$$\bar{N}_A = \frac{\{AX_u\}}{\{AX_u\} + \{BX_v\}}$$

and

$$\bar{N}_B = \frac{\{BX_v\}}{\{AX_u\} + \{BX_v\}} \, . \tag{6.5}$$

Vanselow (1932) assumed that K_V was equal to K_{eq}. However, he failed to realize one very important point. The activity of a "component of a homogeneous mixture is equal to its mole fraction only if the mixture is ideal" (Guggenheim, 1967), i.e., $f_A = f_B = 1$, where f_A and f_B are the exchanger phase activity coefficients for cations A and B, respectively. If the mixture is not ideal, then the activity is a product of $\bar{N}$ and f. Thus, K_{eq} is correctly written as

$$K_{eq} = \left(\frac{\gamma_B^u \, C_B^u \, \bar{N}_A^v \, f_A^v}{\gamma_A^v \, C_A^v \, \bar{N}_B^u \, f_B^u} \right) = K_V \left(\frac{f_A^v}{f_B^u} \right) , \tag{6.6}$$

where

$$f_A \equiv (AX_u)/\bar{N}_A \text{ and } f_B \equiv (BX_v)/\bar{N}_B . \tag{6.7}$$

Thus,

$$K_V = K_{eq} \, (f_B^u / f_A^v) \tag{6.8}$$

and K_V is an apparent equilibrium exchange constant or a cation exchange selectivity coefficient.

Other Empirical Exchange Equations

A number of other cation exchange selectivity coefficients have also been employed in environmental soil chemistry. Krishnamoorthy and Overstreet (1949) used a statistical mechanics approach and included a factor for valence of the ions, 1 for monovalent ions, 1.5 for divalent ions, and 2 for trivalent ions, to obtain a selectivity coefficient K_{KO}. Gaines and Thomas (1953) and Gapon (1933) also introduced exchange equations that yielded selectivity coefficients (K_{GT} and K_G, respectively). For K–Ca exchange on a soil, the Gapon Convention would be written as

$$\text{Ca}_{1/2}\text{-soil} + \text{K}^+ \rightleftharpoons \text{K-soil} + {}^1/_2\text{Ca}^{2+}, \tag{6.9}$$

where there are chemically equivalent quantities of the exchanger phases and the exchangeable cations. The Gapon selectivity coefficient for K–Ca exchange would be expressed as

$$K_G = \frac{\{\text{K-soil}\}\,[\text{Ca}^{2+}]^{1/2}}{\{\text{Ca}_{1/2}\text{-soil}\}\,[\text{K}^+]}, \tag{6.10}$$

where brackets represent the concentration in the aqueous phase, expressed as mol liter^{-1}, and braces represent the concentration in the exchanger phase, expressed as mol kg^{-1}. The selectivity coefficient obtained from the Gapon equation has been the most widely used in soil chemistry and appears to vary the least as exchanger phase composition changes. The various cation exchange selectivity coefficients for homovalent and heterovalent exchange are given in Table 6.1.

Thermodynamics of Ion Exchange

Theoretical Background

Thermodynamic equations that provide a relationship between exchanger phase activity coefficients and the exchanger phase composition were independently derived by Argersinger *et al.* (1950) and Hogfeldt (Ekedahl *et al.*, 1950; Hogfeldt *et al.*, 1950). These equations, as shown later, demonstrated that the calculation of an exchanger phase activity coefficient, f, and the thermodynamic equilibrium constant, K_{eq}, were reduced to the measurement of the Vanselow selectivity coefficient, K_V, as a function of the exchanger phase composition (Sposito, 1981a). Argersinger *et al.* (1950) defined f as $f = a/\bar{N}$, where a is the activity of the exchanger phase.

Before thermodynamic parameters for exchange equilibria can be calculated, standard states for each phase must be defined. The choice of standard state affects the value of the thermodynamic parameters and their physical interpretation (Goulding, 1983a). Table 6.2 shows the different standard states and the effects of using them. Normally, the standard state for the adsorbed phase is the homoionic exchanger in equilibrium with a solution of the saturating cation at constant ionic strength.

TABLE 6.1. *Cation Exchange Selectivity Coefficients for Homovalent (K–Na) and Heterovalent (K–Ca) Exchange*

Selectivity coefficient	Homovalent exchange[a]	Heterovalent exchange[b]
Kerr	$K_K = \dfrac{\{K\text{-soil}\}\,[Na^+]^c}{\{Na\text{-soil}\}\,[K^+]}$	$K_K = \dfrac{\{K\text{-soil}\}^2\,[Ca^{2+}]}{\{Ca\text{-soil}\}\,[K^+]^2}$
Vanselow[d]	$K_V = \dfrac{\{K\text{-soil}\}\,[Na^+]}{\{Na\text{-soil}\}\,[K^+]}$, or $K_V = K_K$	$K_V = \left[\dfrac{\{K\text{-soil}\}^2\,[Ca^{2+}]}{\{Ca\text{-soil}\}\,[K^+]^2}\right]$ $\left[\dfrac{1}{\{K\text{-soil}\}+[Ca\text{-soil}]}\right]$ or $K_K\left[\dfrac{1}{\{K\text{-soil}\}+[Ca\text{-soil}]}\right]$
Krishnamoorthy–Overstreet	$K_{KO} = \dfrac{\{K\text{-soil}\}\,[Na^+]}{\{Na\text{-soil}\}\,[K^+]}$, or $K_{KO} = K_K$	$K_{KO} = \left[\dfrac{\{K\text{-soil}\}^2\,[Ca^{2+}]}{\{Ca\text{-soil}\}\,[K^+]^2}\right]$ $\left[\dfrac{1}{\{K\text{-soil}\}+1.5\,\{Ca\text{-soil}\}}\right]$
Gaines–Thomas[d]	$K_{GT} = \dfrac{\{K\text{-soil}\}\,[Na^+]}{\{Na\text{-soil}\}\,[K^+]}$, or $K_{GT} = K_K$	$K_{GT} = \left[\dfrac{\{K\text{-soil}\}^2\,[Ca^{2+}]}{\{Ca\text{-soil}\}\,[K^+]^2}\right]$ $\left[\dfrac{1}{2[2\{Ca\text{ soil}\}+\{K\text{ soil}\}]}\right]$
Gapon	$K_G = \dfrac{\{K\text{-soil}\}\,[Na^+]}{\{Na\text{-soil}\}\,[K^+]}$, or $K_G = K_K$	$K_G = \dfrac{\{K\text{ soil}\}\,[Ca^{2+}]^{1/2}}{\{Ca_{1/2}\text{ soil}\}\,[K^+]}$

[a] The homovalent exchange reaction (K–Na exchange) is Na-soil + K$^+$ $\rightleftharpoons$ K-soil + Na$^+$.
[b] The heterovalent exchange reaction (K–Ca exchange) is Ca-soil + 2K$^+$ $\rightleftharpoons$ 2K-soil + Ca^{2+}, except for the Gapon convention where the exchange reaction would be Ca$_{1/2}$-soil + K$^+$ $\rightleftharpoons$ K-soil + $^1/_2$ Ca^{2+}.
[c] Brackets represent the concentration in the aqueous phase, which is expressed in mol liter^{-1}; braces represent the concentration in the exchanger phase, which is expressed in mol kg^{-1}.
[d] Vanselow (1932) and Gaines and Thomas (1953) originally expressed both aqueous and exchanger phases in terms of activity. For simplicity, they are expressed here as concentrations.

Argersinger *et al.* (1950), based on Eq. (6.8), assumed that any change in K_V with regard to exchanger phase composition occurred because of a variation in exchanger phase activity coefficients. This is expressed as

$$v \ln f_A - u \ln f_B = \ln K_{eq} - \ln K_V. \tag{6.11}$$

Taking differentials of both sides, realizing that K_{eq} is a constant, results in

$$vd \ln f_A - ud \ln f_B = - d \ln K_V. \tag{6.12}$$

Any change in the activity of BX_v (s) must be accounted for by a change in the activity of AX_u (s), such that the mass in the exchanger is conserved. This necessity, an application of the Gibbs–Duhem equation (Guggenheim, 1967), results in

$$\bar{N}_A d \ln f_A + \bar{N}_B d \ln f_B = 0. \tag{6.13}$$

Equations (6.12) and (6.13) can be solved, resulting in

$$vd \ln f_A = \left(\frac{-v\bar{N}_B}{u\bar{N}_A + v\bar{N}_B} \right) d \ln K_V \tag{6.14}$$

$$ud \ln f_B = \left(\frac{-v\bar{N}_A}{u\bar{N}_A + v\bar{N}_B} \right) d \ln K_V, \tag{6.15}$$

where $(u\bar{N}_A/(u\bar{N}_A + v\bar{N}_B))$ is equal to $\bar{E}_A$ or the equivalent fraction of AX_u (s) and $\bar{E}_B$ is $(v\bar{N}_B/(u\bar{N}_A + v\bar{N}_B))$ or the equivalent fraction of BX_v (s) and the identity $\bar{N}_A$ and $\bar{N}_B = 1$.

TABLE 6.2. *Some of the Standard States Used in Calculating the Thermodynamic Parameters of Cation-Exchange Equilibria[a]*

Standard state			
Adsorbed phase	Solution phase	Implications	Reference
Activity = mole fraction when the latter = 1	Activity = molarity as concentration $\rightarrow 0$	Can calculate f, K_V, etc., but all depend on ionic strength	Argersinger *et al.* (1950)
Homoionic exchanger in equilibrium with an infinitely dilute solution of the ion	Activity = molarity as concentration $\rightarrow 0$	ΔG_{ex}^{o} expresses relative affinity of exchanger for cations	Gaines and Thomas (1953)
Activity = mole fraction when the latter = 0.5. Components not in equilibrium	Activity = molarity as concentration $\rightarrow 0$	ΔG_{ex}^{o} expresses relative affinity of exchanger for cations when mole fraction = 0.5	Babcock (1963)

[a] From Goulding (1983b), with permission.

In terms of $\bar{E}_A$, Eqs. (6.14) and (6.15) become

$$vd \ln f_A = -(1 - \bar{E}_A) d \ln K_V \tag{6.16}$$

$$ud \ln f_B = \bar{E}_A d \ln K_V. \tag{6.17}$$

Integrating Eqs. (6.16) and (6.17) by parts, noting that $\ln f_A = 0$ at $\bar{N}_A = 1$, or $\bar{E}_A = 1$, and similarly $\ln f_B = 0$ at $\bar{N}_A = 0$, or $\bar{E}_A = 0$,

$$-v \ln f_A = + (1 - \bar{E}_A) \ln K_V - \int_{\bar{E}_A}^{1} \ln K_V d\bar{E}_A, \tag{6.18}$$

$$-u \ln f_B = -\bar{E}_A \ln K_V + \int_{0}^{\bar{E}_A} \ln K_V d\bar{E}_A. \tag{6.19}$$

Substituting these into Eq. (6.11) leads to

$$\ln K_{eq} = \int_{0}^{1} \ln K_V d\bar{E}_A, \tag{6.20}$$

which provides for calculation of the thermodynamic equilibrium exchange constant. Thus, by plotting $\ln K_V$ vs $\bar{E}_A$ and integrating under the curve, from $\bar{E}_A = 0$ to $\bar{E}_A = 1$, one can calculate K_{eq}, or in ion exchange studies,

K_{ex}, the equilibrium exchange constant. Other thermodynamic parameters can then be determined as given below,

$$\Delta G^0_{ex} = -RT \ln K_{ex}, \tag{6.21}$$

where ΔG^0_{ex} is the standard Gibbs free energy of exchange. Examples of how exchanger phase activity coefficients and K_{ex} and ΔG^0_{ex} values can be calculated for binary exchange processes are provided in Boxes 6.2 and 6.3, respectively.

Using the van't Hoff equation one can calculate the standard enthalpy of exchange, ΔH^0_{ex}, as

$$\ln \frac{K_{ex_{T_2}}}{K_{ex_{T_1}}} = \left(\frac{-\Delta H^0_{ex}}{R} \right) \left(\frac{1}{T_2} - \frac{1}{T_1} \right), \tag{6.22}$$

where subscripts 1 and 2 denote temperatures 1 and 2. From this relationship,

$$\Delta G^0_{ex} = \Delta H^0_{ex} - T \Delta S^0_{ex}. \tag{6.23}$$

The standard entropy of exchange, ΔS^0_{ex}, can be calculated, using

$$\Delta S^0_{ex} = (\Delta H^0_{ex} - \Delta G^0_{ex})/T. \tag{6.24}$$

BOX 6.2. *Calculation of Exchanger Phase Activity Coefficients*

It would be instructive at this point to illustrate how exchanger phase activity coefficients would be calculated for the homovalent and heterovalent exchange reactions in Table 6.1. For the homovalent reaction, K–Na exchange, the f_K and f_{Na} values would be calculated as (Argersinger *et al.*, 1950)

$$-\ln f_K = (1 - \bar{E}_K) \ln K_V - \int_{\bar{E}_K}^{1} \ln K_V d\bar{E}_K, \tag{6.2a}$$

$$-\ln f_{Na} = -\bar{E}_K \ln K_V + \int_{0}^{\bar{E}_K} \ln K_V d\bar{E}_K, \tag{6.2b}$$

and

$$\ln K_{ex} = \int_{0}^{1} \ln K_V d\bar{E}_K. \tag{6.2c}$$

For the heterovalent exchange reaction, K–Ca exchange, the f_K and f_{Ca} values would be calculated as (Ogwada and Sparks, 1986a)

$$2 \ln f_K = -(1 - \bar{E}_K) \ln K_V + \int_{\bar{E}_K}^{1} \ln K_V d\bar{E}_K, \tag{6.2d}$$

$$\ln f_{Ca} = \bar{E}_K \ln K_V - \int_{0}^{\bar{E}_K} \ln K_V d\bar{E}_K, \tag{6.2e}$$

and

$$\ln K_{ex} = \int_{0}^{1} \ln K_V d\bar{E}_K. \tag{6.2f}$$

BOX 6.3.   *Calculation of Thermodynamic Parameters for K–Ca Exchange on a Soil*

Consider the general binary exchange reaction in Eq. (6.1).

$$vACl_u \text{ (aq)} + uBX_v \text{ (s)} \rightleftharpoons uBCl_v \text{ (aq)} + vAX_u \text{ (s)}. \tag{6.3a}$$

If one is studying K–Ca exchange where A is K^+, B is Ca^{2+}, v is 2, and u is 1, then Eq. (6.3a) can be rewritten as

$$2KCl + Ca\text{-soil} \rightleftharpoons CaCl_2 + 2K\text{-soil}. \tag{6.3b}$$

Using the experimental methodology given in the text, one can calculate K_v, K_{ex}, and ΔG_{ex}^0 parameters for the K–Ca exchange reaction in Eq. (6.3b) as shown in the calculations below. Assume the ionic strength (I) was 0.01 and the temperature at which the experiment was conducted is 298 K.

Exchanger test	Solution (aq.) phase concentration (mol liter^{-1})		Exchanger phase concentration (mol kg^{-1})		Mole fractions^a		$K_V{}^b$	$\ln K_V$	$\bar{E}_K{}^c$
	K^+	Ca^{2+}	K^+	Ca^{2+}	$\bar{N}_K$	$\bar{N}_{Ca}$			
1	0	3.32×10^{-3}	0	1.68×10^{-2}	0	1.000	—	5.11^d	0
2	1×10^{-3}	2.99×10^{-3}	2.95×10^{-3}	1.12×10^{-2}	0.2086	0.7914	134.20	4.90	0.116
3	2.5×10^{-3}	2.50×10^{-3}	7.88×10^{-3}	1.07×10^{-2}	0.4232	0.5768	101.36	4.62	0.268
4	4.0×10^{-3}	1.99×10^{-3}	8.06×10^{-3}	5.31×10^{-3}	0.6030	0.3970	92.95	4.53	0.432
5	7.0×10^{-3}	9.90×10^{-4}	8.63×10^{-3}	2.21×10^{-3}	0.7959	0.2041	51.16	3.93	0.661
6	8.5×10^{-3}	4.99×10^{-4}	1.17×10^{-2}	1.34×10^{-3}	0.8971	0.1029	44.07	3.79	0.813
7	9.0×10^{-3}	3.29×10^{-4}	1.43×10^{-2}	1.03×10^{-3}	0.9331	0.0669	43.13	3.76	0.875
8	1.0×10^{-2}	0	1.45×10^{-2}	0	1.000	0.0000	—	3.70^d	1

^{a} $\bar{N}_K = \dfrac{\{K^+\}}{\{K^+\} + \{Ca^{2+}\}}$; $\bar{N}_{Ca} = \dfrac{\{Ca^{2+}\}}{\{K^+\} + \{Ca^{2+}\}}$,

where braces indicate the exchanger phase composition, in mol kg^{-1}; e.g., for exchanger test 2,

$$\bar{N}_K = \frac{(2.95\times10^{-3})}{(2.95\times10^{-3}) + (1.12\times10^{-2})} = 0.2086.$$

^{b} $K_V = \dfrac{\gamma_{Ca^{2+}} C_{Ca^{2+}} (\bar{N}_K)^2}{(\gamma_{K^+})^2 (C_{K^+})^2 (\bar{N}_{Ca})}$,

where γ is the solution phase activity coefficient calculated according to Eq. (4.15) and C is the solution concentration; e.g., for exchanger test 2,

$$K_V = \frac{(0.6653)(2.99\times10^{-3} \text{ mol liter}^{-1})(0.2086)^2}{(0.9030)^2 (1\times10^{-3} \text{ mol liter}^{-1})^2 (0.7914)} = 134.20.$$

^{c} $\bar{E}_K$ is the equivalent fraction of K^+ on the exchanger,

$$\bar{E}_K = \frac{u\bar{N}_K}{u\bar{N}_K + v\bar{N}_{Ca}} = \frac{\bar{N}_K}{\bar{N}_K + 2\bar{N}_{Ca}} ;$$

e.g., for exchanger test 2,

$$\frac{0.2086}{0.2086 + (0.7914)(2)} = \frac{0.2086}{1.7914} = 0.116.$$

^{d} Extrapolated $\ln K_V$ values.

Using Eq. (6.20),

$$\ln K_{ex} = \int_0^1 \ln K_V \, d\bar{E}_K ,$$

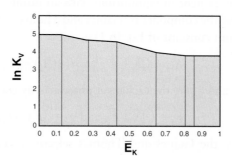

FIGURE 6.B1.

one can determine $\ln K_{ex}$ by plotting $\ln K_V$ vs $\bar{E}_K$ (Fig. 6.B1) and integrating under the curve by summing the areas of the trapezoids using the relationship

$$\frac{1}{2} \sum_{i=1}^{8} (\bar{E}_K^{i+1} - \bar{E}_K^i) \, (y^i + y^{i+1}) , \tag{6.3c}$$

where $\bar{E}_K^1 \ldots \bar{E}_K^8$ are the experimental values of $\bar{E}_K$, $(\bar{E}_K^{i+1} - \bar{E}_K^i)$ gives the width of the ith trapezoid, and $y^1 \ldots y^8$ represent the corresponding $\ln K_V$ values.

Accordingly, $\ln K_{ex}$ for the exchange reaction in Eq. (6.3b) would be

$$\ln K_{ex} = \frac{1}{2} \; [(0.116 - 0) \, (5.11 + 4.90) + (0.268 - 0.116)$$

$$\times \, (4.90 + 4.62) + (0.432 - 0.268) \, (4.62 + 4.53)$$

$$+ \, (0.661 - 0.432) \, (4.53 + 3.93) + (0.813 - 0.661)$$

$$\times \, (3.93 + 3.79) + (0.875 - 0.813) \, (3.79 + 3.76)$$

$$+ \, (1 - 0.875) \, (3.76 + 3.70)],$$

where $\ln K_{ex} = 4.31$ and $K_{ex} = 74.45$. From this value one can then calculate ΔG_{ex}^0 using Eq. (6.21):

$$\Delta G_{ex}^0 = -RT \ln K_{ex}.$$

Substituting 8.314 J mol^{-1} K^{-1} for R and assuming $T = 298$ K, $\Delta G_{ex}^0 = -(8.314$ J mol^{-1} K^{-1}) (298 K) (4.31) = $-10,678$ J mol^{-1} = -10.68 kJ mol^{-1}.

Since ΔG_{ex}^0 is negative, this would indicate that K$^+$ is preferred over Ca^{2+} on the soil.

Gaines and Thomas (1953) also described the thermodynamics of cation exchange and made two contributions. They included a term in the Gibbs–Duhem equation for the activity of water that may be adsorbed on the exchanger. This activity may change as the exchanger phase composition changes. Some workers later showed that changes in water activity with exchanger composition variations had little effect on K_{ex} calculations (Laudelout and Thomas, 1965) but can affect calculation of f values for zeolites (Barrer and Klinowski, 1974). Gaines and Thomas (1953) also defined the reference state of a component of the exchanger as the homoionic exchanger made up of the component in equilibrium with an infinitely dilute aqueous solution containing the components. Gaines and Thomas (1953) defined the exchange equilibrium constant of Eq. (6.1) as

$$K_{ex} = (BCl_v)^u \, g_A^v \bar{E}_A^v / (ACl_u)^v g_B^u \bar{E}_B^u ,\qquad (6.25)$$

where g_A and g_B are the exchanger phase activity coefficients and are defined as

$$g_A \equiv (AX_u)/\bar{E}_A \text{ and } g_B \equiv (BX_v)/\bar{E}_B . \qquad (6.26)$$

Thus, the Gaines and Thomas selectivity coefficient, K_{GT}, would be defined as

$$K_{GT} = (BCl_v)^u \, \bar{E}_A^v / (ACl_u)^v \, \bar{E}_B^u . \qquad (6.27)$$

Hogfeldt et al. (1950) also defined the exchanger phase activity coefficients in terms of the equivalent fraction rather than the Vanselow (1932) convention of mole fraction. Of course, for homovalent exchange, mole and equivalent fractions are equal.

There has been some controversy as to whether the Argersinger et al. (1950) or the Gaines and Thomas (1953) approach should be used to calculate thermodynamic parameters, particularly exchanger phase activity coefficients. Sposito and Mattigod (1979) and Babcock and Doner (1981) have questioned the use of the Gaines and Thomas (1953) approach. They note that except for homovalent exchange, the g values are not true activity coefficients, since the activity coefficient is the ratio of the actual activity to the value of the activity under those limiting conditions when Raoult's law applies (Sposito and Mattigod, 1979). Thus, for exchanger phases, an activity coefficient is the ratio of an actual activity to a mole fraction. Equivalents are formal quantities not associated with actual chemical species except for univalent ions.

Goulding (1983b) and Ogwada and Sparks (1986a) compared the two approaches for several exchange processes and concluded that while there were differences in the magnitude of the selectivity coefficients and adsorbed phase activity coefficients, the overall trends and conclusions concerning ion preferences were the same. Ogwada and Sparks (1986a) studied K–Ca exchange on soils at several temperatures and compared the Argersinger et al. (1950) and Gaines and Thomas (1953) approaches. The difference in the exchanger phase activity coefficients with the two approaches was small at low fractional K^+ saturation values but increased as fractional K^+ saturation increased (Fig. 6.1). However, as seen in Fig. 6.1 the minima, maxima, and inflexions occurred at the same fractional K^+ saturations with both approaches.

Experimental Interpretations

In conducting an exchange study to measure selectivity coefficients, exchanger phase activity coefficients, equilibrium exchange constants, and standard free energies of exchange, the exchanger is first saturated to make it homoionic (one ion predominates on the exchanger). For example, if one wanted to study K–Ca exchange, i.e., the conversion of the soil from Ca^{2+} to K^+, one would equilibrate the soil several times with 1 M $CaCl_2$ or $Ca(ClO_4)_2$ and then remove excess salts with water and perhaps an organic solvent such as methanol. After the soil is in a homoionic form, i.e., Ca-soil, one would equilibrate the soil with a series of salt solutions containing a range of Ca^{2+} and K^+ concentrations (Box 6.3). For example, in the K–Ca exchange experiment described in Box 6.3 the Ca-soil would be reacted by shaking with or leaching with the varying solutions until equilibrium had been obtained; i.e., the concentrations of K^+ and Ca^{2+} in the equilibrium (final) solutions were equal to the initial solution concentrations of K^+ and Ca^{2+}. To calculate the quantity of ions adsorbed on the exchanger (exchanger phase concentration in Box 6.3) at equilibrium, one would exchange the ions from the soil using a different electrolyte solution, e.g., ammonium acetate, and measure the exchanged ions using inductively coupled plasma (ICP) spectrometry or some other analytical technique. Based on such an exchange experiment, one could then calculate the mole fractions of the adsorbed ions, the selectivity coefficients, and K_{ex} and ΔG^0_{ex} as shown in Box 6.3.

From the data collected in an exchange experiment, exchange isotherms that show the relationship between the equivalent fraction of an ion on the exchanger phase ($\bar{E}_i$) versus the equivalent fraction of that ion in solution (E_i) are often presented. In homovalent exchange, where the equivalent fraction in the exchanger phase is not affected by the ionic strength and exchange equilibria are also not affected by valence effects, a diagonal line through the exchange isotherm can be used as a nonpreference exchange isotherm ($\Delta G^0_{ex} = 0$; $E_i = \bar{E}_i$ where i refers to ion i (Jensen and Babcock, 1973)). That

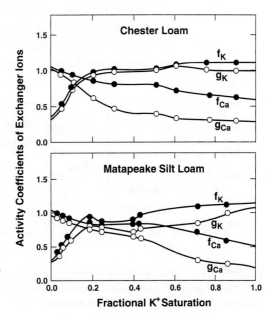

FIGURE 6.1. *Exchanger phase activity coefficients for K^+ and Ca^{2+} calculated according to the Argersinger et al. (1950) approach (f_K and f_{Ca}, respectively) and according to the Gaines and Thomas (1953) approach (g_K and g_{Ca}, respectively) versus fractional K^+ saturation (percentage of exchanger phase saturated with K^+). From Ogwada and Sparks (1986a), with permission.*

is, if the experimental data lie on the diagonal, there is no preference for one ion over the other. If the experimental data lie above the nonpreference isotherm, the final ion or product is preferred, whereas if the experimental data lie below the diagonal, the reactant is preferred. In heterovalent exchange, however, ionic strength affects the course of the isotherm and the diagonal cannot be used (Jensen and Babcock, 1973). By using Eq. (6.28) below, which illustrates divalent–univalent exchange, e.g., Ca–K exchange, nonpreference exchange isotherms can be calculated (Sposito, 2000),

$$\bar{E}_{Ca} = 1 - \left\{1 + \frac{2}{\Gamma I}\left[\frac{1}{(1 - E_{Ca})^2} - \frac{1}{(1 - E_{Ca})}\right]\right\}^{-1/2}, \qquad (6.28)$$

where I = ionic strength of the solution, $\bar{E}_{Ca}$ = equivalent fraction of Ca^{2+} on the exchanger phase, E_{Ca} = equivalent fraction of Ca^{2+} in the solution phase, and $\Gamma = \gamma_K^2/\gamma_{Ca}$. If the experimental data lie above the curvilinear nonpreference isotherm calculated using Eq. (6.28), then $K_V > 1$ and the final ion or product is preferred (in this case, Ca^{2+}). If the data lie below the non-preference isotherm, the initial ion or reactant is preferred (in this case, K^+). Thus, from Fig. 6.2 (Jensen and Babcock, 1973) one sees that K^+ is preferred over Na^+ and Mg^{2+} and Ca^{2+} is preferred over Mg^{2+}.

Table 6.3, from Jensen and Babcock (1973), shows the effect of ionic strength on thermodynamic parameters for several binary exchange systems of a Yolo soil from California. The K_{ex} and ΔG_{ex}^0 values are not affected by ionic strength. Although not shown in Table 6.3, the K_V was dependent on ionic strength with the K exchange systems (K–Na, K–Mg, K–Ca), and there was a selectivity of K^+ over Na^+, Mg^{2+}, and Ca^{2+} which decreased with increasing K^+ saturation. For Mg–Ca exchange, the K_V values were independent of ionic strength and exchanger composition. This system behaved ideally.

It is often observed, particularly with K^+, that K_V values decrease as the equivalent fraction of cation on the exchanger phase or fractional cation saturation increases (Fig. 6.3). Ogwada and Sparks (1986a) ascribed the decrease in the K_V with increasing equivalent fractions to the heterogeneous

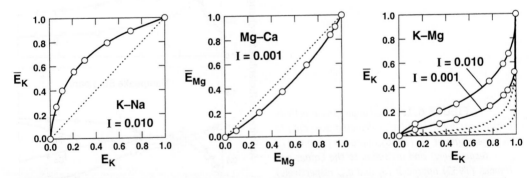

FIGURE 6.2. *Cation exchange isotherms for several cation exchange systems. E = equivalent fraction in the solution phase, $\bar{E}$ = equivalent fraction on the exchanger phase. The broken lines represent nonpreference exchange isotherms. From Jensen and Babcock (1973), with permission.*

TABLE 6.3. *Effect of Ionic Strength on Thermodynamic Parameters for Several Cation-Exchange Systems[a]*

Ionic strength (I)	Standards Gibbs free energy of exchange (ΔG^0_{ex}) (kJ mol^{-1})				Equilibrium exchange constant (K_{ex})			
	K–Na exchange	Mg–Ca exchange	K–Mg exchange	K–Ca exchange	K–Na exchange	Mg–Ca exchange	K–Mg exchange	K–Ca exchange
0.001	—	1.22	–7.78	—	—	0.61	22.93	—
0.010	–4.06	1.22	–7.77	–6.18	5.12	0.61	22.85	12.04
0.100	–4.03	—	—	—	5.08	—	—	—

[a] From Jensen and Babcock (1973), with permission. The exchange studies were conducted on a Yolo loam soil.

exchange sites and a decreasing specificity of the surface for K$^+$ ions. Jardine and Sparks (1984a,b) had shown earlier that there were different sites for K$^+$ exchange on soils.

One also observes with K$^+$ as well as other ions that the exchanger phase activity coefficients do not remain constant as exchanger phase composition changes (Fig. 6.1). This indicates nonideality since if ideality existed f_{Ca} and f_K would both be equal to 1 over the entire range of exchanger phase composition. A lack of ideality is probably related to the heterogeneous sites and the heterovalent exchange. Exchanger phase activity coefficients correct the equivalent or mole fraction terms for departures from ideality. They thus reflect the change in the status, or fugacity, of the ion held at exchange sites and the heterogeneity of the exchange process. Fugacity is the degree of freedom an ion has to leave the adsorbed state, relative to a standard state of maximum freedom of unity. Plots of exchanger phase activity coefficients versus equivalent fraction of an ion on the exchanger phase show how this freedom changes during the exchange process, which tells something about the exchange heterogeneity. Selectivity changes during the exchange process can also be gleaned (Ogwada and Sparks, 1986a).

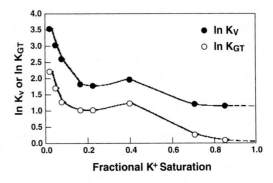

FIGURE 6.3. *Natural logarithm of Vanselow selectivity coefficients (K_V, •) and Gaines and Thomas selectivity coefficients (K_{GT}, 0) as a function of fractional K$^+$ saturation (percentage of exchanger phase saturated with K$^+$) on Chester loam soil at 298 K. From Ogwada and Sparks (1986a), with permission.*

The ΔG^0_{ex} values indicate the overall selectivity of an exchanger at constant temperature and pressure, and independently of ionic strength. For K–Ca exchange a negative ΔG^0_{ex} would indicate that the product or K^+ is preferred. A positive ΔG^0_{ex} would indicate that the reactant, i.e., Ca^{2+} is preferred. Some ΔG^0_{ex} values as well as ΔH^0_{ex} parameters for exchange on soils and soil components are shown in Table 6.4.

TABLE 6.4. *Standard Gibbs Free Energy of Exchange (ΔG^0_{ex}) and Standard Enthalpy of Exchange (ΔH^0_{ex}) Values for Binary Exchange Processes on Soils and Soil Components[a]*

Exchange process	Exchanger	ΔG^0_{ex} (kJ mol^{-1})	ΔH^0_{ex} (kJ mol^{-1})	Reference
Ca–Na	Soils	2.15 to 7.77		Mehta *et al.* (1983)
Ca–Na	Calcareous soils	2.38 to 6.08		Van Bladel and Gheyi (1980)
Ca–Na	World vermiculite	0.06	39.98	Wild and Keay (1964)
Ca–Na	Camp Berteau montmorillonite	0.82		
Ca–Mg	Calcareous soils	0.27 to 0.70		Van Bladel and Gheyi (1980)
Ca–Mg	Camp Berteau montmorillonite	0.13		Van Bladel and Gheyi (1980)
Ca–K	Chambers montmorillonite	7.67	16.22	Hutcheon (1966)
Ca–NH$_4$	Camp Berteau montmorillonite	8.34	23.38	Laudelout *et al.* (1967)
Ca–Cu	Wyoming bentonite	0.11	−18.02	El-Sayed *et al.* (1970)
Na–Ca	Soils (304 K)	0.49 to 4.53		Gupta *et al.* (1984)
Na–Li	World vermiculite	−6.04	−23.15	Gast and Klobe (1971)
Na–Li	Wyoming bentonite	−0.20	−0.63	Gast and Klobe (1971)
Na–Li	Chambers montmorillonite	−0.34	−0.47	Gast and Klobe (1971)
Mg–Ca	Soil	1.22		Jensen and Babcock (1973)
Mg–Ca	Kaolinitic soil clay (303 K)	1.07	6.96	Udo (1978)
Mg–Na	Soils	0.72 to 7.27		Mehta *et al.* (1983)
Mg–Na	World vermiculite	−1.36	40.22	Wild and Keay (1964)
Mg–NH$_4$	Camp Berteau montmorillonite	8.58	23.38	Laudelout *et al.* (1967)
K–Ca	Soils	−4.40 to −14.33		Deist and Talibudeen (1967a)
K–Ca	Soil	−6.18		Jensen and Babcock (1973)
K–Ca	Soils	−7.42 to −14.33	−3.25 to −5.40	Deist and Talibudeen (1967b)
K–Ca	Soil	1.93	−15.90	Jardine and Sparks (1984b)

TABLE 6.4. *Standard Gibbs Free Energy of Exchange (ΔG^0_{ex}) and Standard Enthalpy of Exchange (ΔH^0_{ex}) Values for Binary Exchange Processes on Soils and Soil Components (contd)*

Exchange process	Exchanger	ΔG^0_{ex} (kJ mol^{-1})	ΔH^0_{ex} (kJ mol^{-1})	Reference
K–Ca	Soils	1.10 to –4.70	–3.25 to –5.40	Goulding and Talibudeen (1984)
K–Ca	Soils	–4.61 to –4.74	–16.28	Ogwada and Sparks (1986b)
K–Ca	Soil silt	0.36		Jardine and Sparks (1984b)
K–Ca	Soil clay	–2.84		Jardine and Sparks (1984b)
K–Ca	Kaolinitic soil clay (303 K)	–6.90	–54.48	Udo (1978)
K–Ca	Clarsol montmorillonite	–6.26		Jensen (1972)
K–Ca	Danish kaolinite	–8.63		Jensen (1972)
K–Mg	Soil	–4.06		Jensen and Babcock (1973)
K–Na	Soils	–3.72 to –4.54		Deist and Talibudeen (1967a)
K–Na	Wyoming bentonite	–1.28	–2.53	Gast (1972)
K–Na	Chambers montmorillonite	–3.04	–4.86	Gast (1972)

[a] Unless specifically noted, the exchange studies were conducted at 298 K.

Binding strengths of an ion on a soil or soil component exchanger can be determined from ΔH^0_{ex} values. Enthalpy expresses the gain or loss of heat during the reaction. If the reaction is exothermic, the enthalpy is negative and heat is lost to the surroundings. If it is endothermic, the enthalpy change is positive and heat is gained from the surroundings. A negative enthalpy change implies stronger bonds in the reactants. Enthalpies can be measured using the van't Hoff equation (Eq. (6.22)) or one can use calorimetry.

Relationship Between Thermodynamics and Kinetics of Ion Exchange

Another way that one can obtain thermodynamic exchange parameters is to employ a kinetic approach (Ogwada and Sparks, 1986b; Sparks, 1989). We know that if a reaction is reversible, then $k_1/k_{-1} = K_{ex}$, where k_1 is the forward reaction rate constant and k_{-1} is the backward reaction rate constant. However, this relationship is valid only if mass transfer or diffusion processes are not rate-limiting; i.e., one must measure the actual chemical exchange reaction (CR) process (see Chapter 7 for discussion of mass transfer and CR processes).

Ogwada and Sparks (1986b) found that this assumption is not valid for most kinetic techniques. Only if mixing is very rapid does diffusion become insignificant, and at such mixing rates one must be careful not to alter the surface area of the adsorbent. Calculation of energies of activation for the forward and backward reactions, E_1 and E_{-1}, respectively, using a kinetics approach, are given below:

$$d \ln k_1 / dT = E_1 / RT^2 \tag{6.29}$$

$$d \ln k_{-1} / dT = E_{-1} / RT^2. \tag{6.30}$$

Substituting,

$$d \ln k_1 / dT - d \ln k_{-1} / dT = d \ln K_{ex} / dT. \tag{6.31}$$

From the van't Hoff equation, ΔH_{ex}° can be calculated,

$$d \ln K_{ex} / dT = \Delta H_{ex}^{\circ} / RT^2, \tag{6.32}$$

or

$$E_1 - E_{-1} = \Delta H_{ex}^{\circ}, \tag{6.33}$$

and ΔG_{ex}° and ΔS_{ex}° can be determined as given in Eqs. (6.21) and (6.24), respectively.

Suggested Reading

Argersinger, W. J., Jr., Davidson, A. W., and Bonner, O. D. (1950). Thermodynamics and ion exchange phenomena. *Trans. Kans. Acad. Sci.* **53**, 404–410.

Babcock, K. L. (1963). Theory of the chemical properties of soil colloidal systems at equilibrium. *Hilgardia* **34**, 417–452.

Gaines, G. L., and Thomas, H. C. (1953). Adsorption studies on clay minerals. II. A formulation of the thermodynamics of exchange adsorption. *J. Chem. Phys.* **21**, 714–718.

Goulding, K. W. T. (1983). Thermodynamics and potassium exchange in soils and clay minerals. *Adv. Agron.* **36**, 215–261.

Jensen, H. E., and Babcock, K. L. (1973). Cation exchange equilibria on a Yolo loam. *Hilgardia* **41**, 475–487.

Sposito, G. (1981). Cation exchange in soils: An historical and theoretical perspective. *In* "Chemistry in the Soil Environment" (R. H. Dowdy, J. A. Ryan, V. V. Volk, and D. E. Baker, Eds.), Spec. Publ. 40, pp. 13–30. Am. Soc. Agron./Soil Sci. Soc. Am., Madison, WI.

Sposito, G. (1981). "The Thermodynamics of the Soil Solution." Oxford Univ. Press (Clarendon), Oxford.

Sposito, G. (2000). Ion exchange phenomena. *In* "Handbook of Soil Science" (M. E. Sumner, Ed.), pp. 241–263. CRC Press, Boca Raton, FL.

Sumner, M. E., and Miller, W. P. (1996). Cation exchange capacity and exchange coefficients. *In* "Methods of Soil Analysis, Part 3. Chemical Methods" (D. L. Sparks, Ed.), Soil Sci. Soc. Am. Book Ser. 5, pp. 1201–1229. Soil Sci. Soc. Am., Madison, WI.

7 Kinetics of Soil Chemical Processes

Many soil chemical processes are time-dependent. To fully understand the dynamic interactions of metals, oxyanions, radionuclides, pesticides, industrial chemicals, and plant nutrients with soils and to predict their fate with time, a knowledge of the kinetics of these reactions is important. This chapter will provide an overview of this topic, with applications to environmentally important reactions. The reader is referred to several sources for more definitive discussions on the topic (Sparks, 1989; Sparks and Suarez, 1991; Sposito, 1994).

Rate-Limiting Steps and Time Scales of Soil Chemical Reactions

A number of transport and chemical reaction processes can affect the rate of soil chemical reactions. The slowest of these will limit the rate of a particular reaction. The actual chemical reaction (CR) at the surface, for example, adsorption, is usually very rapid and not rate-limiting. Transport processes (Fig. 7.1) include: (1) transport in the solution phase, which is rapid, and in the laboratory, can be eliminated by rapid mixing; (2) transport across a liquid film at the particle/liquid interface (film diffusion (FD)); (3) transport in liquid-filled macropores (>2 nm), all of which are nonactivated diffusion processes and occur in mobile regions; (4) diffusion of a sorbate along pore wall surfaces (surface diffusion); (5) diffusion of sorbate occluded in micropores (<2 nm) (pore diffusion); and (6) diffusion processes in the bulk of the solid, all of which are activated diffusion processes. Pore and surface diffusion can be referred to as interparticle diffusion while diffusion in the solid is intraparticle diffusion.

Soil chemical reactions occur over a wide time scale (Fig. 7.2), ranging from microseconds and milliseconds for ion association (ion pairing, complexation, and chelation-type reactions in solution), ion exchange, and some sorption reactions to years for mineral solution (precipitation/dissolution reactions including discrete mineral phases) and mineral crystallization reactions (Amacher, 1991). These reactions can occur simultaneously and consecutively.

The type of soil component can drastically affect the reaction rate. For example, sorption reactions are often more rapid on clay minerals such as kaolinite and smectites than on vermiculitic and micaceous minerals. This is

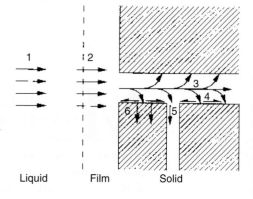

FIGURE 7.1. *Transport processes in solid–liquid soil reactions: nonactivated processes, (1) transport in the soil solution, (2) transport across a liquid film at the solid–liquid interface, and (3) transport in a liquid-filled macropore; activated processes, (4) diffusion of a sorbate at the surface of the solid, (5) diffusion of a sorbate occluded in a micropore, and (6) diffusion in the bulk of the solid. From Aharoni, C., and Sparks, D. L. (1991). Soil Sci. Soc. Am. Spec. Pub. 27. Reproduced with permission of the Soil Science Society of America.*

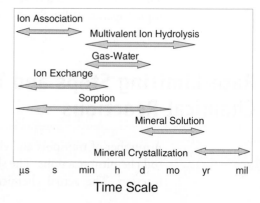

FIGURE 7.2. *Time ranges required to attain equilibrium by different types of reactions in soil environments. From Amacher (1991), with permission.*

in large part due to the availability of sites for sorption. For example, kaolinite has readily available planar external sites and smectites have primarily internal sites that are also quite available for retention of sorbates. Thus, sorption reactions on these soil constituents are often quite rapid, even occurring on time scales of seconds and milliseconds (Sparks, 1989).

On the other hand, vermiculite and micas have multiple sites for retention of metals and organics, including planar, edge, and interlayer sites, with some of the latter sites being partially to totally collapsed. Consequently, sorption and desorption reactions on these sites can be slow, tortuous, and mass transfer controlled. Often, an apparent equilibrium may not be reached even after several days or weeks. Thus, with vermiculite and mica, sorption can involve two to three different reaction rates: high rates on external sites, intermediate rates on edge sites, and low rates on interlayer sites (Jardine and Sparks, 1984a; Comans and Hockley, 1992).

Metal sorption reactions on oxides, hydroxides, and humic substances depend on the type of surface and metal being studied, but the CR appears to be rapid. For example, CR rates of metals and oxyanions on goethite occur on millisecond time scales (Sparks and Zhang, 1991; Grossl et al., 1994, 1997). Half-times for divalent Pb, Cu, and Zn sorption on peat ranged from 5 to 15 sec (Bunzl et al., 1976). A number of studies have shown that heavy metal sorption on oxides (Barrow, 1986; Brummer et al., 1988; Ainsworth et al., 1994; Scheidegger et al., 1997, 1998) and clay minerals (Lövgren et al., 1990) increases with longer residence times (contact time between metal and sorbent). The mechanisms for these lower reaction rates are not well understood, but have been ascribed to diffusion phenomena, sites of lower reactivity, and surface nucleation/precipitation (Scheidegger and Sparks, 1997; Sparks, 1998, 1999). More detail on metal and oxyanion retention rates and mechanisms at the soil mineral/water interface will be discussed later.

Sorption/desorption of metals, oxyanions, radionuclides, and organic chemicals on soils can be very slow, and may demonstrate a residence time effect, which has been attributed to diffusion into micropores of inorganic minerals and into humic substances, retention on sites of varying reactivity, and surface nucleation/precipitation (Scheidegger and Sparks, 1997; Sparks, 1998, 1999, 2000; Strawn and Sparks, 1999; Alexander, 2000; Pignatello, 2000).

It would be instructive at this point to define two important terms – *chemical kinetics* and *kinetics*. Chemical kinetics can be defined as "the investigation of chemical reaction rates and the molecular processes by which reactions occur where transport is not limiting" (Gardiner, 1969). Kinetics is the study of time-dependent processes.

The study of chemical kinetics in homogeneous solutions is difficult, and when one studies heterogeneous systems such as soil components and, particularly, soils, the difficulties are magnified. It is extremely difficult to eliminate transport processes in soils because they are mixtures of several inorganic and organic adsorbates. Additionally, there is an array of different particle sizes and porosities in soils that enhance their heterogeneity.

Thus, when dealing with soils and soil components, one usually studies the kinetics of the reactions.

Rate Laws

There are two important reasons for investigating the rates of soil chemical processes (Sparks, 1989): (1) to determine how rapidly reactions attain equilibrium, and (2) to infer information on reaction mechanisms. One of the most important aspects of chemical kinetics is the establishment of a rate law. By definition, a rate law is a differential equation. For the reaction (Bunnett, 1986)

$$aA + bB \rightarrow yY + zZ, \tag{7.1}$$

the rate of the reaction is proportional to some power of the concentrations of reactants A and B and/or other species (C, D, etc.) in the system. The terms a, b, y, and z are stoichiometric coefficients, and are assumed to be equal to one in the following discussion. The power to which the concentration is raised may equal zero (i.e., the rate is independent of that concentration), even for reactant A or B. Rates are expressed as a decrease in reactant concentration or an increase in product concentration per unit time. Thus, the rate of reactant A above, which has a concentration $[A]$ at any time t, is $(-d[A]/(dt))$ while the rate with regard to product Y having a concentration $[Y]$ at time t is $(d[Y]/(dt))$.

The rate expression for Eq. (7.1) is

$$d[Y]/dt = -d[A]/dt = k[A]^\alpha [B]^\beta \ldots, \tag{7.2}$$

where k is the rate constant, α is the order of the reaction with respect to reactant A and can be referred to as a partial order, and β is the order with respect to reactant B. These orders are experimentally determined and not necessarily integral numbers. The sum of all the partial orders (α, β, etc.) is the overall order (n) and may be expressed as

$$n = \alpha + \beta + \ldots. \tag{7.3}$$

Once the values of α, β, etc., are determined experimentally, the rate law is defined. Reaction order provides only information about the manner in which rate depends on concentration. Order does not mean the same as "molecularity," which concerns the number of reactant particles (atoms, molecules, free radicals, or ions) entering into an elementary reaction. One can define an elementary reaction as one in which no reaction intermediates have been detected or need to be postulated to describe the chemical reaction on a molecular scale. An elementary reaction is assumed to occur in a single step and to pass through a single transition state (Bunnett, 1986).

To prove that a reaction is elementary, one can use experimental conditions different from those employed in determining the law. For example, if one conducted a kinetic study using a flow technique (see later discussion on this technique) and the rate of influent solution (flow rate) was 1 ml min⁻¹, one could study several other flow rates to see whether reaction rate and rate constants change. If they do, one is not determining mechanistic rate laws.

Rate laws serve three purposes: they assist one in predicting the reaction rate, mechanisms can be proposed, and reaction orders can be ascertained. There are four types of rate laws that can be determined for soil chemical processes (Skopp, 1986): mechanistic, apparent, transport with apparent, and transport with mechanistic. Mechanistic rate laws assume that only chemical kinetics are operational and transport phenomena are not occurring. Consequently, it is difficult to determine mechanistic rate laws for most soil chemical systems due to the heterogeneity of the system caused by different particle sizes, porosities, and types of retention sites. There is evidence that with some kinetic studies using relaxation techniques (see later discussion) mechanistic rate laws are determined since the agreement between equilibrium constants calculated from both kinetics and equilibrium studies are comparable (Tang and Sparks, 1993). This would indicate that transport processes in the kinetics studies are severely limited (see Chapter 5). Apparent rate laws include both chemical kinetics and transport-controlled processes. Apparent rate laws and rate coefficients indicate that diffusion and other microscopic transport processes affect the reaction rate. Thus, soil structure, stirring, mixing, and flow rate all would affect the kinetics. Transport with apparent rate laws emphasizes transport phenomena. One often assumes first- or zero-order reactions (see discussion below on reaction order). In determining transport with mechanistic rate laws one attempts to describe simultaneously transport-controlled and chemical kinetics phenomena. One is thus trying to accurately explain both the chemistry and the physics of the system.

Determination of Reaction Order and Rate Constants

There are three basic ways to determine rate laws and rate constants (Bunnett, 1986; Skopp, 1986; Sparks, 1989): (1) using initial rates, (2) directly using integrated equations and graphing the data, and (3) using nonlinear least-squares analysis.

Let us assume the following elementary reaction between species A, B, and Y,

$$A + B \underset{k_{-1}}{\overset{k_1}{\rightleftharpoons}} Y. \tag{7.4}$$

A forward reaction rate law can be written as

$$d[A]/dt = -k_1[A][B], \tag{7.5}$$

where k_1 is the forward rate constant and α and β (see Eq. (7.2)) are each assumed to be 1.

The reverse reaction rate law for Eq. (7.4) is

$$d[A]/dt = +k_{-1}[Y],\qquad(7.6)$$

where k_{-1} is the reverse rate constant.

Equations (7.5) and (7.6) are only applicable far from equilibrium where back or reverse reactions are insignificant. If both these reactions are occurring, Eqs. (7.5) and (7.6) must be combined such that

$$d[A]/dt = -k_1[A][B] + k_{-1}[Y].\qquad(7.7)$$

Equation (7.7) applies the principle that the net reaction rate is the difference between the sum of all reverse reaction rates and the sum of all forward reaction rates.

One way to ensure that back reactions are not important is to measure initial rates. The initial rate is the limit of the reaction rate as time reaches zero. With an initial rate method, one plots the concentration of a reactant or product over a short reaction time period during which the concentrations of the reactants change so little that the instantaneous rate is hardly affected. Thus, by measuring initial rates, one could assume that only the forward reaction in Eq. (7.4) predominates. This would simplify the rate law to that given in Eq. (7.5), which as written would be a second-order reaction, first-order in reactant A and first-order in reactant B. Equation (7.4), under these conditions, would represent a second-order irreversible elementary reaction. To measure initial rates, one must have available a technique that can measure rapid reactions such as a relaxation method (see detailed discussion on this later) and an accurate analytical detection system for determining product concentrations.

Integrated rate equations can also be used to determine rate constants. If one assumes that reactant B in Eq. (7.5) is in large excess of reactant A, which is an example of the "method of isolation" to analyze kinetic data, and $Y_0 = 0$, where Y_0 is the initial concentration of product Y, Eq. (7.5) can be simplified to

$$d[A]/dt = -k_1[A].\qquad(7.8)$$

The first-order dependence of $[A]$ can be evaluated using the integrated form of Eq. (7.8) using the initial conditions at $t = 0$, $A = A_0$,

$$\log [A]_t = \log [A]_0 - \frac{k_1 t}{2.303}.\qquad(7.9)$$

The half-time $(t_{1/2})$ for the above reaction is equal to $0.693/k_1$ and is the time required for half of reactant A to be consumed.

If a reaction is first-order, a plot of $\log[A]_t$ vs t should result in a straight line with a slope $= -k_1/2.303$ and an intercept of $\log[A]_0$. An example of first-order plots for Mn^{2+} sorption on δ-MnO_2 at two initial Mn^{2+} concentrations, $[Mn^{2+}]_0$, 25 and 40 μM, is shown in Fig. 7.3. One sees that the plots are

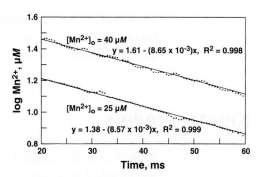

FIGURE 7.3. *Initial reaction rates depicting the first-order dependence of Mn^{2+} sorption as a function of time for initial Mn^{2+} concentrations ($[Mn^{2+}]_0$) of 25 and 40 μM. From Fendorf et al. (1993), with permission.*

linear at both concentrations, which would indicate that the sorption process is first-order. The $[Mn^{2+}]_0$ values, obtained from the intercept of Fig. 7.3, were 24 and 41 μM, in good agreement with the two $[Mn^{2+}]_0$ values. The rate constants were 3.73×10^{-3} and 3.75×10^{-3} s^{-1} at $[Mn^{2+}]_0$ of 25 and 40 μM, respectively. The findings that the rate constants are not significantly changed with concentration is a very good indication that the reaction in Eq. (7.8) is first-order under the experimental conditions that were imposed.

It is dangerous to conclude that a particular reaction order is correct, based simply on the conformity of data to an integrated equation. As illustrated above, multiple initial concentrations that vary considerably should be employed to see that the rate is independent of concentration. One should also test multiple integrated equations. It may be useful to show that reaction rate is not affected by species whose concentrations do not change considerably during an experiment; these may be substances not consumed in the reaction (i.e., catalysts) or present in large excess (Bunnett, 1986; Sparks, 1989).

Least-squares analysis can also be used to determine rate constants. With this method, one fits the best straight line to a set of points linearly related as $y = mx + b$, where y is the ordinate and x is the abscissa datum point, respectively. The slope, m, and the intercept, b, can be calculated by least-squares analysis using Eqs. (7.10) and (7.11), respectively (Sparks, 1989),

$$m = \frac{n\Sigma xy - \Sigma x \Sigma y}{n\Sigma x^2 - (\Sigma x)^2} \, , \tag{7.10}$$

$$b = \frac{\Sigma y \Sigma x^2 - \Sigma x \Sigma xy}{n\Sigma x^2 - (\Sigma x)^2} \, , \tag{7.11}$$

where n is the number of data points and the summations are for all data points in the set.

Curvature may result when kinetic data are plotted. This may be due to an incorrect assumption of reaction order. If first-order kinetics is assumed and the reaction is really second-order, downward curvature is observed. If second-order kinetics is assumed but the reaction is first-order, upward curvature is observed. Curvature can also be due to fractional, third, higher, or mixed reaction order. Nonattainment of equilibrium often results in downward curvature. Temperature changes during the study can also cause curvature; thus, it is important that temperature be accurately controlled during a kinetic experiment.

Kinetic Models

While first-order models have been used widely to describe the kinetics of soil chemical processes, a number of other models have been employed. These include various ordered equations such as zero-, second-, and fractional-order, and Elovich, power function or fractional power, and parabolic diffusion models. A brief discussion of some of these will be given; the final forms of the equations are given in Table 7.1. For more complete details and applications of these models one should consult Sparks (1989, 1998, 1999, 2000).

Elovich Equation

The Elovich equation was originally developed to describe the kinetics of heterogeneous chemisorption of gases on solid surfaces (Low, 1960). It seems to describe a number of reaction mechanisms, including bulk and surface diffusion and activation and deactivation of catalytic surfaces.

TABLE 7.1.   *Linear Forms of Kinetic Equations Commonly Used in Environmental Soil Chemistry*[a]

Zero order[b]
$$[A]_t = [A]_0 - kt$$

First order[b]
$$\log [A]_t = \log [A]_0 - \frac{kt}{2.303}[c]$$

Second order
$$\frac{1}{[A]_t} = \frac{1}{[A]_0} + kt$$

Elovich
$$q_t = (1/\beta) \ln (\alpha\beta) + (1/\beta) \ln t$$

Parabolic diffusion
$$\frac{q_t}{q_\infty} = R_D t^{1/2}$$

Power function
$$\ln q = \ln k + v \ln t$$

---

[a] Terms are defined in the text.
[b] Describing the reaction $A \rightarrow Y$.
[c] $\ln x = 2.303 \log x$ is the conversion from natural logarithms (ln) to base 10 logarithms (log).

In soil chemistry, the Elovich equation has been used to describe the kinetics of sorption and desorption of various inorganic materials on soils (see Sparks, 1989). It can be expressed as (Chien and Clayton, 1980).

$$q_t = (1/\beta) \ln (\alpha\beta) + (1/\beta) \ln t, \tag{7.12}$$

where q_t is the amount of sorbate per unit mass of sorbent at time t and α and β are constants during any one experiment. A plot of q_t vs $\ln t$ should give a linear relationship if the Elovich equation is applicable with a slope of $(1/\beta)$ and an intercept of $(1/\beta) \ln (\alpha\beta)$.

An application of Eq. (7.12) to phosphate sorption on soils is shown in Fig. 7.4.

Some investigators have used the α and β parameters from the Elovich equation to estimate reaction rates. For example, it has been suggested that a decrease in β and/or an increase in α would increase reaction rate. However, this is questionable. The slope of plots using Eq. (7.12) changes with the concentration of the adsorptive and with the solution to soil ratio (Sharpley, 1983). Therefore, the slopes are not always characteristic of the soil but may depend on various experimental conditions.

Some researchers have also suggested that "breaks" or multiple linear segments in Elovich plots could indicate a changeover from one type of binding site to another (Atkinson et al., 1970). However, such mechanistic suggestions may not be correct (Sparks, 1989).

Parabolic Diffusion Equation

The parabolic diffusion equation is often used to indicate that diffusion-controlled phenomena are rate-limiting. It was originally derived based on radial diffusion in a cylinder where the ion concentration throughout the cylinder is uniform. It is also assumed that ion diffusion through the upper and lower faces of the cylinder is negligible. Following Crank (1976), the parabolic diffusion equation, as applied to soils, can be expressed as

$$(q_t/q_\infty) = \frac{4}{\pi^{1/2}} Dt^{1/2}/r^2 - Dt/r^2, \tag{7.13}$$

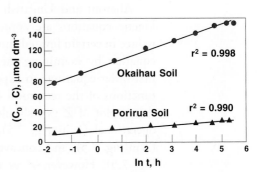

FIGURE 7.4. *Plot of Elovich equation for phosphate sorption on two soils where C_0 is the initial phosphorus concentration added at time 0 and C is the phosphorus concentration in the soil solution at time t. The quantity (C_0-C) can be equated to q_t, the amount sorbed at time t. From Chien and Clayton (1980), with permission.*

where r is the average radius of the soil particle, q_t was defined earlier, q_∞ is the corresponding quantity of sorbate at equilibrium, and D is the diffusion coefficient.

Equation (7.13) can be simply expressed as

$$q_t/q_\infty = R_D t^{1/2} + \text{constant}, \tag{7.14}$$

where R_D is the overall diffusion coefficient. If the parabolic diffusion law is valid, a plot of q_t/q_∞ versus $t^{1/2}$ should yield a linear relationship.

The parabolic diffusion equation has successfully described metal reactions on soils and soil constituents (Chute and Quirk, 1967; Jardine and Sparks, 1984a), feldspar weathering (Wollast, 1967), and pesticide reactions (Weber and Gould, 1966).

Fractional Power or Power Function Equation

This equation can be expressed as

$$q = kt^v, \tag{7.15}$$

where q is the amount of sorbate per unit mass of sorbent, k and v are constants, and v is positive and <1. Equation (7.15) is empirical, except for the case where $v = 0.5$, when Eq. (7.15) is similar to the parabolic diffusion equation.

Equation (7.15) and various modified forms have been used by a number of researchers to describe the kinetics of soil chemical processes (Kuo and Lotse, 1974; Havlin and Wesfall, 1985).

Comparison of Kinetic Models

In a number of studies it has been shown that several kinetic models describe the rate data well, based on correlation coefficients and standard errors of the estimate (Chien and Clayton, 1980; Onken and Matheson, 1982; Sparks and Jardine, 1984). Despite this, there often is not a consistent relation between the equation that gives the best fit and the physicochemical and mineralogical properties of the sorbent(s) being studied. Another problem with some of the kinetic equations is that they are empirical and no meaningful rate parameters can be obtained.

Aharoni and Ungarish (1976) and Aharoni (1984) noted that some kinetic equations are approximations to which more general expressions reduce in certain limited time ranges. They suggested a generalized empirical equation by examining the applicability of power function, Elovich, and first-order equations to experimental data. By writing these as the explicit functions of the reciprocal of the rate Z, which is $(dq/dt)^{-1}$, one can show that a plot of Z vs t should be convex if the power function equation is operational (1 in Fig. 7.5), linear if the Elovich equation is appropriate (2 in Fig. 7.5), and concave if the first-order equation is appropriate (3 in Fig. 7.5). However, Z vs t plots for soil systems (Fig. 7.6) are usually

S-shaped, convex at small t, concave at large t, and linear at some intermediate t. These findings suggest that the reaction rate can best be described by the power function equation at small t, by the Elovich equation at an intermediate t, and by a first-order equation at large t. Thus, the S-shaped curve indicates that the above equations may be applicable, each at some limited time range.

One of the reasons a particular kinetic model appears to be applicable may be that the study is conducted during the time range when the model is most appropriate. While sorption, for example, decreases over many orders of magnitude before equilibrium is approached, with most methods and experiments, only a portion of the entire reaction is measured, and over this time range the assumptions associated with a particular equation are valid. Aharoni and Suzin (1982a,b) showed that the S-shaped curves could be well described using homogeneous and heterogeneous diffusion models. In homogeneous diffusion situations, the initial and final portions of the S-shaped curves (conforming to the power function and first-order equations, respectively) predominated (see Fig. 7.6 showing data conformity to a homogeneous diffusion model), whereas in instances where the heterogeneous diffusion model was operational, the linear portion of the S-shaped curve, which conformed to the Elovich equation, predominated.

The fact that diffusion models describe a number of soil chemical processes is not surprising since in most cases, mass transfer and chemical kinetics phenomena are occurring simultaneously and it is difficult to separate them.

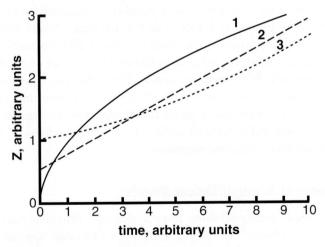

FIGURE 7.5. *Plots of Z vs time implied by (1) power function model, (2) Elovich model, and (3) first-order model. The equations for the models were differentiated and expressed as explicit functions of the reciprocal of the rate, Z. From Aharoni and Sparks (1991), with permission.*

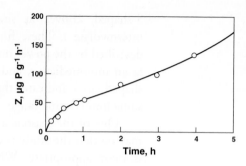

FIGURE 7.6.   *Sorption of phosphate by a Typic Dystrochrept soil plotted as Z vs time. The circles represent the experimental data of Polyzopoulos et al. (1986). The solid line is a curve calculated according to a homogeneous diffusion model. From Aharoni and Sparks (1991), with permission.*

Multiple Site Models

Based on the previous discussion, it is evident that simple kinetic models such as ordered reaction, power function, and Elovich models may not be appropriate to describe reactions in heterogeneous systems such as soils, sediments, and soil components. In these systems where there is a range of particle sizes and multiple retention sites, both chemical kinetics and transport phenomena are occurring simultaneously, and a fast reaction is often followed by a slower reaction(s). In such systems, nonequilibrium models that describe both chemical and physical nonequilibrium and that consider multiple components and sites are more appropriate. Physical nonequilibrium is ascribed to some rate-limiting transport mechanism such as FD or interparticle diffusion, while chemical nonequilibrium is due to a rate-limiting mechanism at the particle surface (CR). Nonequilibrium models include two-site, multiple site, radial diffusion (pore diffusion), surface diffusion, and multiprocess models (Table 7.2). Emphasis here will be placed on the use of these models to describe sorption phenomena.

The term sites can have a number of meanings (Brusseau and Rao, 1989): (1) specific, molecular scale reaction sites; (2) sites of differing degrees of accessibility (external, internal); (3) sites of differing sorbent type (organic matter and inorganic mineral surfaces); and (4) sites with different sorption mechanisms. With chemical nonequilibrium sorption processes, the sorbate may undergo two or more types of sorption reactions, one of which is rate-limiting. For example, a metal cation may sorb to organic matter by one mechanism and to mineral surfaces by another mechanism, with one of the mechanisms being time-dependent.

Chemical Nonequilibrium Models

Chemical nonequilibrium models describe time-dependent reactions at sorbent surfaces. The one-site model is a first-order approach that assumes that the reaction rate is limited by only one process or mechanism on a single class of sorbing sites and that all sites are of the time-dependent type. In many cases, this model appears to describe soil chemical reactions quite well. However, often it does not. This model would seem not appropriate for most heterogeneous systems since multiple sorption sites exist.

The two-site (two-compartment, two-box) or bicontinuum model has been widely used to describe chemical nonequilibrium (Leenheer and Ahlrichs, 1971; Hamaker and Thompson, 1972; Karickhoff, 1980; McCall and Agin, 1985; Jardine *et al.*, 1992) and physical nonequilibrium (Nkedi-Kizza *et al.*, 1984; Lee *et al.*, 1988; van Genuchten and Wagenet, 1989) (Table 7.2). This model assumes that there are two reactions occurring, one that is fast and reaches equilibrium quickly and a slower reaction that can continue for long time periods. The reactions can occur either in series or in parallel (Brusseau and Rao, 1989).

In describing chemical nonequilibrium with the two-site model, two types of sorbent sites are assumed. One site involves an instantaneous equilibrium reaction and the other, the time-dependent reaction. The former is described by an equilibrium isotherm equation while a first-order equation is usually employed for the latter.

With the two-site model, there are two adjustable or fitting parameters, the fraction of sites at local equilibrium (X_1) and the rate constant (k). A distribution (K_d) or partition coefficient (K_p) is determined independently from a sorption/desorption isotherm.

TABLE 7.2. *Comparison of Chemical and Physical Nonequilibrium Sorption Kinetic Models[a,b]*

Conceptual model	Fitting parameter(s)	Model limitations
Chemical Nonequilibrium Models		
One-site model (Coates and Elzerman, 1986) $S \xrightarrow{k_d} C$	k_d	Cannot describe biphasic sorption/ desorption
Two-site model (Coates and Elzerman, 1986) $S_1 \xleftrightarrow{X_1 K_p} C \xleftarrow{k_d} S_2$	$k_d, K_p^{\,c}, X_1$	Cannot describe the "bleeding" or slow, reversible, nonequilibrium desorption for residual sorbed compounds (Karickhoff, 1980)
Multisite continuum compartment model (Connaughton *et al.*, 1993) $F(t) = 1 - \dfrac{M(t)}{M} = 1 - \left(\dfrac{\beta}{\beta + t}\right)^{a}$	α, β	Assumption of homogeneous, spherical particles and diffusion only in aqueous phase
Physical Nonequilibrium Models		
Radial diffusion penetration retardation (pore diffusion) model (Wu and Gschwend, 1986) $S' \xleftrightarrow{K_p} C' \xrightarrow{D_{eff}} C$	$D_{eff}^{\,d} = f(n,t)D_m n/(1-n)\rho_s K_p$	Cannot describe instantaneous uptake without additional correction factor (Ball, 1989); did not describe kinetic data for times greater than 10^3 min (Wu and Gschwend, 1986)
Dual-resistance surface diffusion model (Miller and Pedit, 1992) $S' \xrightarrow{D_s} C_s' \xrightarrow{k_b} C$	D_s, k_b	Model calibrated with sorption data predicted more desorption than occurred in the desorption experiments (Miller and Pedit, 1992)

TABLE 7.2. *Comparison of Chemical and Physical Nonequilibrium Sorption Kinetic Models^{a,b} (contd)*

Conceptual model	Fitting parameter(s)	Model limitations
Pore space diffusion model (Fuller *et al.*, 1993) $$\left(\varepsilon + \frac{S_a}{n} K_s C(r)^{(1-1/n)}\right) \frac{\partial C(r)}{\partial t}$$ $$= D_e \left(\frac{\partial^2 C(r)}{\partial r^2} + \frac{2\partial C(r)}{r\partial r}\right)$$	$D_e, \varepsilon, K_s, 1/n, F_{eq}$	
Multiple particle class pore diffusion model (Pedit and Miller, 1995) $$\left(\theta_p^i + \rho_a^i \frac{\partial q_r^i(r,t)}{\partial C_p^i(r,t)}\right) \frac{\partial C_p^i(r,t)}{\partial t}$$ $$= \frac{\theta_p^i D_p^i}{r^2} \frac{\partial}{\partial r}\left(\frac{r^2 \partial C_p^i(r,t)}{\partial r}\right)$$ $$-\theta_p^i \lambda_p^i C_p^i(r,t) - \rho_a^i \lambda_r^i q_r^i(r,t)$$	$\theta_p^i, \rho_a^i, D_p^i, \lambda_p^i, \lambda_r^i$	Multiple fitting parameters; variations in sorption equilibrium and rates that might occur within a particle class or an individual particle grain are not addressed.

^{a} Reprinted with permission from Connaughton *et al.* (1993). Copyright 1993 American Chemical Society.
^{b} Abbreviations used are as follows: S, concentration of the bulk sorbed contaminant (g g^{-1}); C, concentration of the bulk aqueous-phase contaminant (g ml^{-1}); k_d, first-order desorption rate coefficient (min^{-1}); S_2, concentration of the sorbed contaminant that is rate-limited (g g^{-1}); S_1, concentration of the contaminant that is in equilibrium with the bulk aqueous concentration (g g^{-1}); X_1, fraction of the bulk sorbed contaminant that is in equilibrium with the aqueous concentration; K_p, sorption equilibrium partition coefficient (ml g^{-1}); $F(t)$, fraction of mass released through time t; $M(t)$, mass remaining after time t; M, total initial mass; β, scale parameter necessary for determination of mean and standard deviation of k_s; α, shape parameter; D_{eff}, effective diffusivity of sorbate molecules or ions in the particles (cm^2 s^{-1}); S', concentration of contaminant in immobile bound state (mol g^{-1}); C', concentration of contaminant free in the pore fluid (mol cm^{-3}); n, porosity of the sorbent (cm^3 of fluid cm^{-3}); D_m, pore fluid diffusivity of the sorbate (cm^2 s^{-1}); ρ_s, specific gravity of the sorbent (g cm^{-3}); $f(n,t)$, pore geometry factor; k_b, boundary layer mass transfer coefficient (m s^{-1}); r, radius of the spherical solid particle, assumed constant (m); ρ, macroscopic particle density of the solid phase (g m^{-3}); C_s', solution-phase solute concentration corresponding to an equilibrium with the solid-phase solute concentration at the exterior of the particle (g L^{-1}); D_s, surface diffusion coefficient (m s^{-1}).
^{c} K_p can be determined independently.
^{d} K_p, D_m, and ρ can be determined independently; ε, internal porosity of sorbent; $C(r)$, concentration of sorptive in the aqueous phase in the pore fluid at radial distance r; S_s is the surface of sorbent per unit volume of solid; $1/n$, the adsorption isotherm intercept; D_e, effective diffusion coefficient; a, radius of the aggregate; F_{eq}, equilibrium fraction of adsorption sites; θ_p^i, intraparticle porosity of particle class i; ρ_a^i, apparent particle density of particle class i; r, radial distance; C_p^i (r, t), intraparticle fluid-phase solute concentration of the particle class i; D_p^i, pore diffusion coefficient for particle class i; λ_p^i, intraparticle fluid-phase first-order reaction rate coefficient for particle class i; λ_r^i, intraparticle solid-phase first-order reaction rate coefficient for particle class i; q_r^i (r, t), intraparticle solid-phase solute concentration of particle class i.

To account for the multiple sites that may exist in heterogeneous systems, Connaughton *et al.* (1993) developed a multisite compartment (continuum) model (Γ) that incorporates a continuum of sites or compartments with a distribution of rate coefficients that can be described by a gamma density function. A fraction of the sorbed mass in each compartment is at equilibrium with a desorption rate coefficient or distribution coefficient for each compartment or site (Table 7.2). The multisite model has two fitting parameters, α, a shape parameter, and $1/\beta$, which is a scale parameter that determines the mean standard deviation of the rate coefficients.

Physical Nonequilibrium Models

A number of models can be used to describe physical nonequilibrium reactions. Since transport processes in the mobile phase are not usually rate-limiting, physical nonequilibrium models focus on diffusion in the immobile phase or interparticle/diffusion processes such as pore and/or surface diffusion. The transport between mobile and immobile regions is accounted for in physical nonequilibrium models in three ways (Brusseau and Rao, 1989): (1) explicitly with Fick's law to describe the physical mechanism of diffusive transfer, (2) explicitly by using an empirical first-order mass transfer expression to approximate solute transfer, and (3) implicitly by using an effective or lumped dispersion coefficient that includes the effects of sink/source differences and hydrodynamic dispersion and axial diffusion.

A pore diffusion model (Table 7.2) has been used by a number of investigators to study sorption processes using batch systems (Wu and Gschwend, 1986; Steinberg *et al.*, 1987). The sole fitting parameter in this model is the effective diffusivity (D_e), which may be estimated *a priori* from chemical and colloidal properties. However, this estimation is only valid if the sorbent material has a narrow particle size distribution so that an accurate, average particle size can be defined. Moreover, in the pore diffusion model, an average representative D_e is assumed, which means there is a continuum in properties across an entire pore size spectrum. This is not a valid assumption for micropores in which there are higher adsorption energies of sorbates causing increased sorption. The increased sorption reduces diffusive transport rates in nonlinear isotherms for sorbents with pores less than several sorbate diameters in size. Other factors including steric hindrance, which increases as the pore size approaches the solute size, and greatly increased surface-area to-pore-volume ratios, which occur as pore size decreases, can cause reduced transport rates in micropores.

Another problem with the pore diffusion model is that sorption and desorption kinetics may have been measured over a narrow concentration range. This is a problem since a sorption/desorption mechanism in micropores at one concentration may be insignificant at another concentration.

Fuller *et al.* (1993) used a pore space diffusion model (Table 7.2) to describe arsenate adsorption on ferrihydrite that included a subset of sites whereby sorption was at equilibrium. A Freundlich model was used to describe sorption on these sites. Intraparticle diffusion was described by Fick's second law of diffusion; homogeneous, spherical aggregates, and diffusion only in the solution phase were assumed. Figure 7.7 shows the fit of the model when sorption at all sites was controlled by intraparticle diffusion. The fit was better when sites that had attained sorption equilibrium were included based on the assumption that there was an initial rapid sorption on external surface sites before intraparticle diffusion.

Pedit and Miller (1995) have developed a general multiple particle class pore diffusion model that accounts for differences in physical and sorptive properties for each particle class (Table 7.2). The model includes both instantaneous equilibrium sorption and time-dependent pore diffusion for each particle class. The pore diffusion portion of the model assumes that solute transfer between the intraparticle fluid and the solid phases is fast vis-à-vis interparticle pore diffusion processes.

Surface diffusion models, assuming a constant surface diffusion coefficient, have been used by a number of researchers (Weber and Miller, 1988; Miller and Pedit, 1992). The dual resistance model (Table 7.2) combines both pore and surface diffusion.

Kinetic Methodologies

A number of methodologies can be used to study the rates of soil chemical processes. These can be broadly classified as methods for slower reactions (>15 sec), which include batch and flow techniques, and rapid techniques that can measure reactions on milli- and microsecond time scales. It should be recognized that none of these methods is a panacea for kinetic analyses. They all have advantages and disadvantages. For comprehensive discussions on kinetic methodologies one should consult Sparks (1989), Amacher (1991), Sparks and Zhang (1991), and Sparks *et al.* (1995).

Batch Methods

Batch methods have been the most widely used kinetic techniques. In the simplest traditional batch technique, an adsorbent is placed in a series of vessels such as centrifuge tubes with a particular volume of adsorptive. The tubes are then mixed by shaking or stirring. At various times a tube is sacrificed for analysis; i.e., the suspension is either centrifuged or filtered to obtain a clear supernatant for analysis. A number of variations of batch methods exist and these are discussed in Amacher (1991).

There are a number of disadvantages to traditional batch methods. Often the reaction is complete before a measurement can be made,

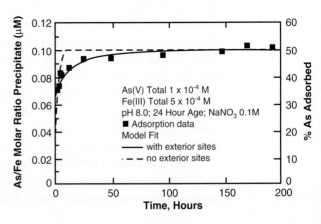

FIGURE 7.7. *Comparison of pore space diffusion model fits of As(V) sorption with experimental data (dashed curve represents sorption where all surface sites are diffusion-limited and the solid curve represents sorption on equilibrium sites plus diffusion-limited sites). From Fuller et al. (1993), with permission.*

particularly if centrifugation is necessary, and the solid:solution ratio may be altered as the experiment proceeds. Too much mixing may cause abrasion of the adsorbent, altering the surface area, while too little mixing may enhance mass transfer and transport processes. Another major problem with all batch techniques, unless a resin or chelate material such as Na-tetraphenylboron is used, is that released species are not removed. This can cause inhibition in further adsorbate release and promotion of secondary precipitation in dissolution studies. Moreover, reverse reactions are not controlled, which makes the calculation of rate coefficients difficult and perhaps inaccurate.

Many of the disadvantages listed above for traditional batch techniques can be eliminated by using a method like that of Zasoski and Burau (1978), shown in Fig. 7.8. In this method an adsorbent is placed in a vessel containing the adsorptive, pH and suspension volume are adjusted, and the suspension is vigorously mixed with a magnetic stirrer. At various times, suspension aliquots are withdrawn using a syringe containing N_2 gas. The N_2 gas prevents CO_2 and O_2 from entering the reaction vessel. The suspension is rapidly filtered and the filtrates are then weighed and analyzed. With this apparatus a constant pH can be maintained, reactions can be measured at 15-sec intervals, excellent mixing occurs, and a constant solid-to-solution ratio is maintained.

Flow Methods

Flow methods can range from continuous flow techniques (Fig. 7.9), which are similar to liquid-phase chromatography, to stirred-flow methods (Fig. 7.10) that combine aspects of both batch and flow methods. Important attributes of flow techniques are that one can conduct studies at realistic soil-

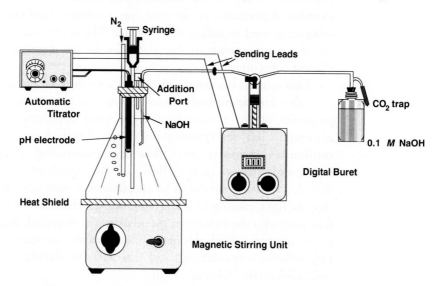

FIGURE 7.8. *Schematic diagram of equipment used in batch technique of Zasoski and Burau (1978), with permission.*

to-solution ratios that better simulate field conditions, the adsorbent is exposed to a greater mass of ions than in a static batch system, and the flowing solution removes desorbed and detached species.

With continuous flow methods, samples can be injected as suspensions or spread dry on a membrane filter. The filter is attached to its holder by securely capping it, and the filter holder is connected to a fraction collector and peristaltic pump, the latter maintaining a constant flow rate. Influent solution then passes through the filter and reacts with the adsorbent, and at various times, effluents are collected for analysis. Depending on flow rate and the amount of effluent needed for analysis, samples can be collected about every 30–60 sec. One of the major problems with this method is that the colloidal particles may not be dispersed; i.e., the time necessary for an adsorptive to travel through a thin layer of colloidal particles is not equal at all locations of the layer. This plus minimal mixing promotes significant transport effects. Thus, apparent rate laws and rate coefficients are measured, with the rate coefficients changing with flow rate. There can also be dilution of the incoming adsorptive solution by the liquid used to load the adsorbent on the filter, particularly if the adsorbent is placed on the filter as a suspension, or if there is washing out of remaining adsorptive solution during desorption. This can cause concentration changes not due to adsorption or desorption.

A more preferred method for measuring soil chemical reaction rates is the stirred-flow method. The experimental setup is similar to the continuous flow method (Fig. 7.9) except there is a stirred-flow reaction chamber rather than a membrane filter. A schematic of this method is shown in Fig. 7.10. The sorbent is placed into the reaction chamber where a magnetic stir bar or an overhead stirrer keeps it suspended during the experiment. There is a filter placed in the top of the chamber that keeps the solids in the reaction chamber. A peristaltic pump maintains a constant flow rate, and a fraction collector is used to collect the leachates. The stirrer effects perfect mixing; i.e., the concentration of the adsorptive in the chamber is equal to the effluent concentration.

This method has several advantages over the continuous flow technique and other kinetic methods. Reaction rates are independent of the physical properties of the porous media, the same apparatus can be used for adsorption and desorption experiments, desorbed species are removed, continuous measurements allow for monitoring reaction progress, experimental factors such as flow rate and adsorbent mass can be easily altered, a variety of solids can be used (however, sometimes fine particles can clog the filter, causing a buildup in pressure, which results in a nonconstant flow rate) with the technique, the adsorbent is dispersed, and dilution errors can be measured. With this method, one can also use stopped-flow tests and vary influent concentrations and flow rates to elucidate possible reaction mechanisms (Bar-Tal *et al.*, 1990).

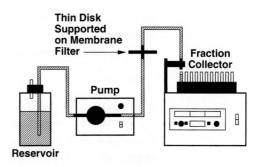

FIGURE 7.9. *Thin-disk flow (continuous flow) method experimental setup. Background solution and solute are pumped from the reservoir through the thin disk and are collected as aliquots by the fraction collector. From Amacher (1991), with permission.*

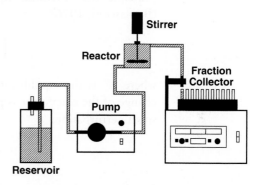

FIGURE 7.10. *Stirred-flow reactor method experimental setup. Background solution and solute are pumped from the reservoir through the stirred reactor containing the solid phase and are collected as aliquots by the fraction collector. Separation of solid and liquid phases is accomplished by a membrane filter at the outlet end of the stirred reactor. From Amacher (1991), with permission.*

Relaxation Techniques

As noted earlier, many soil chemical reactions are very rapid, occurring on milli- and microsecond time scales. These include metal and organic sorption–desorption reactions, ion exchange processes, and ion association reactions. Batch and flow techniques, which measure reaction rates of >15 sec, cannot be employed to measure these reactions. Chemical relaxation methods must be used to measure very rapid reactions. These include pressure-jump (p-jump), electric field pulse, temperature-jump (t-jump), and concentration-jump (c-jump) methods. These methods are fully outlined in other sources (Sparks, 1989; Sparks and Zhang, 1991). Only a brief discussion of the theory of chemical relaxation and a description of p-jump methods will be given here. The theory of chemical relaxation can be found in a number of sources (Eigen, 1954; Takahashi and Alberty, 1969; Bernasconi, 1976). It should be noted that relaxation techniques are best used with soil components such as oxides and clay minerals and not whole soils. Soils are heterogeneous, which complicates the analyses of the relaxation data.

All chemical relaxation methods are based on the theory that the equilibrium of a system can be rapidly perturbed by some external factor such as pressure, temperature, or electric field strength. Rate information can then be obtained by measuring the approach from the perturbed equilibrium to the final equilibrium by measuring the relaxation time, τ (the time that it takes for the system to relax from one equilibrium state to another, after the perturbation pulse), by using a detection system such as conductivity. The relaxation time is related to the specific rates of the elementary reactions involved. Since the perturbation is small, all rate expressions reduce to first-order equations regardless of reaction order or molecularity (Bernasconi, 1976). The rate equations are then linearized such that

$$\tau^{-1} = k_1(C_A + C_B) + k_{-1},\qquad(7.16)$$

where k_1 and k_{-1} are the forward and backward rate constants and C_A and C_B are the concentrations of reactants A and B at equilibrium. From a linear plot of τ^{-1} vs $(C_A + C_B)$ one could calculate k_1 and k_{-1} from the slope and intercept, respectively. Pressure-jump relaxation is based on the principle that chemical equilibria depend on pressure as shown below (Bernasconi, 1976),

$$\left(\frac{\partial \ln K^\circ}{\partial \ln p}\right)_T = -\Delta V/RT,\qquad(7.17)$$

where K° is the equilibrium constant, ΔV is the standard molar volume change of the reaction, p is pressure, and R and T were defined earlier. For a small perturbation,

$$\frac{\Delta K^\circ}{K^\circ} = \frac{-\Delta V \Delta p}{RT}.\qquad(7.18)$$

Details on the experimental protocol for a p-jump study can be found in several sources (Sparks, 1989; Zhang and Sparks, 1989; Grossl *et al.*, 1994).

Fendorf *et al.* (1993) used an electron paramagnetic resonance stopped-flow (EPR-SF) method (an example of a c-jump method) to study reactions in colloidal suspensions *in situ* on millisecond time scales. If one is studying an EPR active species (paramagnetic) such as Mn, this technique has several advantages over other chemical relaxation methods. With many relaxation methods, the reactions must be reversible and reactant species are not directly measured. Moreover, in some relaxation studies, the rate constants are calculated from linearized rate equations that are dependent on equilibrium parameters. Thus, the rate parameters are not directly measured.

With the EPR-SF method of Fendorf *et al.* (1993) the mixing can be done in <10 msec and EPR digitized within a few microseconds. A diagram of the EPR-SF instrument is shown in Fig. 7.11. Dual 2-ml in-port syringes feed a mixing cell located in the EPR spectrometer. This allows for EPR detection of the cell contents. A single outflow port is fitted with a 2-ml effluent collection syringe equipped with a triggering switch. The switch activates the data acquisition system. Each run consists of filling the in-port syringes with the desired reactants, flushing the system with the reactants several times, and initiating and monitoring the reaction. Fendorf *et al.* (1993) used this system to study the kinetics of Mn^{2+} sorption on γ-MnO_2. The sorption reaction was complete in 200 msec. Data were taken every 50 μsec and 100 points were averaged to give the time-dependent sorption of Mn(II).

Choice of Kinetic Method

The method that one chooses to study the kinetics of soil chemical reactions depends on several factors. The reaction rate will certainly dictate the choice of method. With batch and flow methods, the most rapid measurements one can make require about 15 sec. For more rapid reactions, one must use relaxation techniques where millisecond time scales can be measured.

Another factor in deciding on a kinetic method is the objective of one's experiments. If one wishes to measure the chemical kinetics of a reaction where transport is minimal, most batch and flow techniques are unsuitable and a relaxation technique should be employed. On the other hand, if one wants to simulate time-dependent reactions in the field, perhaps a flow technique would be more realistic than a batch method.

Effect of Temperature on Reaction Rates

Temperature has a marked effect on reaction rate. Arrhenius noted the following relationship between k and T,

$$k = A_f e^{-E_a/RT} , \qquad (7.19)$$

where A_f is a frequency factor and E_a is the energy of activation. Converting Eq. (7.19) to a linear form results in

$$\ln k = \ln A_f - E_a/RT. \qquad (7.20)$$

A plot of $\ln k$ vs $1/T$ would yield a linear relationship with the slope equal to $-E_a/R$ and the intercept equal to $\ln A_f$. Thus, by measuring k values at several temperatures, one could determine the E_a value.

The magnitude of E_a can reveal important information on the reaction process or mechanism. Some E_a values for various reactions or processes are given in Table 7.3. Low E_a values usually indicate diffusion-controlled transport and physical adsorption processes, whereas higher E_a values would indicate chemical reaction or surface-controlled processes (Sparks, 1985).

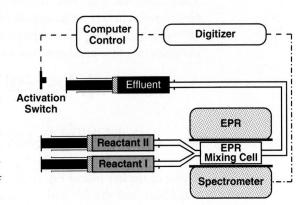

FIGURE 7.11. *Schematic diagram of the electron paramagnetic resonance monitored stopped-flow kinetic apparatus. From Fendorf et al. (1993), with permission.*

For example, E_a values of 6.7–26.4 kJ mol^{-1} were observed for pesticide sorption on soils and soil components, which appeared to be diffusion-controlled (Haque et al., 1968; Leenheer and Ahlrichs, 1971; Khan, 1973), while gibbsite dissolution in acid solutions, which appeared to be a surface-controlled reaction, was characterized by E_a values ranging from 59 ± 4.3 to 67 ± 0.6 kJ mol^{-1} (Bloom and Erich, 1987).

Kinetics of Important Soil Chemical Processes

Sorption–Desorption Reactions

HEAVY METALS AND OXYANIONS

The chemical reaction (CR) rate of heavy metal (e.g., Cu^{2+}, Pb^{2+}) and oxyanion (e.g., arsenate, chromate, and phosphate) sorption on soil components is rapid, occurring on millisecond time scales. For such rapid reactions, chemical relaxation techniques, e.g., pressure-jump relaxation must be employed (Sparks and Zhang, 1991; Sparks, 2000).

Zhang and Sparks (1990b) studied the kinetics of selenate adsorption on goethite using pressure-jump relaxation and found that adsorption occurred mainly under acidic conditions. The dominant species was $(SeO_4)^{2-}$. As pH increased $(SeO_4)^{2-}$ adsorption decreased. Selenate was described using the modified triple-layer model (see Chapter 5). A single relaxation was observed and the mechanism proposed was

$$XOH + H^+ + SeO_4^{-2} \rightleftharpoons XOH_2^+ - SeO_4^{2-}, \qquad (7.21)$$

where XOH is 1 mol of reactive surface hydroxyl bound to an Fe ion in goethite.

A linearized rate equation given below was developed and tested,

$$\tau^{-1} = k_1 \, ([XOH][SeO_4^{-2}] + [XOH][H^+] + [SeO_4^{2-}][H^+]) + k_{-1}, \quad (7.22)$$

where the terms in the brackets are the concentrations of species at equilibrium. Since the reaction was conducted at the solid/liquid interface, the electrostatic effect must be considered to calculate the intrinsic rate constants (k_1^{int} and k_{-1}^{int}). Using the modified triple-layer model to obtain electrostatic parameters, a first-order reaction was derived (Zhang and Sparks, 1990b):

$$\tau^{-1} \exp\left(\frac{-F(\psi_\alpha - 2\psi_\beta)}{2RT}\right) = k_1^{int} \left[\exp\left(\frac{-F(\psi_\alpha - 2\psi_\beta)}{RT}\right)\right.$$

$$\left. \times \, ([XOH][SeO_4^{2-}] + [XOH][H^+] + [SeO_4^{2-}][H^+])\right] + k_{-1}^{int}. \quad (7.23)$$

A plot of the left side of Eq. (7.23) vs the terms in brackets on the right side was linear and the k_1^{int} and k_{-1}^{int} values were calculated from the slope and

TABLE 7.3. *Energy of Activation (E_a) Values for Various Reactions or Processes[a]*

Reaction or process	Typical range of E_a values (kJ mol^{-1})
Physical adsorption	8 to 25
Aqueous diffusion	<21
Pore diffusion	20 to 40
Cellular and life-related reactions	21 to 84
Mineral dissolution or precipitation	34 to 151
Mineral dissolution via surface reaction control	42 to 84
Polymer diffusion	60 to 70
Ion exchange	>84
Isotopic exchange in solution	75 to 201
Solid-state diffusion in minerals at low temperatures	84 to 503

[a] From Langmuir (1997), with permission.

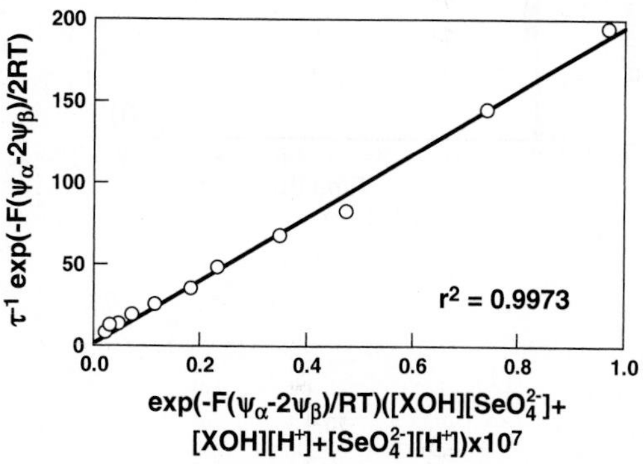

FIGURE 7.12. *Plot of relationship between τ^{-1} with exponential and concentration terms in Eq. (7.23). Reprinted with permission from Zhang and Sparks (1990b). Copyright 1990 American Chemical Society.*

intercept, respectively (Fig. 7.12). The linear relationship would indicate that the outer-sphere complexation mechanism proposed in Eq. (7.21) was plausible. Of course, one would need to use spectroscopic approaches to definitively determine the mechanism. This was done earlier with X-ray absorption fine structure spectroscopy (XAFS) to prove that selenate is adsorbed as an outer-sphere complex on goethite (Hayes *et al.*, 1987).

The kinetics of heavy metal and oxyanion sorption on soil components and soils is typically characterized by a biphasic process in which sorption is initially rapid followed by slow reactions (Fig. 7.13). The rapid step, which occurs over milliseconds to hours, can be ascribed to CR and film diffusion processes. During this rapid reaction process, a large portion of the sorption may occur. For example, in Fig. 7.13a one sees that ~90% of the total Ni sorbed on kaolinite and pyrophyllite occurred within the first 24 hr. For Pb sorption on a Matapeake soil, 78% of the total Pb sorption occurred in 8 min. Following the initial fast reaction, slow sorption continued, but only about 1% additional Pb was sorbed after 800 hr (Fig. 7.13b). Figure 7.13c shows a biphasic reaction for As(V) sorption on ferrihydrite. Within 5 min, a majority of the total sorption had occurred. Slow sorption continued for at least 192 hr.

However, in some cases, the magnitude of sorption can greatly increase with longer reaction times. For example, Bruemmer *et al.* (1988) studied Ni^{2+}, Zn^{2+}, and Cd^{2+} sorption on goethite, a porous Fe oxide that has defect structures in which metals can be incorporated to satisfy charge imbalances. It was found at pH 6 that as reaction time increased from 2 hr to 42 days (at 293 K), adsorbed Ni^{2+} increased from 12 to 70% of total sorption, and total Zn^{2+} and Cd^{2+} sorption over the same time increased by 33 and 21%, respectively. Metal uptake was hypothesized to occur by a three-step

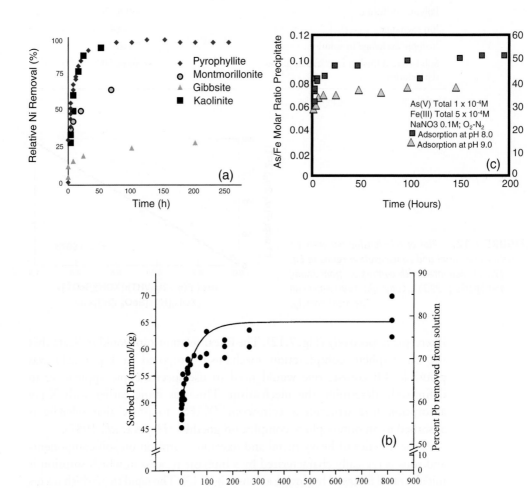

FIGURE 7.13.   *Kinetics of metal and oxyanion sorption on soil minerals and soil. (a) Kinetics of Ni sorption (%) on pyrophyllite (◆), kaolinite (■), gibbsite (▲), and montmorillonite (○) from a 3 mM Ni solution, an ionic strength I = 0.1 M NaNO₃, and a pH of 7.5 (from Scheidegger et al. (1997), with permission); (b) kinetics of Pb sorption on a Matapeake soil from a 12.25 mM Pb solution, an ionic strength I = 0.05 M, and a pH of 5.5 (from Strawn and Sparks (2000), with permission); (c) kinetics of As(V) sorption on ferrihydrite at pH 8.0 and 9.0 (from Fuller et al. (1993), with permission from Elsevier Science).*

mechanism: (1) adsorption of metals on external surfaces, (2) solid-state diffusion of metals from external to internal sites, and (3) metal binding and fixation at positions inside the goethite particle.

The amount of contact time between metals and soil sorbents (residence time) can dramatically affect the degree of desorption, depending on the metal and sorbent. Examples of this are shown in Fig. 7.14 and Table 7.4. In Fig. 7.14 the effect of residence time on Pb^{2+} and Co^{2+} desorption from hydrous Fe oxide (HFO) was studied. With Pb, between pH 3 and 5.5 there was a minor effect of residence time (over 21 weeks) on Pb^{2+} desorption, with only minor hysteresis occurring (hysteresis varied from <2% difference between sorption and desorption to ~10%). At pH 2.5, Pb^{2+} desorption was complete within a 16-hr period and was not affected by residence time (Fig. 7.14a). In a soil where 2.1% SOM was present, residence time had little

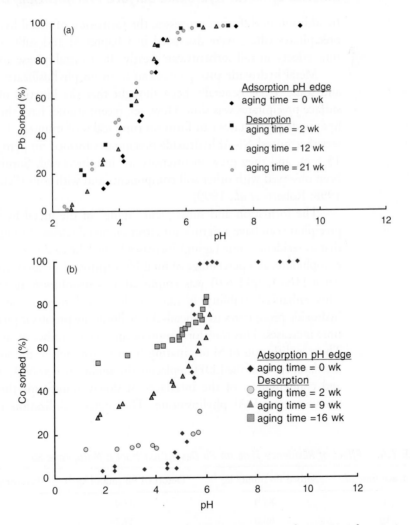

FIGURE 7.14. *Fractional sorption–desorption of (a) Pb^{2+} and (b) Co^{2+} to hydrous Fe(III) oxide (HFO) as a function of pH and HFO–Pb^{2+} (a) and HFO–Co^2 (b) aging time. From Ainsworth et al. (1994), with permission.*

effect on the amount of Pb desorbed, but marked hysteresis was observed at all residence times (Table 7.4). This could be ascribed to the strong metal–soil complexes that occur and perhaps to diffusion processes.

With Co^{2+}, extensive hysteresis was observed over a 16-week residence time (Fig. 7.14b), and the hysteresis increased with residence time. After a 16-week residence time, 53% of the Co^{2+} was not desorbed, and even at pH 2.5, hysteresis was observed. The extent of Co reversibility with residence time was attributed to Co incorporation into a recrystallizing solid by isomorphic substitution and not to micropore diffusion. Similar residence time effects on Co desorption have been observed both with Fe and Mn oxides (Backes et al., 1995) and with soil clays (McLaren et al., 1998).

Kinetics of Metal Hydroxide Surface Precipitation/Dissolution

In addition to diffusion processes, the formation of metal hydroxide surface precipitates (these were discussed in Chapter 5) and subsequent residence time effects on soil sorbents can greatly affect metal release and hysteresis.

Metal hydroxide precipitates can form on phyllosilicates, metal oxides, and soils. It has generally been thought that the kinetics of formation of surface precipitates was slow. However, recent studies have shown that metal hydroxide precipitates can form on time scales of minutes. In Fig. 7.15 one sees that mixed Ni–Al hydroxide precipitates formed on pyrophyllite within 15 min, and they grew in intensity as time increased. Similar results have been observed with other soil components and with soils (Scheidegger et al., 1998; Roberts et al., 1999).

The formation and subsequent "aging" of the metal hydroxide surface precipitate can have a significant effect on metal release. In Fig. 7.16 one sees that as residence time (aging) increases from 1 hr to 2 years, Ni release from pyrophyllite, as a percentage of total Ni sorption, decreased from 23 to ~0%, when HNO_3 (pH 6.0) was employed as a dissolution agent for 14 days. This enhanced stability is due to the transformation of the metal–Al hydroxide precipitates to a metal–Al phyllosilicate precursor phase as residence time increases. This transformation occurs via a number of steps (Fig. 7.17). There is diffusion of Si originating from weathering of the sorbent into the interlayer space of the LDH, replacing the anions such as NO_3^-. Polymerization and condensation of the interlayer Si slowly transforms the LDH into a precursor metal–Al phyllosilicate. The metal stabilization that occurs in

TABLE 7.4. *Effect of Residence Time on Pb Desorption from a Matapeake Soil[a]*

Residence time (d)	Sorbed Pb (mmol kg^{-1})	Desorbed Pb (mmol kg^{-1})	Percentage Pb desorbed
1	54.9	27.9	50.8
10	60.0	28.7	47.
32	66.1	30.5	46.1

[a] From Strawn and Sparks (2000), with permission.

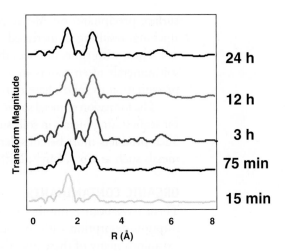

FIGURE 7.15. *Radial structure functions (derived from XAFS analyses) for Ni sorption on pyrophyllite for reaction times up to 24 hr, demonstrating the appearance and growth of the second shell (peak at ≈2.8 Å) contributions due to surface precipitation and growth of a mixed Ni–Al hydroxide phase. From Scheidegger et al. (1998), with permission from Elsevier Science.*

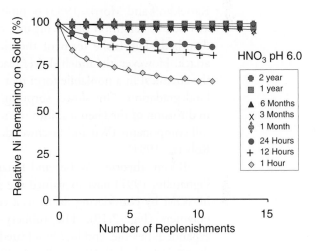

FIGURE 7.16. *Dissolution of Ni from surface precipitates formed on pyrophyllite at residence times of 1 hr to 2 years. The figure shows the relative amount of Ni^{2+} remaining on the pyrophyllite surface following extraction for 24-hr periods (each replenishment represents a 24-hr extraction) with HNO_3 at pH 6.0. From Scheckel et al. (2001), with permission.*

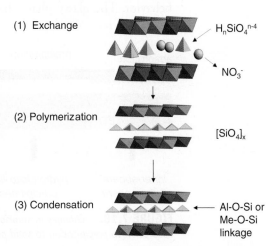

FIGURE 7.17. *Hypothetical reaction process illustrating the transformation of an initially precipitated Ni–Al LDH into a phyllosilicate-like phase during aging. The initial step involves the exchange of dissolved silica for nitrate within the LDH interlayer followed by polymerization and condensation of silica onto the octahedral Ni–Al layer. The resultant solid possesses structural features common to 1:1 and 2:1 phyllosilicates. From Ford et al. (2001), with permission from Academic Press, Orlando, FL.*

surface precipitates on Al-free sorbents (e.g., talc) may be due to Ostwald ripening, resulting in increased crystallization (Scheckel and Sparks, 2001).

Thus, with time, one sees that metal (e.g., Co, Ni, and Zn) sorption on soil minerals often results in a continuum of processes from adsorption to precipitation to solid phase transformation (Fig. 7.18).

The formation of metal surface precipitates could be an important mechanism for sequestering metals in soils such that they are less mobile and bioavailable. Such products must be considered when modeling the fate and mobility of metals such as Co^{2+}, Mn^{2+}, Ni^{2+}, and Zn^{2+} in soil and water environments.

ORGANIC CONTAMINANTS

There have been a number of studies on the kinetics of organic chemical sorption/desorption with soils and soil components. Similarly to metals and oxyanions many of these investigations have shown that sorption/desorption is characterized by a rapid, reversible stage followed by a much slower, nonreversible stage (Karickhoff et al., 1979; DiToro and Horzempa, 1982; Karickhoff and Morris, 1985) or biphasic kinetics. The rapid phase has been ascribed to retention of the organic chemical in a labile form that is easily desorbed. The labile form of the chemical is also available for microbial attack. However, the much slower reaction phase involves the entrapment of the chemical in a nonlabile form that is difficult to desorb and is resistant to biodegradation. This slower sorption/desorption reaction has been ascribed to diffusion of the chemical into micropores of organic matter and inorganic soil components (Wu and Gschwend, 1986; Steinberg et al., 1987; Ball and Roberts, 1991).

Recent theories (Weber and Huang, 1996; Pignatello, 2000; Xing and Pignatello, 1997) have explained the slow diffusion of the organic chemical in SOM by considering SOM as a combination of "rubbery" and "glassy" polymers (Fig. 7.19). The rubbery-like phases are characterized by an expanded, flexible, and highly solvated structure with pores of subnanometer dimensions (holes) (Pignatello, 1998, 2000; Xing and Pignatello, 1997). Sorption in the rubbery phase results in linear, noncompetitive, and reversible behavior. The glassy phases have pores that are of subnanometer size and

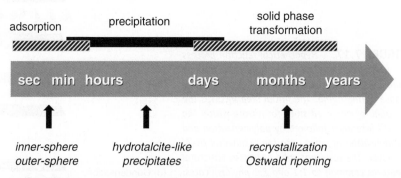

FIGURE 7.18. *Changes in sorption processes with time showing a continuum from adsorption to precipitation to solid phase transformation.*

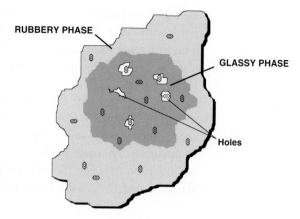

FIGURE 7.19. *Rubbery–glassy polymer concept of soil organic matter. The perspective is intended to be three-dimensional. The rubbery and glassy phases both have dissolution domains in which sorption is linear and noncompetitive. The glassy phase, in addition, has pores of subnanometer dimension ("holes") in which adsorption-like interactions occur with the walls, giving rise to nonlinearity and competitive sorption. From Xing and Pignatello (1997), with permission. Copyright 1997 American Chemical Society.*

sorption in this phase is characterized by nonlinearity and it is competitive (Xing and Pignatello, 1997).

The above theories, relating slow diffusion into organic matter to diffusion in polymers, are partially validated in recent studies measuring E_a values for organic chemical sorption. Cornelissen *et al.* (1997) studied the temperature dependence of slow adsorption and desorption kinetics of some chlorobenzenes, polychlorinated biphenyls (PCBs), and polycyclic aromatic hydrocarbons (PAHs) in laboratory and field contaminated sediments. E_a values of 60–70 kJ mol^{-1}, which are in the range for diffusion in polymers (Table 7.3), were determined. These values are much higher than those for pore diffusion (20–40 kJ mol^{-1}), suggesting that intraorganic matter diffusion may be a more important mechanism for slow organic chemical sorption than interparticle pore diffusion.

An example of the biphasic kinetics observed for many organic chemical reactions in soils/sediments is shown in Fig. 7.20. In this study 55% of the labile PCBs was desorbed from sediments in a 24-hr period, while little of the remaining 45% nonlabile fraction was desorbed in 170 hr (Fig. 7.20a). Over another 1-year period about 50% of the remaining nonlabile fraction desorbed (Fig. 7.20b).

In another study with volatile organic compounds (VOCs), Pavlostathis and Mathavan (1992) observed a biphasic desorption process for field soils contaminated with trichloroethylene (TCE), tetrachloroethylene (PCE), toluene (TOL), and xylene (XYL). A fast desorption reaction occurred in 24 hr, followed by a much slower desorption reaction beyond 24 hr. In 24 hr, 9–29, 14–48, 9–40, and 4–37% of the TCE, PCE, TOL, and XYL, respectively, were released.

A number of studies have also shown that with increased residence time the nonlabile portion of the organic chemical in the soil/sediment becomes more resistant to release (McCall and Agin, 1985; Steinberg *et al.*, 1987; Pignatello and Huang, 1991; Pavlostathis and Mathavan, 1992; Alexander, 1995). One of the first studies to clearly demonstrate this aging or residence time effect was in the research of Steinberg *et al.* (1987). They showed that 1,2-dibromoethane (EDB) release from soils reacted in the laboratory over a short period of time was much more rapid than EDB release from field soils that had been contaminated with EDB for many years. This difference in release was related to a greater diffusion into micropores of clay minerals and humic components that occurred at longer times (Steinberg et al., 1987).

One way to gauge the effect of time on organic contaminant retention in soils is to compare K_d (sorption distribution coefficient) values for "freshly aged" and "aged" soil samples (see Chapter 5 for discussion of these coefficients). In most studies, K_d values are measured based on a 24-hr equilibration between the soil and the organic chemical. When these values are compared to K_d values for field soils previously reacted with the organic

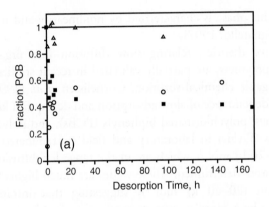

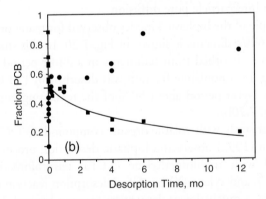

FIGURE 7.20. *(a) Short-term PCB desorption in hours (h) from Hudson River sediment contaminated with 25 mg kg^{-1} PCB. Distribution of the PCB between the sediment (■) and XAD-4 resin (○) is shown, as well as the overall mass balance (△). The resin acts as a sink to retain the PCB that is desorbed. (b) Long-term PCB desorption in months (mo) from Hudson River sediment contaminated with 25 mg kg^{-1} PCB. Distribution of the PCB between the sediment (■) and XAD-4 resin (•) is shown. The line represents a nonlinear regression of the data by the two-box (site) model. (Reprinted with permission from Carroll et al. (1994). Copyright © 1994 American Chemical Society.)*

chemical (aged samples) the latter have much higher K_d values, indicating that much more of the organic chemical is in a sorbed state. For example, Pignatello and Huang (1991) measured K_d values in freshly aged (K_d) and aged soils (K_{app}, apparent sorption distribution coefficient) reacted with atrazine and metolachlor, two widely used herbicides. The aged soils had been treated with the herbicides 15–62 months before sampling. The K_{app} values were 2.3–42 times higher than the K_d values (Table 7.5).

Scribner *et al.* (1992), studying simazine (a widely used triazine herbicide for broadleaf and grass control in crops) desorption and bioavailability in aged soils, found that K_{app} values were 15 times higher than K_d values. Scribner *et al.* (1992) also showed that 48% of the simazine added to the freshly aged soils was biodegradable over a 34-day incubation period while none of the simazine in the aged soil was biodegraded.

One of the implications of these results is that while many transport and degradation models for organic contaminants in soils and waters assume that the sorption process is an equilibrium process, the above studies clearly show that kinetic reactions must be considered when making predictions about the mobility and fate of organic chemicals. Moreover, K_d values determined based on a 24-hr equilibration period and commonly used in fate and risk assessment models can be inaccurate since 24-hr K_d values often overestimate the amount of organic chemical in the solution phase.

The finding that many organic chemicals are quite persistent in the soil environment has both good and bad features. The beneficial aspect is that the organic chemicals are less mobile and may not be readily transported in groundwater supplies. The negative aspect is that their persistence and inaccessibility to microbes may make decontamination more difficult, particularly if *in situ* remediation techniques such as biodegradation are employed.

Ion Exchange Kinetics

Ion exchange kinetics are greatly dependent on the adsorbent and the ion. Figure 7.21 clearly shows how the type of clay mineral affects K–Ca exchange. Reaction rates are much more rapid on kaolinite where the exchange sites are external than on vermiculite, which contains both external and internal exchange sites. The internal sites may be fully expanded or partially collapsed.

The type of ion also has a pronounced effect on the rate of exchange. Exchange of ions like K^+, NH_4^+, and Cs^+ is often slower than that of ions such as Ca^{2+} and Mg^{2+}. This is related to the smaller hydrated radius of the former ions. The smaller ions fit well in the interlayer spaces of clay minerals, which causes partial or total interlayer space collapse. The exchange is thus slow and interparticle diffusion-controlled. However, with the exception of K^+, NH_4^+, and Cs^+ exchange on 2:1 clay minerals like vermiculite and mica, ion exchange kinetics are usually very rapid, occurring on millisecond time scales (Tang and Sparks, 1993). Figure 7.22 shows that Ca–Na exchange on montmorillonite was complete in <100 msec.

TABLE 7.5. *Sorption Distribution Coefficients for Herbicides in "Freshly Aged" and "Aged" Soils[a]*

Herbicide	Soil	K_d[b]	K_{app}[c]
Metolachlor	CVa	2.96	39
	CVb	1.46	27
	W1	1.28	49
	W2	0.77	33
Atrazine	CVa	2.17	28
	CVb	1.32	29
	W3	1.75	4

[a] Adapted from Pignatello and Huang (1991) with permission; herbicides had been added to soils 31 months prior to sampling for CVa and CVb soils, 15 months for the W1 and W2 soils, and 62 months for the W3 soil.
[b] Sorption distribution coefficient (liter kg^{-1}) of "freshly aged" soil based on a 24-hr equilibration period.
[c] Apparent sorption distribution coefficient (liter kg^{-1}) in contaminated soil ("aged" soil) determined using a 24-hr equilibration period.

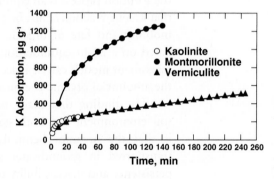

FIGURE 7.21. *Potassium adsorption versus time for clay minerals: ○, kaolinite; •, montmorillonite; ▲, vermiculite. From Sparks and Jardine (1984), with permission.*

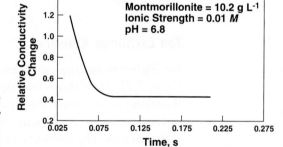

FIGURE 7.22. *Typical pressure-jump relaxation curve for Ca–Na exchange on montmorillonite showing relative change in conductivity vs. time. From Tang and Sparks (1993), with permission.*

Kinetics of Mineral Dissolution

RATE-LIMITING STEPS

Dissolution of minerals involves several steps (Stumm and Wollast, 1990): (1) mass transfer of dissolved reactants from the bulk solution to the mineral surface, (2) adsorption of solutes, (3) interlattice transfer of reacting species, (4) surface chemical reactions, (5) removal of reactants from the surface, and (6) mass transfer of products into the bulk solution. Under field conditions mineral dissolution is slow and mass transfer of reactants or products in the aqueous phase (Steps 1 and 6) is not rate-limiting. Thus, the rate-limiting steps are either transport of reactants and products in the solid phase (Step 3) or surface chemical reactions (Step 4) and removal of reactants from the surface (Step 5).

Transport-controlled dissolution reactions or those controlled by mass transfer or diffusion can be described using the parabolic rate law (Stumm and Wollast, 1990)

$$r = \frac{dC}{dt} = kt^{-1/2}, \tag{7.24}$$

where r is the reaction rate, C is the concentration in solution, t is time, and k is the reaction rate constant. Integrating, C increases with $t^{1/2}$,

$$C = C_0 + 2kt^{1/2} \tag{7.25}$$

where C_0 is the initial concentration in solution.

If the surface reactions are slow compared to the transport reactions, dissolution is surface-controlled, which is the case for most dissolution reactions of silicates and oxides. In surface-controlled reactions the concentrations of solutes next to the surface are equal to the bulk solution concentrations and the dissolution kinetics are zero-order if steady state conditions are operational on the surface. Thus, the dissolution rate, r, is

$$r = \frac{dC}{dt} = kA, \tag{7.26}$$

and r is proportional to the mineral's surface area, A. Thus, for a surface-controlled reaction the relationship between time and C should be linear. Figure 7.23 compares transport- and surface-controlled dissolution mechanisms.

Intense arguments concerning the mechanism for mineral dissolution have ensued over the years. Those that supported a transport-controlled mechanism believed that a leached layer formed as mineral dissolution proceeded and that subsequent dissolution took place via diffusion through the leached layer (Petrovic et al., 1976). Advocates of this theory found that dissolution was described by the parabolic rate law (Eq. (7.24)). However, the "apparent" transport-controlled kinetics may be an artifact caused by dissolution of hyperfine particles formed on the mineral surfaces after grinding, which are highly reactive sites, or by use of batch methods that cause reaction products to accumulate, causing precipitation of secondary minerals. These experimental artifacts can cause incongruent reactions and pseudoparabolic kinetics. Studies employing surface spectroscopies such as X-ray photoelectron spectroscopy and nuclear resonance profiling (Schott and Petit, 1987; Casey et al., 1989) have demonstrated that although some incongruency may occur in the initial dissolution process, which may be diffusion-controlled, the overall reaction is surface-controlled. An illustration of the surface-controlled dissolution of γ-Al_2O_3 resulting in a linear release of Al^{3+} with time is shown in Fig. 7.24. The dissolution rate, r, can be obtained from the slope of Fig. 7.24.

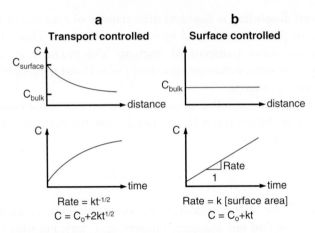

FIGURE 7.23. *Transport- vs surface-controlled dissolution. Schematic representation of concentration in solution, C, as a function of distance from the surface of the dissolving mineral. In the lower part of the figure, the change in concentration is given as a function of time. From W. Stumm, "Chemistry of the Solid–Water Interface." Copyright © 1992 John Wiley & Sons, Inc. Reprinted by permission of John Wiley & Sons, Inc.*

FIGURE 7.24. *Linear dissolution kinetics observed for the dissolution of γ-Al₂O₃. Representative of processes whose rates are controlled by a surface reaction and not by transport. Reprinted from Furrer, G., and Stumm, W. (1986), The coordination chemistry of weathering, Geochim. Cosmochim. Acta **50**, 1847–1860. Copyright © 1986, with kind permission from Elsevier Science Ltd., The Boulevard, Langford Lane, Kidlington OX5 1GB, UK.*

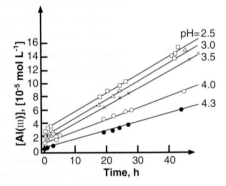

SURFACE-CONTROLLED DISSOLUTION MECHANISMS

Dissolution of oxide minerals via a surface-controlled reaction by ligand- and proton-promoted processes has been described by Stumm and co-workers (Furrer and Stumm, 1986; Zinder *et al.*, 1986; Stumm and Furrer, 1987) using a surface coordination approach. The important reactants in these processes are H_2O, H^+, OH^-, ligands, and reductants and oxidants (see definitions in Chapter 8). The reaction mechanism occurs in two steps (Stumm and Wollast, 1990):

$$\text{Surface sites + reactants (H}^+\text{, OH}^-\text{, or ligands)} \xrightarrow{\text{fast}} \text{Surface species} \quad (7.27)$$

$$\text{Surface species} \xrightarrow[\substack{\text{detachment of} \\ \text{metal (M)}}]{\text{slow}} \text{M (aq).} \quad (7.28)$$

Thus, the attachment of the reactants to the surface sites is fast and detachment of metal species from the surface into solution is slow and rate-limiting.

LIGAND-PROMOTED DISSOLUTION

Figure 7.25 shows how the surface chemistry of the mineral affects dissolution. One sees that surface protonation of the surface ligand increases dissolution by polarizing interatomic bonds close to the central surface ions, which promotes the release of a cation surface group into solution. Hydroxyls that bind to surface groups at higher pH's can ease the release of an anionic surface group into the solution phase.

Ligands that form surface complexes via ligand exchange with a surface hydroxyl add negative charge to the Lewis acid center coordination sphere, and lower the Lewis acid acidity. This polarizes the M–oxygen bonds, causing detachment of the metal cation into the solution phase. Thus, inner-sphere surface complexation plays an important role in mineral dissolution. Ligands such as oxalate, salicylate, F^-, EDTA, and NTA increase dissolution but others, e.g., SO_4^{2-}, CrO_4^{2-}, and benzoate, inhibit dissolution. Phosphate and arsenate enhance dissolution at low pH and dissolution is inhibited at pH >4 (Stumm, 1992).

The reason for these differences may be that bidentate species that are mononuclear promote dissolution while binuclear bidentate species inhibit dissolution. With binuclear bidentate complexes, more energy may be needed to remove two central atoms from the crystal structure. With phosphate and arsenate, at low pH mononuclear species are formed, while at higher pH (around pH 7) binuclear or trinuclear surface complexes form. Mononuclear bidentate complexes are formed with oxalate while binuclear bidentate complexes form with CrO_4^{2-}. Additionally, the electron donor properties of CrO_4^{2-} and oxalate are also different. With CrO_4^{2-} a high redox potential is maintained at the oxide surface, which restricts reductive dissolution (Stumm and Wollast, 1990; Stumm, 1992).

Dissolution can also be inhibited by cations such as VO^{2+}, Cr(III), and Al(III) that block surface functional groups.

One can express the rate of the ligand-promoted dissolution, R_L, as

$$R_L = k'_L (\equiv ML) = k'_L C_L^s, \tag{7.29}$$

where k'_L is the rate constant for ligand-promoted dissolution (time^{-1}), $\equiv ML$ is the metal–ligand complex, and C_L^s is the surface concentration of the ligand complex (mol m^{-2}). Figure 7.26 shows that Eq. (7.29) adequately described ligand-promoted dissolution of γ-Al$_2$O$_3$.

PROTON-PROMOTED DISSOLUTION

Under acid conditions, protons can promote mineral dissolution by binding to surface oxide ions, causing bonds to weaken. This is followed by detachment of metal species into solution. The proton-promoted dissolution rate, R_H, can be expressed as (Stumm, 1992)

$$R_H = k'_H (\equiv MOH_2^+)^j = k'_H (C_H^s)^j, \tag{7.30}$$

where k'_H is the rate constant for proton-promoted dissolution, $\equiv MOH_2^+$ is the metal–proton complex, C_H^s is the concentration of the surface-adsorbed

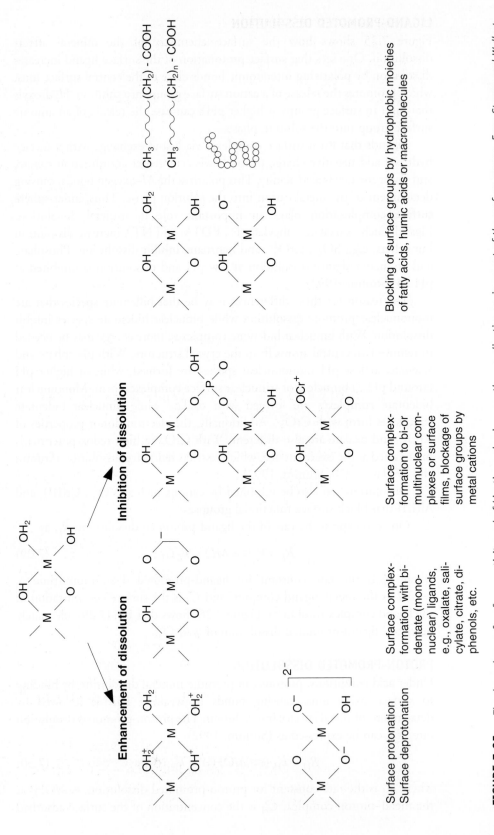

FIGURE 7.25. *The dependence of surface reactivity and of kinetic mechanisms on the coordinative environment of the surface groups. From Stumm and Wollast (1990), Rev. Geophys.* **28**, *53–69. Copyright by the American Geophysical Union.*

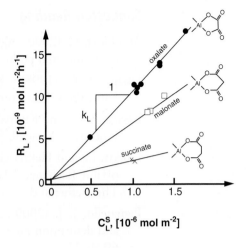

FIGURE 7.26. *The rate of ligand-catalyzed dissolution of γ-Al$_2$O$_3$ by the aliphatic ligands oxalate, malonate, and succinate, R_L (nmol m^{-2} h^{-1}), can be interpreted as a linear dependence on the surface concentrations of the ligand complexes, C_L^s. In each case the individual values for C_L^s were determined experimentally. Reprinted from Furrer, G., and Stumm, W. (1986), The coordination chemistry of weathering, Geochim. Cosmochim. Acta **50**, 1847–1860. Copyright © 1986, with kind permission from Elsevier Science Ltd., The Boulevard, Langford Lane, Kidlington OX5 1GB, UK.*

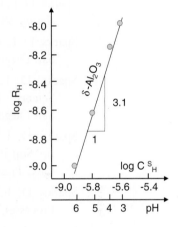

FIGURE 7.27. *The dependence of the rate of proton-promoted dissolution of γ-Al$_2$O$_3$, R_H (mol m^{-2} h^{-1}), on the surface concentration of the proton complexes. C_H^s (mol m^{-2}). Reprinted from Furrer, G., and Stumm, W. (1986), The coordination chemistry of weathering, Geochim. Cosmochim. Acta **50**, 1847–1860. Copyright © 1986, with kind permission from Elsevier Science Ltd., The Boulevard, Langford Lane, Kidlington OX5 1GB, UK.*

proton complex (mol m^{-2}), and j corresponds to the oxidation state of the central metal ion in the oxide structure (i.e., $j = 3$ for Al(III) and Fe(III) in simple cases). If dissolution occurs by only one mechanism j is an integer. Figure 7.27 shows an application of Eq. (7.30) for the proton-promoted dissolution of γ-Al$_2$O$_3$.

OVERALL DISSOLUTION MECHANISMS

The rate of mineral dissolution, which is the sum of the ligand-, proton-, and deprotonation-promoted (or bonding of OH$^-$ ligands) dissociation $(R_{OH} = k'_{OH} (C_{OH}^s)^i)$ rates, along with the pH-independent portion of the dissolution rate (k'_{H_2O}), which is due to hydration, can be expressed as (Stumm, 1992)

$$R = k'_L (C_L^s) + k'_H (C_H^s)^j + k'_{OH} (C_{OH}^s)^i + k'_{H_2O}. \qquad (7.31)$$

Equation (7.31) is valid if dissolution occurs in parallel at varying metal centers (Furrer and Stumm, 1986).

Suggested Reading

Alexander, M. (2000). Aging, bioavailability, and overestimation of risk from environmental pollutants. *Environ. Sci. Technol.* **34**, 4259–4265.

Lasaga, A. C. (1998). "Kinetic Theory in the Earth Sciences." Princeton Univ. Press, Princeton, NJ.

Lasaga, A. C., and Kirkpatrick, R. J., Eds. (1981). "Kinetics of Geochemical Processes." Mineral. Soc. Am., Washington, DC.

Liberti, L., and Helfferich, F. G., Eds. (1983). "Mass Transfer and Kinetics of Ion Exchange," NATO ASI Ser. E, No. 71. Nijhoff, The Hague, The Netherlands.

Pignatello, J. J. (2000). The measurement and interpretation of sorption and desorption rates for organic compounds in soil media. *Adv. Agron.* **69**, 1–73.

Sparks, D. L. (1985). Kinetics of ion reactions in clay minerals and soils. *Adv. Agron.* **38**, 231–266.

Sparks, D. L. (1989). "Kinetics of Soil Chemical Processes." Academic Press, San Diego, CA.

Sparks, D. L. (1992). Soil kinetics. *In* "Encyclopedia of Earth Systems Science" (W. A. Nirenberg, Ed.), Vol. 4, pp. 219–229. Academic Press, San Diego, CA.

Sparks, D. L. (1999). Kinetics of reactions in pure and mixed systems. *In* "Soil Physical Chemistry" (D. L. Sparks, Ed.), 2nd ed., pp. 83–145. CRC Press, Boca Raton, FL.

Sparks, D. L., and Suarez, D. L., Eds. (1991). "Rates of Soil Chemical Processes," SSSA Spec. Publ. 27. Soil Sci. Soc. Am., Madison, WI.

Sposito, G. (1994). "Chemical Equilibria and Kinetics in Soils." Wiley, New York.

Stumm, W., Ed. (1990). "Aquatic Chemical Kinetics." Wiley, New York.

8 Redox Chemistry of Soils

Oxidation–Reduction Reactions and Potentials

S oil chemical reactions involve some combination of proton and electron transfer. Oxidation occurs if there is a loss of electrons in the transfer process while reduction occurs if there is a gain of electrons. The oxidized component or oxidant is the electron acceptor and the reduced component or reductant is the electron donor. Table 8.1 lists oxidants and reductants found in natural environments. The electrons are not free in the soil solution; thus the oxidant must be in close contact with the reductant. Both oxidation and reduction must be considered to completely describe oxidation–reduction (redox) reactions (Bartlett and James, 1993; Patrick *et al.*, 1996).

To determine if a particular reaction will occur (i.e., the Gibbs free energy for the reaction, ΔG_r <0), one can write reduction and oxidation half-reactions (a half-reaction or half-cell reaction can be referred to as a redox couple) and calculate equilibrium constants for the half-reactions. Redox reactions of soil oxidants can be defined conventionally by the general half-reduction reaction (Patrick *et al.*, 1996)

245

$$Ox + mH^+ + ne^- \rightarrow Red, \tag{8.1}$$

where Ox is the oxidized component or the electron acceptor, Red is the reduced component or electron donor, m is the number of hydrogen ions participating in the reaction, and n is the number of electrons involved in the reaction. The electrons in Eq. (8.1) must be supplied by an accompanying oxidation half-reaction. For example, in soils, soil organic matter is the primary source of electrons. Thus, to completely describe a redox reaction, an oxidation reaction must balance the reduction reaction.

Let us illustrate these concepts for the redox reaction of $Fe(OH)_3$ reduction (Patrick *et al.*, 1996):

$$4Fe(OH)_3 + 12H^+ + 4e^- \rightarrow 4\ Fe^{2+} + 12H_2O \text{ (reduction)} \tag{8.2}$$

$$CH_2O + H_2O \rightarrow CO_2 + 4H^+ + 4e^- \text{ (oxidation)} \tag{8.3}$$

$$4\ Fe(OH)_3 + CH_2O + 8H^+ \rightarrow 4\ Fe^{2+} + CO_2 + 11H_2O \text{ (net reaction), } \tag{8.4}$$

where CH_2O is soil organic matter. Equation (8.2) represents the reduction half-reaction and Eq. (8.3) represents the oxidation half-reaction. The reduction (Eq. (8.2)) reaction can also be described by calculating ΔG_r, the Gibbs free energy for the reaction,

$$\Delta G_r = \Delta G_r^\circ + RT \ln (Red)/(Ox)(H^+)^m, \tag{8.5}$$

where ΔG_r° is the standard free energy change for the reaction. The Nernst equation can be employed to express the reduction reaction in terms of electrochemical energy (millivolts) using the expression $\Delta G_r = -nFE$ such that (Patrick *et al.*, 1996)

$$Eh = E^\circ - RT/nF \ln(Red)/(Ox) + mRT/nF \ln H^+, \tag{8.6}$$

where Eh is the electrode potential, or in the case of the reduction half-reaction in Eq. (8.2), a reduction potential, E° is the standard half-reaction reduction potential (with each half-reaction, for example, Eqs. (8.2) and (8.3), there is a standard potential; the standard potential means the activities of all reactants and products are unity), F is the Faraday constant, n is the number of electrons exchanged in the half-cell reaction, m is the number of protons exchanged, and the activities of the oxidized and reduced species are in parentheses. Determination of Eh will provide quantitative information on electron availability and can be either an oxidation or reduction potential depending on how the reaction is written (see Eqs. (8.2)–(8.3)). Oxidation potentials are more often used in chemistry, while in soil chemistry reduction potentials are more frequently used to describe soil and other natural systems (Patrick *et al.*, 1996). It should also be pointed out that the Nernst equation is valid for predicting the activity of oxidized and reduced species only if the system is at equilibrium, which is seldom the case for soils and sediments. As noted in Chapter 7, the heterogeneity of soils which promotes transport processes causes many soil chemical reactions to be very slow. Thus, it is difficult to use Eh values to quantitatively measure the activities of oxidized and reduced species for such heterogeneous systems (Bohn, 1968).

TABLE 8.1. *Selected Reduction Half-Reactions Pertinent to Soil, Natural Water, Plant, and Microbial Systems[a]*

Half-reaction	log $K^{\circ b}$	pe^c pH 5	pe^c pH 7
Nitrogen species			
$1/2N_2O + e^- + H^+ = 1/2N_2 + 1/2H_2O$	29.8	22.9	20.9
$NO + e^- + H^+ = 1/2N_2O + 1/2H_2O$	26.8	19.8	17.8
$1/2NO_2^- + e^- + 3/2H^+ = 1/4N_2O + 3/4H_2O$	23.6	15.1	12.1
$1/5\,NO_3^- + e^- + 6/5H^+ = 1/10N_2 + 3/5H_2O$	21.1	14.3	11.9
$NO_2^- + e^- + 2H^+ = NO + H_2O$	19.8	9.8	5.8
$1/4NO_3^- + e^- + 5/4H^+ = 1/8N_2O + 5/8H_2O$	18.9	12.1	9.6
$1/6NO_2^- + e^- + 4/3H^+ = 1/6NH_4^+ + 1/3H_2O$	15.1	8.4	5.7
$1/8NO_3^- + e^- + 5/4H^+ = 1/8NH_4^+ + 3/8H_2O$	14.9	8.6	6.1
$1/2NO_3^- + e^- + H^+ = 1/2NO_2^- + 1/2H_2O$	14.1	9.1	7.1
$1/6NO_3^- + e^- + 7/6H^+ = 1/6NH_2OH + 1/3H_2O$	11.3	5.4	3.1
$1/6N_2 + e^- + 4/3H^+ = 1/3NH_4^+$	4.6	−0.7	−3.3
Oxygen species			
$1/2O_3 + e^- + H^+ = 1/2O_2 + 1/2H_2O$	35.1	28.4	26.4
$OH\cdot + e^- = OH^-$	33.6	33.6	33.6
$O_2^- + e^- + 2H^+ = H_2O_2$	32.6	22.6	18.6
$1/2H_2O_2 + e^- + H^+ = H_2O$	30.0	23.0	21.0
$1/4O_2 + e^- + H^+ = 1/2H_2O$	20.8	15.6	13.6
$1/2O_2 + e^- + H^+ = 1/2H_2O_2$	11.6	8.2	6.2
$O_2 + e^- = O_2^-$	−9.5	−6.2	−6.2
Sulfur species			
$1/8SO_4^{2-} + e^- + 5/4H^+ = 1/8H_2S + 1/2H_2O$	5.2	−1.0	−3.5
$1/2SO_4^{2-} + e^- + 2H^+ = 1/2SO_2 + H_2O$	2.9	−7.1	−11.1
Iron and manganese compounds			
$1/2Mn_3O_4 + e^- + 4H^+ = 3/2Mn^{2+} + 2H_2O$	30.7	16.7	8.7
$1/2Mn_2O_3 + e^- + 3H^+ = Mn^{2+} + 3/2H_2O$	25.7	14.7	8.7
$Mn^{3+} + e^- = Mn^{2+}$	25.5	25.5	25.5
$\gamma MnOOH + e^- + 3H^+ = Mn^{2+} + 2H_2O$	25.4	14.4	8.4
$0.62MnO_{1.8} + e^- + 2.2H^+ = 0.62Mn^{2+} + 1.1H_2O$	22.1	13.4	8.9
$1/2Fe_3(OH)_8 + e^- + 4H^+ = 3/2Fe^{2+} + 4H_2O$	21.9	7.9	−0.1
$1/2MnO_2 + e^- + 2H^+ = 1/2Mn^{2+} + H_2O$	20.8	12.8	8.8
$[Mn^{3+}(PO_4)_2]^{3-} + e^- = [Mn^{2+}(PO_4)_2]^{4-}$	20.7	20.7	20.7
$Fe(OH)_2^+ + e^- + 2H^+ = Fe^{2+} + 2H_2O$	20.2	10.2	6.2
$1/2Fe_3O_4 + e^- + 4H^+ = 3/2Fe^{2+} + 2H_2O$	17.8	3.9	−4.1
$MnO_2 + e^- + 4H^+ = Mn^{3+} + 2H_2O$	16.5	0.54	−7.5
$Fe(OH)_3 + e^- + 3H^+ = Fe^{2+} + 3H_2O$	15.8	4.8	−1.2
$Fe(OH)^{2+} + e^- + H^+ = Fe^{2+} + H_2O$	15.2	10.2	8.2
$1/2Fe_2O_3 + e^- + 3H^+ = Fe^{2+} + 3/2H_2O$	13.4	2.4	−3.6
$FeOOH + e^- + 3H^+ = Fe^{2+} + 2H_2O$	13.0	2.0	−4.0
$Fe^{3+} + e^- = Fe^{2+}$ phenanthroline	18.0	—[d]	—
$Fe^{3+} + e^- = Fe^{2+}$	13.0	13.0	13.0
$Fe^{3+} + e^- = Fe^{2+}$ acetate	—	5.8	—
$Fe^{3+} + e^- = Fe^{2+}$ malonate	—	4.4 (pH 4)	—
$Fe^{3+} + e^- = Fe^{2+}$ salicylate	—	4.4 (pH 4)	—
$Fe^{3+} + e^- = Fe^{2+}$ hemoglobin	—	—	2.4
$Fe^{3+} + e^- = Fe^{2+}$ cyt b_3 (plants)	—	—	0.68
$Fe^{3+} + e^- = Fe^{2+}$ oxalate	—	—	0.034
$Fe^{3+} + e^- = Fe^{2+}$ pyrophosphate	−2.4	—	—
$Fe^{3+} + e^- = Fe^{2+}$ peroxidase	—	—	−4.6
$Fe^{3+} + e^- = Fe^{2+}$ ferredoxin (spinach)	—	—	−7.3
$1/3KFe_3(SO_4)_2(OH)_6 + e^- + 2H^+ = Fe^{2+} + 2H_2O + 2/3SO_4^{2-}$ $+ 1/3K^+$	8.9	6.9	2.9
$[Fe(CN)_6]^{3-} + e^- = [Fe(CN)_6]^{4-}$	—	—	6.1

TABLE 8.1. *Selected Reduction Half-Reactions Pertinent to Soil, Natural Water, Plant, and Microbial Systems[a] (contd)*

Half-reaction	log $K^{∘b}$	pe^{c} pH 5	pH 7
Carbon species			
$1/2CH_3OH + e^- + H^+ = 1/2CH_4 + 1/2H_2O$	9.9	4.9	2.9
$1/2o\text{-quinone} + e^- + H^+ = 1/2\text{diphenol}$	—	—	5.9
$1/2p\text{-quinone} + e^- + H^+ = 1/2\text{hydroquinone}$	—	—	4.7
$1/12C_6H_{12}O_6 + e^- + H^+ = 1/4C_2H_5OH + 1/4H_2O$	4.4	0.1	-1.9
$\text{Pyruvate} + e^- + H^+ = \text{lactate}$	—	—	-3.1
$1/8CO_2 + e^- + H^+ = 1/8CH_4 + 1/4H_2O$	2.9	-2.1	-4.1
$1/2CH_2O + e^- + H^+ = 1/2CH_3OH$	2.1	-2.9	-4.9
$1/2HCOOH + e^- + H^+ = 1/2CH_2O + 1/2H_2O$	1.5	-3.5	-5.5
$1/4CO_2 + e^- + H^+ = 1/24C_6H_{12}O_6 + 1/4H_2O$	-0.21	-5.9	-7.9
$1/2\text{deasc} + e^- + H^+ = 1/2\text{asc}$	1.0	-3.5	-5.5
$1/4CO_2 + e^- + H^+ = 1/4CH_2O + 1/4H_2O$	-1.2	-6.1	-8.1
$1/2CO_2 + e^- + H^+ = 1/2HCOOH$	-1.9	-6.7	-8.7
Pollutant/nutrient group			
$Co^{3+} + e^- = Co^{2+}$	30.6	30.6	30.6
$1/2NiO_2 + e^- + 2H^+ = 1/2Ni^{2+} + H_2O$	29.8	21.8	17.8
$PuO_2^+ + e^- = PuO_2$	26.0	22.0	22.0
$1/2PbO_2 + e^- + 2H^- = 1/2Pb^{2+} + H_2O$	24.8	16.8	12.8
$PuO_2 + e^- + 4H^+ = Pu^{3+} + 2H_2O$	9.9	-6.1	-14.1
$1/3HCrO_4^- + e^- + 4/3H^+ = 1/3Cr(OH)_3 + 1/3H_2O$	18.9	10.9	8.2
$1/2AsO_4^{3-} + e^- + 2H^+ = 1/2AsO_2^- + H_2O$	16.5	6.5	2.5
$Hg^{2+} + e^- = 1/2Hg_2^{2+}$	15.4	13.4	13.4
$1/2MoO_4^{2-} + e^- + 2H^+ = 1/2MoO_2 + H_2O$	15.0	3.0	-1.0
$1/2SeO_4^{2-} + e^- + H^+ = 1/2SeO_3^{2-} + 1/2H_2O$	14.9	9.9	7.9
$1/4SeO_3^{2-} + e^- + 3/2H^+ = 1/4Se + 3/4H_2O$	14.8	6.3	3.3
$1/6SeO_3^{2-} + 4/3H^+ = 1/6H_2Se + 1/2H_2O$	7.62	1.0	-1.7
$1/2VO_2^+ + e^- + 1/2\,H_3O^+ = 1/2\,V\,(OH)_3$	6.9	2.4	1.4
$Cu^{2+} + e^- = Cu^+$	2.6	2.6	2.6
$PuO_2 + e^- + 3H^+ = PuOH^{2+} + H_2O$	2.9	-8.1	-14.1
Analytical couples			
$CeO_2 + e^- + 4H^+ = Ce^{3+} + 2H_2O$	47.6	31.6	23.6
$1/2ClO^- + e^- + H^+ = 1/2Cl^- + 1/2H_2O$	29.0	24.0	22.0
$HClO + e^- = 1/2Cl_2 + H_2O$	27.6	20.6	18.6
$1/2Cl_2 + e^- = Cl^-$	23.0	25.0	25.0
$1/6IO_3^- + e^- + H^+ = 1/6I^- + 1/2H_2O$	18.6	13.6	11.6
$1/2Pt(OH)_2 + e^- + H^+ = 1/2Pt + H_2O$	16.6	11.6	9.6
$1/2I_2 + e^- = I^-$	9.1	11.1	11.1
$1/2Hg_2Cl_2 + e^- = Hg + Cl^-$	4.5	3.9	3.9
$e^- + H^+ = 1/2H_2$	0	-5.0	-7
$1/2PtS + e^- + H^+ = 1/2Pt + 1/2H_2S$	-5.0	-10.0	-12.0

[a] From Bartlett and James (1993), with permission.
[b] Calculated for reaction as written according to Eq. (8.14). Free energy of formation data were taken from Lindsay (1979) as a primary source, and when not available from that source, from Garrels and Christ (1965) and Loach (1976).
[c] Calculated using tabulated log $K^{∘}$ values, reductant and oxidant = 10^{-4} M soluble ions and molecules, and activities of solid phases = 1; partial pressures for gases that are pertinent to soils: 1.01×10^{-4} MPa for trace gases, 2.12×10^{-2} MPa for O_2, 7.78×10^{-2} MPa for N_2, and 3.23×10^{-5} MPa for CO_2.
[d] Values not listed by Loach (1976).

Using the values of 8.31 J K^{-1} mol^{-1} for R, 9.65×10^4 C mol^{-1} for F, and 298 K for T and the relationship $\ln(x) = 2.303 \log(x)$, Eq. (8.6) becomes

$$\text{Eh(mV)} = E° - 59/n \log (\text{Red})/(\text{Ox}) - 59 \, m/n \text{ pH}. \qquad (8.7)$$

It is obvious from Eqs. (8.6)–(8.7) that Eh increases as the activity of the oxidized species increases, decreases with increases in the activity of the reduced species, and increases as H$^+$ activity increases or pH decreases. If the ratio of protons to electrons is 1 (i.e., $m/n = 1$), one would predict that Eh would change by 59 mV for every unit change in pH. Thus, one could predict the Eh at various pH values by using the 59-mV factor. However, this relationship assumes that redox controls the pH of the system. This assumption is valid for solutions, but in soils pH buffering is affected by soil components such as silicates, carbonates, and oxides, which are not involved in redox reactions. Thus, it may be inappropriate to apply the 59-mV factor (Patrick et al., 1996).

Eh is positive and high in strongly oxidizing systems while it is negative and low in strongly reducing systems. (There is not a neutral point, as one observes with pH.) Eh, like pH, is an intensity factor. The oxygen–nitrogen range has been defined by Eh values of +250 to +100 mV, the iron range as +100 to 0.0 mV, the sulfate range as 0.0 to –200 mV, and the methane–hydrogen range as <200 mV (Liu and Narasimhan, 1989).

Eh vs pH and pe vs pH Diagrams

Diagrams of the activities of Eh vs pH can be very useful in delineating the redox status of a system. Figure 8.1 shows such a diagram for soils. The pH range was narrower in reduced soils (negative Eh) than in oxidized soils (positive Eh). Based on these results, Baas Becking et al. (1960) divided the soils into three categories: normal (oxidized), wet (seasonally saturated), and waterlogged (semipermanently saturated) (Fig. 8.1).

The reduction half-reaction given in Eq. (8.1) can also be expressed in terms of an equilibrium constant $K°$ (Patrick et al., 1996).

$$K° = (\text{Red})/(\text{Ox})(e^-)^n(\text{H}^+)^m. \qquad (8.8)$$

Expressed in log form Eq. (8.8) becomes

$$\log K° = \log(\text{Red}) - \log(\text{Ox}) - n\log (e^-) - m\log (\text{H}^+). \qquad (8.9)$$

The $-\log (e^-)$ term in Eq. (8.9) is defined as pe in a similar way as pH is expressed as $-\log (\text{H}^+)$. The pe is an intensity factor as it is an index of the electron free energy level per mole of electrons (Ponnamperuma, 1972). Thus, pe and pH are master variables of a soil and must be known to completely understand the equilibrium state of a soil. Moreover, to fully determine the redox status of a soil, pe and pH cannot be separated (Bartlett and James, 1993). In strongly oxidizing systems the e^- activity is low and pe is large and positive. In reducing systems pe is small and negative. Sposito (1989) proposed "oxic" (oxidized) soils as those with pe >7, "suboxic" soils in the pe range between +2 and +7, and "anoxic" (reduced) soils with pe <+2, all at pH 7. These ranges are consistent with redox control by oxygen–nitrogen, manganese–iron, and sulfur couples (James and Bartlett, 2000). The pe range of most soils is –6 to +12 (Lindsay, 1979). Rearranging Eq. (8.9) one arrives at an expression that relates pe to pH,

$$pe = [(\log K^\circ - \log (\text{Red}) + \log (\text{Ox}))/n] - m/n \text{ pH}, \qquad (8.10)$$

which represents a straight line with a slope of m/n and an intercept given in brackets. The intercept is a function of $\log K^\circ$ for the half-reaction and the activities of the oxidized and reduced species. When there is a one-electron transfer (i.e., $n = 1$) and consumption of one proton (i.e., $m = 1$), and when (Red) = (Ox), Eq. (8.10) is simplified to

$$pe + \text{pH} = \log K^\circ. \qquad (8.11)$$

At pH = 0,

$$pe = \log K^\circ. \qquad (8.12)$$

One can relate $\log K^\circ$ to ΔG_r° using the equation

$$\Delta G_r^\circ = - RT \ln K^\circ. \qquad (8.13)$$

At 298 K and converting to log,

$$-\Delta G_r^\circ/5.71 = \log K^\circ, \qquad (8.14)$$

where 5.71 is derived from the product of $(RT)(2.303)$, R is $(0.008314 \text{ kJ mol}^{-1} \text{ K}^{-1})$, and $T = 298.15$ K. Therefore, $\log K^\circ$ could be estimated by knowing the free energies of formation (ΔG_f°) of H_2O and the Red and Ox species since those for H^+ and e^- are zero by convention (Bartlett and James, 1993).

Information in Box 8.1 shows how one would calculate $\log K^\circ$ and pe for a reduction half-reaction at pH 5 and 7 using Eqs. (8.11)–(8.14).

The values of $\log K^\circ$ can be used to predict whether a reduction and oxidation reaction will combine to effect the transfer of electrons from reductant to oxidant. Table 8.1 lists a number of reduction half-reactions important in natural systems. The $\log K^\circ$ values are given in descending order and are pe values at pH 0, when the activities of oxidant and reductant are 1, and are standard reference pe values for the reactions. The larger the values

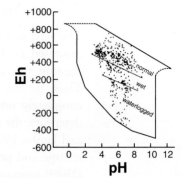

FIGURE 8.1.   *Eh–pH characteristics of soils.*
From Baas Becking et al. (1960),
with permission from the University of Chicago Press.

of log $K°$ or pe, the greater the tendency for an oxidant (left side of the half-reaction equation) to be reduced (converted to the right side of the half-reaction equation). Therefore, an oxidant in a given reduction half-reaction can oxidize the reductant in another half-reaction with a lower pe, at a particular pH. As an example, Mn(III,IV) oxides could oxidize Cr(III) to Cr(VI) at pH 5 because the range of pe values for reduction of Mn (12.8–16.7) is higher than that for Cr(VI) reduction (10.9) (Bartlett and James, 1993). In field moist soils over a pH range of 4–7 it has indeed been observed that Mn(III,IV) oxides can oxidize Cr(III) to Cr(VI) (Bartlett and James, 1979; James and Bartlett, 1983).

The pe–pH relationship expressed in Eq. (8.10) can be used to determine whether an oxidation–reduction reaction can occur spontaneously, i.e., $\Delta G_r < 0$. Figure 8.2 shows pe vs pH stability lines between oxidized and reduced species for several redox couples. If thermodynamic equilibrium is present, the oxidized form of the couple would be preferred if the pe and pH region was above a given line and the reduced form would be favored below a given line (Bartlett, 1986). The line for Fe is often considered the dividing point between an aerobic (oxidized) and an anaerobic (reduced) soil. In aerobic soils oxidized species stay oxidized even though the thermodynamic tendency is toward reduction, as indicated by the high pe. Below the iron line, reduced species are prevalent, even though the thermodynamic tendency is toward oxidation. Sulfide is easily oxidized and nitrite is easily reduced (Bartlett and James, 1993).

BOX 8.1 *Calculation of log $K°$ and pe*

The reduction half-reaction below (see Table 8.1) shows the reduction of Fe^{3+} to Fe^{2+},

$$Fe(OH)^{2+} + e^- + H^+ = Fe^{2+} + H_2O. \qquad (8.1a)$$

In this reaction there is one electron transfer, i.e., n in Eq. (8.8) is 1, there is consumption of one proton, i.e., $m = 1$, and $(Fe^{3+}) = (Fe^{2+})$ is an imposed condition. Thus Eq. (8.10) reduces to Eq. (8.11) and at pH 0, Eq. (8.12) results. Relating $\ln K°$ to $\Delta G_r^°$, one can employ Eq. (8.13),

$$\Delta G_r^° = - RT \ln K°.$$

We know from Eq. (4.7) that

$$\Delta G_r^° = \Sigma \Delta G_f^° \text{ products} - \Sigma \Delta G_f^° \text{ reactants.}$$

Solving $\Delta G_r^°$ for Eq. (8.1a) above,

$$\Delta G_r^° = [(-91.342 \text{ kJ mol}^{-1}) + (-237.52 \text{ kJ mol}^{-1})]$$

$$- [(-241.85 \text{ kJ mol}^{-1}) + (0)] \qquad (8.1b)$$

$$= [-328.86 \text{ kJ mol}^{-1} + 241.85 \text{ kJ mol}^{-1}]$$

$$= -87.01 \text{ kJ mol}^{-1}.$$

Using Eq. (8.14),

$$\log K^{\circ} = \frac{87.01}{5.71} = 15.2. \qquad (8.1c)$$

This value for $\log K^{\circ}$ is the one shown in Table 8.1 for the reaction in Eq. (8.1a).

To calculate pe at pH 5 and pH 7, one would use Eq. (8.11). For pH 5 and substituting in the value of 15.2 for $\log K^{\circ}$,

$$pe = \log K^{\circ} - pH$$

$$pe = 15.2 - 5 = 10.2.$$

For pH 7,

$$pe = 15.2 - 7 = 8.2.$$

These are the pe values shown in Table 8.1 for the reduction half-reaction in Eq. (8.1a).

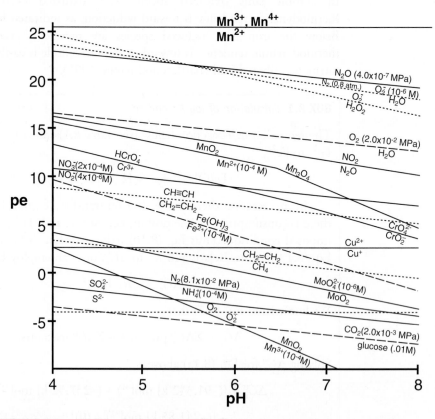

FIGURE 8.2.   *Stability lines between oxidized and reduced species for several redox couples. Solid phases of Mn and Fe are at unit activity and the activities of other species are designated if not equal within a couple. Reprinted with permission from Bartlett (1986). Copyright CRC Press, Boca Raton, FL.*

Figure 8.3 illustrates a pe–pH diagram for several Mn species. One sees that Mn oxides can oxidize Pu(III) to Pu(IV), V(III) to V(V), As(III) to As(V), Se(IV) to Se(VI), and Cr(III) to Cr(VI), because the pe for each of these couples is below the pe for Mn oxides. It has been shown that Mn oxides in soils can indeed effect oxidation of Pu(III), As(III), Se(IV), and, as noted earlier, Cr(III) (Bartlett and James, 1979; Bartlett, 1981; Amacher and Baker, 1982; Moore et al., 1990). The environmental aspects of some of these oxidation processes are discussed later in this chapter.

Another term often used in studying redox chemistry of soils is poise. The poise of a redox system is the resistance to change in redox potential with the addition of small amounts of oxidant or reductant. Poise increases with the total concentration of oxidant plus reductant, and for a fixed total concentration it reaches a maximum when the ratio of oxidant to reductant is 1 (Ponnamperuma, 1955).

Measurement and Use of Redox Potentials

Measurement of redox potentials in soils is usually done with a platinum electrode. This electrode will transfer electrons to or from the medium, but it should not react with the medium. Once the platinum electrode is combined with a half-cell of known potential, reducing systems will transfer electrons to the electrode while oxidizing systems will remove electrons from the electrode. When experimental redox potential measurements are done, there is no electron flow and the potential between the half-cell composed of the platinum in contact with the medium and the known potential of the reference electrode half-cell is determined with a meter that reacts to the electromotive force or potential (Patrick et al., 1996). A number of investigators have noted that measurement of redox potentials in aerated soils is questionable due to the lack of poising of reduction–oxidation systems that are well aerated and have plentiful quantities of oxygen (Ponnamperuma, 1955). Soil atmospheric oxygen measurements are preferred to characterize well-aerated soils. Thus, redox measurements are most reliable for flooded soils and sediments.

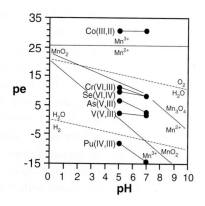

FIGURE 8.3. *A pe–pH diagram for Mn^{3+}, MnO_2, and Mn_3O_4; as compared with reduction between pH 5 and 7 for Co, Cr, Se, As, V, and Pu. Activity for ionic species is 10^{-4} M. From Bartlett and James (1993), with permission.*

Redox potentials can be very useful in characterizing the oxidation–reduction status of a soil. Oxidized soils have redox potentials of +400 to +700 mV. Seasonally saturated soils have redox potentials of +400 to +700 mV (oxidized) to highly reduced (–250 to –300 mV) (Patrick *et al.*, 1996). Redox potentials can help one predict when reducing conditions will begin due to depletion of oxidants such as oxygen and nitrate, and the initiation of oxidizing conditions when oxygen is reintroduced in the soil. Redox potentials can also provide information on conditions that are favorable for increased bioavailability of heavy metals (Gambrell *et al.*, 1977; Reddy and Patrick, 1977), changes in plant metabolism (Mendelssohn *et al.*, 1981), distribution of plant species (Josselyn *et al.*, 1990), and location of wetlands (Faulkner *et al.*, 1989).

If redox potential data are combined with other information such as depth to the water table and oxygen content of the soil, even more accurate information can be gleaned about the wetness of an environment. In nonwetland environments the Eh and oxygen content do not change much during the year. Transitional areas may be either oxidized or reduced as the water table rises and falls. The redox potentials are low until after the water is drained and oxygen moves through the soil. Wetland sites that have low redox have had long periods of flooding and soil saturation (Patrick *et al.*, 1996).

Redox data are also useful in understanding the morphology and genesis of the soil. The color of a soil and the degree of mottling (spots or blotches of different colors or shades of color interspersed with the dominant color; Glossary of Soil Science Terms, 1997) can reveal much about the soil's moisture status. Both color and mottling depend on the redox chemistry of Fe in the soil. When the soil is saturated for long times Fe oxides are reduced under low redox potentials, and the soil will exhibit a gray color. Soils that undergo alternate oxidation and reduction cycles are usually mottled (Patrick *et al.*, 1996).

Submerged Soils

Submerged soils are reduced and they have a low oxidation–reduction potential (Ponnamperuma, 1972). When an aerobic soil is submerged, the Eh decreases for the first few days and reaches a minimum; then it increases, reaches a maximum, and then decreases again to a value characteristic of the soil, usually after 8–12 weeks of submergence (Ponnamperuma, 1955, 1965). The magnitude and rate of the Eh decrease depend on the kind and amount of soil organic matter (SOM), the nature and content of e^- acceptors, temperature, and period of submergence. Native or added SOM enhances the first Eh minimum, while nitrate causes the minimum to disappear. Temperatures above and below 298 K slow the Eh decrease but the retardation varies with the soil. It is greatest in acid soils and not observable in neutral soils with high SOM.

Ponnamperuma and Castro (1964) and Ponnamperuma (1965) have listed the effects of soil properties on Eh changes in submerged soils as follows: soils high in nitrate, with more than 275 mg kg^{-1} NO_3^-, exhibit positive potentials for several weeks after submergence; soils low in SOM (<1.5%) or high in Mn (>0.2%) have positive potentials even 6 months after submergence; soils low in active Mn and Fe (sandy soils) with more than 3% SOM reach Eh values of -0.2 to -0.3 V within 2 weeks of submergence; and stable potentials of 0.2 to -0.3 V are attained after several weeks of submergence.

Redox Reactions Involving Inorganic and Organic Pollutants

As noted in Chapter 2, metal oxides/hydroxides, e.g., Fe(III) and Mn(III/IV), are quite common in soils and sediments as suspended particles and as coatings on clay mineral surfaces. Manganese(III/IV), Fe(III), Co(III), and Pb(IV) oxides/hydroxides are thermodynamically stable in oxygenated systems at neutral pH. However, under anoxic conditions, reductive dissolution of oxides/hydroxides by reducing agents occurs as shown below for MnOOH and MnO_2 (Stone, 1991):

$$Mn(III)OOH(s) + 3H^+ + e^- = Mn^{2+} + 2H_2O \quad E° = +1.50 \text{ V} \quad (8.15)$$

$$Mn(IV)O_2(s) + 4H^+ + 2e^- = Mn^{2+} + 2H_2O \quad E° = +1.23 \text{ V}. \quad (8.16)$$

Changes in the oxidation state of the metals associated with the oxides above can greatly affect their solubility and mobility in soil and aqueous environments. The reductants can be either inorganic or organic.

There are a number of natural and xenobiotic organic functional groups that are good reducers of oxides and hydroxides. These include carboxyl, carbonyl, phenolic, and alcoholic functional groups of SOM. Microorganisms in soils and sediments are also examples of organic reductants. Stone (1987a) showed that oxalate and pyruvate, two microbial metabolites, could reduce and dissolve Mn(III/IV) oxide particles. Inorganic reductants include As(III), Cr(III), and Pu(III).

Table 8.2 gives standard reduction potentials ($E°$) for oxide/hydroxide minerals at a metal concentration of 1.0 M and potentials ($E°'$) determined under more normal environmental conditions of pH 7 and a metal concentration of 1×10^{-6} M. Oxidant strength decreases in the order Mn(III,IV) oxides > Co(III) oxides > Fe(III) oxides. The Fe oxides are more difficult to reduce than Mn(III,IV) oxides. Thus Fe(II) is easier to oxidize than Mn(II). Reduction potentials for oxidation of some important organic half-reduction reactions are provided in Table 8.3.

If the potential of the oxidant half-reaction is higher than the reductant half-reaction potential, then the overall reaction is thermodynamically favored. Thus, comparing $E°'$ values in Tables 8.2 and 8.3, one observes that

TABLE 8.2. *Reduction Half-Reactions Along with Standard Reduction Potentials (E°) and Reduction Potentials Calculated under More Realistic Environmental Conditions (E°′) for Mineral Phases Containing Mn(IV), Mn(III), Fe(III), and Co(III)[a]*

	$E°$ [b] (V)	$E°′$ [c] (V)
Vernadite (Bricker, 1965)		
$1/2\ Mn(IV)\ O_2(s) + 2H^+ + e^- = 1/2Mn^{2+} + H_2O$	+1.29	+0.64
Manganite (Bricker, 1965)		
$Mn(III)\ OOH(s) + 3H^+ + e^- = Mn^{2+} + 2H_2O$	+1.50	+0.61
Goethite (Robie *et al.*, 1978)		
$Fe(III)\ OOH(s) + 3H^+ + e^- = Fe^{2+} + 2H_2O$	+0.67	−0.22
Hematite (Robie *et al.*, 1978)		
$1/2Fe(III)\ O_3(s) + 3H^+ + e^- = Fe^{2+} + 3/2H_2O$	+0.66	−0.23
Magnetite (Robie *et al.*, 1978)		
$1/2Fe_2(III)\ Fe(II)\ O_4(s) + 4H^+ + e^- = 3/2\ Fe^{2+} + 2H_2O$	+0.90	−0.23
Cobalt hydroxide oxide (crystalline) (Hem *et al.*, 1985)		
$Co(III)\ OOH(s) + 3H^+ + e^- = Co^{2+} + 2H_2O$	+1.48	+0.23

[a] From Stone (1991), with permission.
[b] $E°$ = standard reduction potential ([I] = 1.0 M).
[c] $E°′$ = Reduction potential under the following conditions: $[H^+] = 1.0 \times 10^{-7}$ M; $[M^{2+}] = 1.0 \times 10^{-6}$ M.

Mn(III/IV) oxides can oxidize hydroquinone, ascorbate, bisulfide, and oxalate under the specified conditions. Goethite and hematite could only oxidize oxalate; they are not strong enough oxidants to oxidize hydrogen sulfide, ascorbate, or hydroquinone. At pH 4 the Fe(III) oxides can oxidize all of the reductants listed in Table 8.3 (Stone, 1991). Manganese (III/IV) oxides can react with many organic compounds over a wide pH range while reactions with Fe(III) oxides are not thermodynamically stable at neutral and alkaline pH's and with organic compounds that are slightly reduced (Stone, 1991).

TABLE 8.3. *Reduction Half-Reactions Along with Standard Reduction Potentials (E°) and Reduction Potentials Calculated under More Realistic Environmental Conditions (E°′) for Several Organic Compounds[a,b]*

	$E°$ [c] (V)	$E°′$ [d] (V)
Hydroquinone		
p-Benzoquinone + $2H^+ + 2e^-$ = hydroquinone	0.699	0.196
Ascorbate		
Dehydroascorbate + $2H^+ + 2e^-$ = ascorbate	0.40	−0.103
Hydrogen sulfide		
$S°(s) + H^+ + e^-$ = HS^-	−0.062	−0.17
Oxalate		
$2HCO_3^- + 2H^+ + 2e^-$ = $^-OOC–COO^- + 2H_2O$	−0.18	−0.69

[a] From Stone (1991), with permission.
[b] Using thermodynamic data compiled in Latimer (1952) and Stone and Morgan (1984b).
[c] $E°$ = standard reduction potential ([I] = 1.0 M).
[d] $E°′$ = Reduction potential under the following conditions: $[H^+] = 1.0 \times 10^{-7}$ M; reductant concentration = 1.0×10^{-3} M; oxidant concentration = 1.0×10^{-6} M; C_T (total dissolved carbonate) = 1.0×10^{-3} M.

Mechanisms for Reductive Dissolution of Metal Oxides/Hydroxides

The reductive dissolution of metal oxides/hydroxides occurs in the following sequential steps (Stone, 1986, 1991): (1) diffusion of the reductant molecules to the oxide surface, (2) surface chemical reaction, and (3) release of reaction products and diffusion away from the oxide surface. Steps (1) and (3) are transport steps. The rate-controlling step in reductive dissolution of oxides appears to be surface chemical reaction control. Reductive dissolution can be described by both inner- and outer-sphere complex mechanisms that involve (A) precursor complex formation, (B) electron transfer, and (C) breakdown of the successor complex (Fig. 8.4). Inner- and outer-sphere precursor complex formations are adsorption reactions that increase the density of reductant molecules at the oxide surface, which promotes electron transfer (Stone, 1991). In the inner-sphere mechanism, the reductant enters the inner coordination sphere via ligand exchange and bonds directly to the metal center prior to electron transfer. With the outer-sphere complex, the inner coordination sphere is left intact and electron transfer is enhanced by an outer-sphere precursor complex (Stone, 1986). Kinetic studies have shown that high rates of reductive dissolution are favored by high rates of precursor complex formation, i.e., large k_1 and k_{-1} values, high electron transfer rates, i.e., large k_2, and high rates of product release, i.e., high k_3 (Fig. 8.4).

Specifically adsorbed cations and anions may reduce reductive dissolution rates by blocking oxide surface sites or by preventing release of Mn(II) into solution. Stone and Morgan (1984a) showed that PO_4^{3-} inhibited the reductive dissolution of Mn(III/IV) oxides by hydroquinone. Addition of 10^{-2} M PO_4^{3-} at pH 7.68 caused the dissolution rate to be only 25% of the rate when PO_4^{3-} was not present. Phosphate had a greater effect than Ca^{2+}.

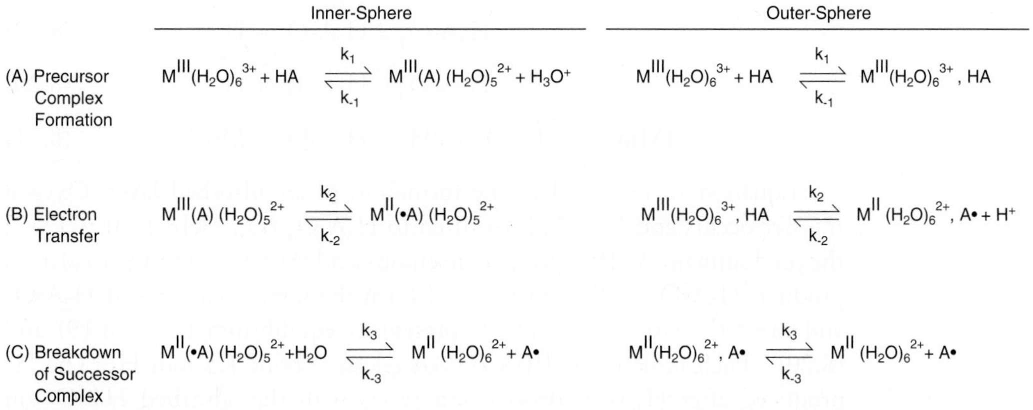

	Inner-Sphere	Outer-Sphere
(A) Precursor Complex Formation	$M^{III}(H_2O)_6^{3+} + HA \underset{k_{-1}}{\overset{k_1}{\rightleftharpoons}} M^{III}(A)(H_2O)_5^{2+} + H_3O^+$	$M^{III}(H_2O)_6^{3+} + HA \underset{k_{-1}}{\overset{k_1}{\rightleftharpoons}} M^{III}(H_2O)_6^{3+}, HA$
(B) Electron Transfer	$M^{III}(A)(H_2O)_5^{2+} \underset{k_{-2}}{\overset{k_2}{\rightleftharpoons}} M^{II}(\bullet A)(H_2O)_5^{2+}$	$M^{III}(H_2O)_6^{3+}, HA \underset{k_{-2}}{\overset{k_2}{\rightleftharpoons}} M^{II}(H_2O)_6^{2+}, A\bullet + H^+$
(C) Breakdown of Successor Complex	$M^{II}(\bullet A)(H_2O)_5^{2+} + H_2O \underset{k_{-3}}{\overset{k_3}{\rightleftharpoons}} M^{II}(H_2O)_6^{2+} + A\bullet$	$M^{II}(H_2O)_6^{2+}, A\bullet \underset{k_{-3}}{\overset{k_3}{\rightleftharpoons}} M^{II}(H_2O)_6^{2+} + A\bullet$

FIGURE 8.4. *Reduction of $M(H_2O)_6^{3+}$ by phenol (HA) in homogeneous solution. Reprinted from Stone (1986). Copyright 1986 American Chemical Society.*

Oxidation of Inorganic Pollutants

As mentioned earlier, Mn oxides can oxidize a number of environmentally important ions that can be toxic to humans and animals. Chromium and plutonium are similar in their chemical behavior in aqueous settings (Rai and Serne, 1977; Bartlett and James, 1979). They can exist in multiple oxidation states and as both cationic and anionic species. Chromium (III) is quite stable and innocuous, and occurs as Cr^{3+} and its hydrolysis products or as CrO_2^-. Chromium(III) can be oxidized to Cr(VI) by Mn(III/IV) oxides (Bartlett and James, 1979; Fendorf and Zasoski, 1992). Chromium(VI) is mobile in the soil environment and is a suspected carcinogen. It occurs as the dichromate, $Cr_2O_7^{2-}$, or chromate, $HCrO_4^-$ and CrO_4^{2-}, anions (Huang, 1991).

Figure 8.5 shows the oxidation kinetics of Cr(III) to Cr(VI) in a soil. Most of the oxidation occurred during the first hour. At higher temperatures, there was a rapid oxidation rate, followed by a slower rate. The decrease in the rate of Cr(III) oxidation has been ascribed to a number of factors including the formation of a surface precipitate that effectively inhibits further oxidation (Fendorf et al., 1992).

Plutonium can exist in the III to VI oxidation states as Pu^{3+}, Pu^{4+}, PuO_2^+, and PuO_2^{2+} in strongly acid solutions (Huang, 1991). Plutonium(VI), which can result from oxidation of Pu(III/IV) by Mn(III/IV) oxides (Amacher and Baker, 1982), is very toxic and mobile in soils and waters.

Arsenic (As) can exist in several oxidation states and forms in soils and waters. In waters, As can exist in the +5, +3, 0, and −3 oxidation states. Arsenite, As(III), and arsine (AsH_3, where the oxidation state of As is −3) are much more toxic to humans than arsenate, As(V). Manganese(III/IV) oxides can oxidize As(III) to As(V) as shown below where As(III) as $HAsO_2$ is added to MnO_2 to produce As(V) as H_3AsO_4 (Oscarson et al., 1983):

$$HAsO_2 + MnO_2 = (MnO_2) \cdot HAsO_2 \tag{8.17}$$

$$(MnO_2) \cdot HAsO_2 + H_2O = H_3AsO_4 + MnO \tag{8.18}$$

$$H_3AsO_4 = H_2AsO_4^- + H^+ \tag{8.19}$$

$$H_2AsO_4^- = HAsO_4^{2-} + H^+ \tag{8.20}$$

$$(MnO_2) \cdot HAsO_2 + 2H^+ = H_3AsO_4 + Mn^{2+}. \tag{8.21}$$

Equation (8.18) involves the formation of an adsorbed layer. Oxygen transfer occurs and $HAsO_2$ is oxidized to H_3AsO_4 (Eq. (8.18)). At pH ≤ 7, the predominant As(III) species is arsenious acid ($HAsO_2$), but the oxidation product, H_3AsO_4, will dissociate and form the same quantities of $H_2AsO_4^-$ and $HAsO_4^{2-}$ with little H_3AsO_4 present at equilibrium (Eqs. (8.19) and (8.20)). Each mole of As(III) oxidized releases about 1.5 mol H^+. The H^+ produced after H_3AsO_4 dissociation reacts with the adsorbed $HAsO_2$ on MnO_2, forming H_3AsO_4, and leads to the reduction and dissolution of Mn(IV) (Eq. (8.21)). Thus, every mole of As(III) oxidized to As(V) results in 1 mol of Mn(IV) in the solid phase being reduced to Mn(II) and partially dissolved in solution (Oscarson *et al.*, 1981).

Oscarson *et al.* (1980) studied the oxidation of As(III) to As(V) in sediments from five lakes in Saskatchewan, Canada. Oxidation of As(III) to As(V) occurred within 48 hr. In general, > 90% of the added As was sorbed on the sediments within 72 hr.

Manganese oxides and hydroxides, e.g., Mn_3O_4 and $MnOOH$, may also catalyze the oxidation of other trace metals such as Co^{2+}, Co^{3+}, Cu^{2+}, Ni^{2+}, Ni^{3+}, Pb^{2+}, and Se^{4+} by disproportionation to Mn^{2+} and MnO_2 (Hem, 1978; Scott and Morgan, 1996). The disproportionation results in vacancies in the Mn oxide structure. Since the Mn^{2+} and Mn^{3+} in the oxides have physical sizes similar to those of Co^{2+}, Co^{3+}, Cu^{2+}, Ni^{2+}, Ni^{3+}, and Pb^{2+}, these metals can occupy the vacancies in the Mn oxide and become part of the structure. With disproportionation or with other redox processes involving the Mn oxides the solubility of the metals can be affected. For example, if during the disproportionation process Co_3O_4, the oxidized form of the metal, forms from Co^{2+}, the reaction can be expressed as (Hem, 1978)

$$2Mn_3O_4(s) + 3Co^{2+} + 4H^+ \rightarrow$$

$$MnO_2(s) + Co_3O_4(s) + 5Mn^{2+} + 2H_2O, \qquad (8.22)$$

and the equilibrium constant ($K°$) is (Hem, 1978)

$$(Mn^{2+})^5/(Co^{2+})^3 (H^+)^4 = 10^{18.73}. \qquad (8.23)$$

Thus, the oxidation of Co(II) to Co(III) reduces its solubility and mobility in the environment. Using X-ray photoelectron spectroscopic analyses (Murray and Dillard, 1979), this reaction has been shown to occur.

Scott and Morgan (1996) studied the oxidation of Se(IV) by synthetic birnessite. Se(IV) was oxidized to Se(VI) with Se(VI) first appearing in the aqueous suspension after 12 hr and was produced at a constant rate over the duration of the experiment (28 days). The following oxidation mechanism was suggested: (1) birnessite directly oxidized Se(IV) through a surface complex mechanism; (2) the rate-limiting step in the production of Se(VI) was the electron transfer step involving a transfer of two electrons from the

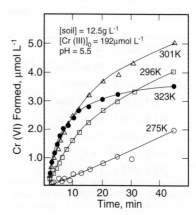

FIGURE 8.5. *Effect of temperature on the kinetics of Cr(III) oxidation in moist Hagerstown silt loam soil. From Amacher and Baker (1982).*

anion to the metal ion, breaking of two Mn–O bonds, and addition of an O from water to Se(VI); and (3) the reaction products Se(VI) and Mn(II) were released from the surface by different steps.

Scott and Morgan (1996) compared their results to those of Eary and Rai (1987), who studied Cr(III) oxidation by pyrolusite (β-MnO$_2$) between pH 3.0 and 4.7, and Scott and Morgan (1995), who studied As(III) oxidation by birnessite (δ-MnO$_2$) at pH values between 4.0 and 8.2 (Table 8.4). The Cr(III) redox transformation on pyrolusite was slowest, which was attributed to unfavorable adsorption on both a positively charged surface and aqueous species and the small thermodynamic driving force. Also, the transfer of three electrons from Cr(III) to Mn(IV) requires the involvement of more than one Mn(IV) atom per Cr(III) atom.

Manganese oxides also appear to play an important role in ligand-facilitated metal transport. Using soil columns that consisted of fractured saprolite coated with amorphous Fe$^-$ and Mn oxides, Jardine et al. (1993) studied the transport of Co(II) EDTA^{2-}, a mixture of Co(II) EDTA^{2-} and Co(III) EDTA$^-$ and Sr EDTA^{2-}. The Mn oxides oxidized Co(II) EDTA^{2-} into Co(III) EDTA$^-$, a very stable complex (log $K°$ value of 41.4, Xue and Traina, 1996). The formation of this complex resulted in enhanced transport of Co.

Xue and Traina (1996) found that an aerobic goethite suspension catalyzed oxidation of Co(II) EDTA^{2-} to Co(III) EDTA$^-$ by dissolved O$_2$. The kinetics were described using a pseudo-first-order rate constant, k_1, of 0.0078 ± 0.0002 h^{-1} at pH 5 and a goethite concentration of 3.09 g liter^{-1}.

Reductive Dissolution of Mn Oxides by Organic Pollutants

A number of investigators have studied the reductive dissolution of Mn oxides by organic pollutants such as hydroquinone (Stone and Morgan, 1984a), substituted phenols (Stone, 1987b), and other organic compounds (Stone and Morgan, 1984b). With substituted phenols the rate of dissolution

TABLE 8.4. *Inorganic Redox Reactions with Manganese Dioxides[a]*

System	Time to oxidize 50%	Driving force at pH 4 $\Delta e°$ (V)[b]	Source
δ-MnO$_2$:As(III) $\rightarrow$ As(V)			
pH 4, 298 K, 14 m^2 liter^{-1}	10 min	+0.529	Scott and Morgan (1995)
δ-MnO$_2$: Se(IV) $\rightarrow$ Se(VI)			
pH 4, 308 K, 14 m^2 liter^{-1}	10 days		
pH 4, 298 K, 28 m^2 liter^{-1}	16 days	+0.092	
pH 4, 298 K, 14 m^2 liter^{-1}	30 days		
β-MnO$_2$: Cr(III) $\rightarrow$ Cr(VI)			
pH 4, 298 K, 71 m^2 liter^{-1}	95 days	+0.011	Eary and Rai (1987)

[a] From Scott and Morgan (1996), with permission. Copyright 1996 American Chemical Society.
[b] The activity ratio for each oxidant/reductant pair is taken as unity.

was proportional to substituted phenol concentration and the rate increased as pH decreased (Stone, 1987b). Phenols containing alkyl, alkoxy, or other electron-donating substituents were more slowly degraded; *para*-nitrophenol reacted slowly with Mn(III/IV) oxides. The increased rate of reductive dissolution at lower pH may be due to more protonation reactions that enhance the creation of precursor complexes or increases in the protonation level of the surface precursor complexes that increase electron transfer rates (Stone, 1987b).

Reduction of Contaminants by Iron and Microbes

Iron(II)-containing oxides such as magnetite [$Fe^{2+} Fe_2^{3+}$]O_4 and ilmenite [$Fe^{2+} Ti$]O_3, and microbes (microbially mediated processes) can play a significant role in the reduction of inorganic and organic contaminants. Chemical reduction of $Cr_2O_7^{2-}$ [Cr(VI)] to Cr(III)] by a generic Fe^{2+} oxide is shown below (White and Peterson, 1998):

$$9[Fe^{2+}]_{oxide} + Cr_2O_7^{2-} + 14\ H^+ \rightarrow 6\ [Fe^{3+}]_{oxide} + 3\ Fe^{2+} + 2\ Cr^{3+} + 7\ H_2O.$$
(8.24)

Figure 8.6 shows changes that occur in the aqueous concentrations of Fe, Cr, and V species with time after reaction with ilmenite at pH 3. As time increases there is a linear increase in the transformation of the oxidized species to the reduced species. That is, Fe(III) is reduced to Fe(II), Cr(VI) is reduced to Cr(III), and V(V) is reduced to V(IV).

However, under field conditions and at longer times, the effectiveness of Fe(II)-containing oxides on reduction of contaminants is affected by the reductive capacity of the oxide minerals, the impact of surface passivation (surface layer that forms, which is different from the bulk oxide structure), and competition effects and poisoning by other aqueous species.

Several classes of organic pollutants can also undergo abiotic reduction in anoxic aqueous environments when ferrous iron and Fe oxides are present (Haderlein and Pecher, 1998). Figure 8.7a shows the reduction of 4-chloro nitrobenzene (4-Cl-NB) with time when both magnetite and Fe(II) in solution were present and when only magnetite or solution Fe(II) was included. Only when both magnetite and solution Fe(II) were present did rapid reduction in 4-Cl-NB occur (Fig. 8.7a). This suggests that continued reduction of the organic contaminant is dependent on the continual replacement of Fe(II) on the mineral surface that has been consumed by oxidation of the organic pollutant as shown in Fig. 8.7b. During the entire period of 4-Cl-NB reduction, the number of electrons transferred to the 4-Cl-NB was similar to the consumption of aqueous Fe(II).

While chemical reduction, for example Cr(VI) reduction by Fe(II) or S(-II), is a major pathway for reduction of contaminants in anaerobic environments, the presence of reductant pools of Fe(II) and S(-II) is dependent on microbial activity (Wielinga *et al.*, 2001). Fe(III) and SO_4^{2-} reduction occurs predominantly through dissimilatory reduction pathways

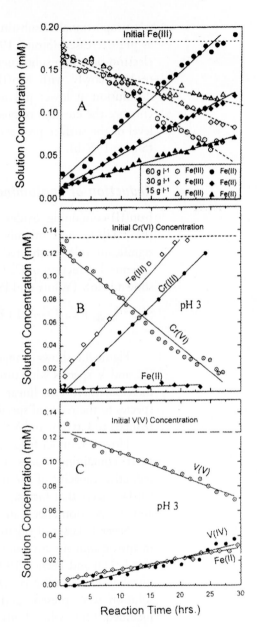

FIGURE 8.6. *Time trends showing the rates of ferric (A), chromate (B), and vanadate (C) reduction in ilmenite suspensions at pH 3 at 298 K. Rates of ferric reduction increase with increasing suspended ilmenite (g liter^{-1}). Solid lines are linear regression fits to data. Dashed lines are initial solution concentrations. Adapted from White and Peterson (1996), with permission. Copyright 1996 American Chemical Society.*

by dissimilatory metal-reducing bacteria (DMRB) (Lovely, 1991). DMRB can reduce solid phase Fe(III) oxides and oxyhydroxides such as ferrihydrite, goethite, hematite, and magnetite (Kostka and Nealson, 1995).

For example, one sees in the following two-step biotic–abiotic reaction the role of DMRB on the reduction of Fe(III) oxide and subsequent reduction of Cr(VI) (Wielinga *et al.*, 2001):

$$^3/_4\ C_3H_5O_3^- + 3\ Fe(OH)_3 \rightarrow$$
$$^3/_4\ C_2H_3O_2^- + 3\ Fe^{2+} + {}^3/_4\ HCO_3^- + 2\ H_2O + 5\ ^1/_4\ OH^- \qquad (8.25)$$

$$3\ Fe^{2+} + H\ CrO_4^- + 8\ H_2O \rightarrow 3\ Fe(OH)_3 + Cr(OH)_3 + 5\ H^+. \quad (8.26)$$

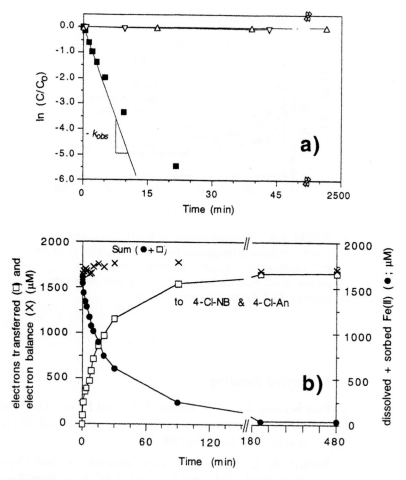

FIGURE 8.7. (a) Reduction of 50 μM 4-chloro nitrobenzene (4-Cl-NB) in the presence of 17 m^2 $liter^{-1}$ magnetite and an initial concentration of 2.3 mM Fe(II) at pH 7.0 and 298 K (■). The rate law deviates from pseudo-first-order behavior for longer observation times. 4-Cl-NB was not reduced significantly in suspensions of magnetite without dissolved Fe(II) (∇) or in solutions of Fe(II) without magnetite (Δ) (adapted from Klausen, 1995). (b) Electron balance (x) for the reduction of 4-Cl-NB repeatedly added at times t = 0, 6, 17 min to a suspension containing 11.2 m^2 $liter^{-1}$ magnetite and an initial concentration of dissolved Fe(II) of 1.6 mM at pH 7.75 and T = 298 K. Adapted from Klausen (1995), with permission. Copyright 1995 American Chemical Society.

Ferrous iron produced in Eq. (8.25) is cycled back to Fe(III) in Eq. (8.26), thus serving as an electron shuttle between the bacteria and Cr.

Figure 8.8 shows microbially induced, ferrous iron-mediated reduction of Cr(VI). *Shewanella alga strain* BrY was the model dissimilatory iron-reducing bacterium (DIRB). BrY is a facultative anaerobic bacterium that can couple the oxidation of organic acids and H_2 to the reduction of Fe(III), Mn(IV), and U(VI) under anaerobic conditions (Caccaro *et al.*, 1992). In Fig. 8.8 one sees the ability of BrY to reduce Cr(VI) via Fe(II) production during iron respiration. When Cr(VI) is introduced, the concentration of Fe(II)$_{(aq)}$ decreases dramatically. Without the bacterium, little Fe^{2+} is produced.

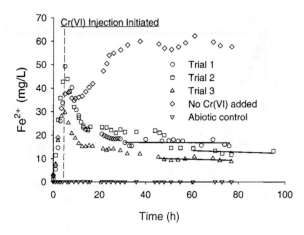

FIGURE 8.8. *Temporal changes in Fe(II)$_{(aq)}$ concentration detected in flow-reactor effluent for goethite. Conditions: 30 mM HEPES buffered medium, pH 7, 10 mM lactate added as electron donor. From Wielinga et al. (2001), with permission. Copyright 2001 American Chemical Society.*

Suggested Reading

Baas Becking, L. G. M., Kaplan, I. R., and Moore, D. (1960). Limits of the natural environment in terms of plant and oxidation–reduction potentials. *J. Geo. Phys.* **68**, 243–284.

Bartlett, R. J. (1986). Soil redox behavior. *In* "Soil Physical Chemistry" (D.L. Sparks, Ed.), pp. 179–207, CRC Press, Boca Raton, FL.

Bartlett, R. J., and James, B. R. (1993). Redox chemistry of soils. *Adv. Agron.* **50**, 151–208.

Haderlein, S. B., and Pecher, K. (1998). Pollutant reduction in heterogeneous Fe(II)–Fe(III) systems. *In* "Mineral–Water Interfacial Reactions" (D. L. Sparks and T. J. Grundl, Eds.), ACS Symp. Ser. 715, pp. 342–357. Am. Chem. Soc., Washington, DC.

Huang, P. M. 1991. Kinetics of redox reactions on manganese oxides and its impact on environmental quality. *In* "Rates of Soil Chemical Processes" (D. L. Sparks and D. L. Suarez, Eds.), SSSA Spec. Publ. 27, pp. 191–230. Soil Sci. Soc. Am., Madison, WI.

James, B. R., and Bartlett, R. J. (2000). Redox phenomena. *In* "Handbook of Soil Science" (M. E. Sumner, Ed.), pp. B169–B194, CRC Press, Boca Raton, FL.

Patrick, W. H., Jr., Gambrell, R. P., and Faulkner, S. P. (1996). Redox measurements of soils. *In* "Methods of Soil Analysis: Part 3—Chemical Methods" (D. L. Sparks, Ed.), Soil Sci. Soc. Am. Book Ser. 5, pp. 1255–1273. Soil Sci. Soc. Am., Madison, WI.

Ponnamperuma, F. N. (1972). The chemistry of submerged soils. *Adv. Agron.* **24**, 29–96.

Stone, A. T. (1986). Adsorption of organic reductants and subsequent electron transfer on metal oxide surfaces. *In* "Rates of Soil Chemical Processes" (D. L. Sparks and D. L. Suarez, eds.), SSSA Spec. Publ. 27, pp. 231–254. Soil Sci. Soc. Am., Madison, WI.

Stumm, W., and Morgan, J. J. (1981). "Aquatic Chemistry." Wiley, New York.

White, A. F., and Peterson, M. L. (1998). The reduction of aqueous metal species on the surfaces of Fe(II)-containing oxides: The role of surface passivation. *In* "Mineral–Water Interfacial Reactions" (D. L. Sparks and T. J. Grundl, Eds.), ACS Symp. Ser. 715, pp. 323—341. Am. Chem. Soc., Washington, DC.

9

The Chemistry of Soil Acidity

Introduction

Soil pH has often been called the master variable of soils and greatly affects numerous soil chemical reactions and processes. It is an important measurement in deciding how acid a soil is, and can be expressed as pH = –log (H^+). Soils that have a pH <7 are acid, those with a pH >7 are considered alkaline, and those with a pH of 7 are assumed to be neutral. Soil pH ranges can be classified as given in Table 9.1. The most important culprits of soil acidity in mineral–organic soils are H and Al, with Al being more important in soils except for those with very low pH values (<4).

Soil pH significantly affects the availability of plant nutrients and microorganisms (Fig. 9.1). At low pH one sees that Al, Fe, and Mn become more soluble and can be toxic to plants. As pH increases, their solubility decreases and precipitation occurs. Plants may suffer deficiencies as pH rises above neutrality.

One of the major problems for plants growing in acid soils is aluminum toxicity. Aluminum in the soil solution causes stunted roots and tops in susceptible plants. The degree of toxicity is dependent on the type of plant

267

and the Al species (Foy, 1984). For example, corn growth was reduced when the Al concentration in solution was >3.6 mg liter^{-1} and soybean growth was depressed at Al concentrations >1.8 mg liter^{-1} (Evans and Kamprath, 1970). While monomeric Al ($Al(H_2O)_6^{3+}$) is particularly toxic to plants, studies have also shown that polymeric Al species in aqueous solutions can be very toxic to plants such as soybeans and wheat (Parker et al., 1988, 1989). For example, Parker et al. (1989), using culture solutions, showed that the Al_{13} polymer (tridecamer species, which will be discussed later) was five to 10 times more rhizotoxic (reduced root growth) to wheat than Al^{3+}. Low pH may also increase the solubility of heavy metals that can also be harmful to plants.

Environmental Aspects of Acidification

Acidity can have a dramatic effect on the soil environment. Two examples of this are the effects of acid rain on soils and the presence of mine spoil and acid sulfate soils.

ACID RAIN

As noted in Chapter 1, acid vapors, primarily sulfuric (H_2SO_4) and nitric (HNO_3), form in the atmosphere as a result of the emission of sulfur dioxide (SO_2) and nitrogen oxides from natural and anthropogenic sources. The largest anthropogenic sources of these gases are from the burning of fossil fuels (source of sulfur gases) and the exhaust from motor vehicles (source of nitrogen oxides). These vapors condense to form aerosol particles and along with basic materials in the atmospheric water determine the pH of precipitation. The major cations in precipitation water are H^+, NH_4^+, Na^+, Ca^{2+}, Mg^{2+}, and K^+ while the major anions are SO_4^{2-}, NO_3^-, and Cl^- (Meszaros, 1992).

As noted in Chapter 1, there has been great concern about the increased acidity of rainfall or acid rain (deposition). An average value for the amount of H^+ produced per year from acid precipitation falling on industrialized areas is 1 kmol H^+ ha^{-1} year^{-1}, but depending on the proximity to the pollution source, it may vary from 0.1 to 6 kmol H^+ ha^{-1} year^{-1} (van Breemen, 1987). In the United States, 60–70% of the acidity in precipitation comes from

TABLE 9.1. Descriptive Terms and Proposed Buffering Mechanisms for Various Soil pH Ranges[a]

Descriptive terms[b]	pH range	Buffering mechanism[c]
Extremely acid	<4.5	Iron range (pH 2.4–3.8)
Very strongly acid	4.5–5.0	Aluminum/iron range (pH 3.0–4.8)
Strongly acid	5.1–5.5	Aluminum range (pH 3.0–4.8)
Moderately acid	5.6–6.0	Cation exchange (pH 4.2–5.0)
Slightly acid to neutral	6.1–7.3	Silicate buffers (all pH values typically >5)
Slightly alkaline	7.4–7.8	Carbonate (pH 6.5–8.3)

[a] From Robarge and Johnson (1992), with permission.
[b] Glossary of Soil Science Terms (1987).
[c] Schwertmann et al. (1987), Ulrich (1987), Reuss and Walthall (1989).

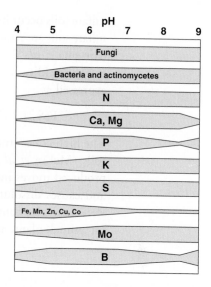

FIGURE 9.1. *Effect of pH on the availability of nutrients important in plant growth and of microorganisms. As the band for a particular nutrient or microbe widens, the availability of the nutrient or activity of the microbes is greater. For example, with K the greatest availability is from pH ~6–9. From Brady (1984), with permission from Pearson Education.*

H_2SO_4 and the remaining 30–40% is derived from HNO_3. While it is well documented that acid rain can deleteriously affect aquatic life by significantly lowering the pH of lakes and streams and can cause damage to buildings, monuments, and plants, such as some types of trees, its effects on agricultural soils appear minimal.

In most cases, the amount of soil acidification that occurs naturally or results from agronomic practices is significantly higher than that occurring from acid rain. For example, if one assumes annual fertilizer application rates of 50–200 kg N ha^{-1} to soils being cropped, soil acidification (due to the reaction $NH_4^+ + 2O_2 \rightarrow NO_3^- + H_2O + 2H^+$) from the fertilizer would be 4–16 times greater than acidification from acid rain in highly industrialized areas (Sumner, 1991). Thus, in most soils used in agriculture, acid rain does not appear to be a problem. This is particularly true for soils that are limed periodically and that have appreciable buffering capacities due to significant clay and organic matter contents. However, on poorly buffered soils, such as many sandy soils, acid rain could increase their acidity over time.

In forests and grasslands, acid rain can have a significant effect not only on the trees but on the chemistry of the soils. Liming of forests is seldom done and acid rain can cause leaching of nutrient cations such as Ca^{2+}, Mg^{2+}, and K^+ from the soil, resulting in low pH's and the solubilization of toxic metals such as Al^{3+} and Mn^{2+}. This can cause reduced soil biological activity such as ammonification (conversion of NH_4^+ to NO_3^-) and reduced fixation of atmospheric N_2 by leguminous plants such as soybeans and can also reduce nutrient cycling. Over time, the productivity of forests and grasslands is decreased due to fewer nutrients and higher levels of toxic metals.

MINE SPOIL AND ACID SULFATE SOILS

Mine spoil and acid sulfate soils have very low pH's due to the oxidation of pyrite. Mine spoil soils are common in surface-mined coal areas, and acid sulfate soils occur in marine flood plains in temperate and tropical areas. Acid sulfate soils are estimated to occupy an area of at least 24 million ha worldwide (Ritsema *et al.*, 2000). When they are drained and pyrite oxidation occurs, extreme acidity is produced. The complete oxidation of pyrite (FeS_2) can be expressed as

$$FeS_2 + 15/4O_2 + 7/2H_2O \rightarrow Fe(OH)_3 + 2H_2SO_4. \tag{9.1}$$

The high concentrations of sulfuric acid cause pH's as low as 2 in mine spoil soils (McFee *et al.*, 1981) and <4 in acid sulfate soils. The extreme acid produced moves into drainage and floodwaters, corrodes steel and concrete, and causes dissolution of clay minerals, releasing soluble Al. The drainage waters can also contain heavy metals and As, which can have profound effects on animal, plant, and human health (Ritsema *et al.*, 2000).

Historical Perspective of Soil Acidity

As previously noted in Chapter 1, one of the great debates in soil chemistry has been the cause of soil acidity. This debate went on for over five decades and there were heated arguments over whether the culprit in soil acidity was H or Al. The history of this debate was described in a lively manner by Thomas (1977). The discussion below is largely taken from the latter review.

As noted in the discussion in Chapter 1 on the history of soil chemistry, Edmund Ruffin was the first person to lime soils for the proper reason, to neutralize acidity, when he applied oyster shells to his soils. It was 70 years after Ruffin's work before research on soil acidity was initiated again. F. P. Veitch (1902) found that titration of soils that had been equilibrated with $Ca(OH)_2$ to a pink endpoint with phenolphthalein was a good test for predicting whether lime (e.g., $CaCO_3$) was needed to neutralize acidity that would be detrimental to crop growth (Thomas, 1977). Hopkins *et al.* (1903) developed a lime requirement test based on the titration of a soil equilibrated with 1 N NaCl. Veitch (1904) showed that a 1 N NaCl extract, while not replacing all the soil's acidity, was a good lime requirement test. A very important finding by Veitch (1904) not recognized as such at the time was that the acidity replaced by 1 N NaCl was $AlCl_3$, not HCl. After Veitch's work a number of soil chemists started to study soil acidity and to debate whether acidity was caused by Al or H. Bradfield (1923, 1925) titrated clays and observed that their pK_a values were similar to those found for weak acids. Kelley and Brown (1926) and Page (1926) hypothesized that "exchangeable Al" was dissolved by exchangeable H^+ during the extraction with salt. Paver and Marshall (1934) believed that the exchangeable H^+ dissolved the clay structure, releasing Al, which in turn became a counterion on the exchange complex. This was indeed an important discovery that was not definitively proved and accepted until the 1950s and early 1960s as we shall see.

Chernov (1947) had shown that electrodialyzed clays and naturally acid clays were primarily Al-saturated. Shortly thereafter, Coleman and Harward (1953) found that H resin-treated clays or clays leached rapidly with 1 M HCl had properties quite different from those of clays that were slowly leached, leached with dilute acid solutions, or electrodialyzed. They concluded, based on their studies, that hydrogen clays were strongly acid. Low (1955), employing potentiometric and conductometric titration analyses (these are discussed later in this chapter), proved that an electrodialyzed clay was Al-saturated. Coleman and Craig (1961) confirmed the earlier finding of Coleman and Harward (1953) that H clays are unstable and rapidly convert to Al clay, with temperature having a dramatic effect on the transformation rate. The research on H vs Al clays was very important in that it showed that Al is more important in soil acidity than H^+.

Also in the 1950s and 1960s there were some important discoveries made about the types of Al found in soils. Rich and Obenshain (1955) showed that in some Virginia soils, formed from mica schist, there was not only exchangeable Al^{3+}, but also nonexchangeable Al, with the latter blocking exchange sites and thus lowering the cation exchange capacity (CEC) of the soils. The nonexchangeable Al also kept vermiculite from collapsing (Rich, 1964; Rich and Black, 1964) and was referred to as interlayer hydroxy-Al.

Solution Chemistry of Aluminum

Monomeric Al Species

Aluminum in aqueous solution rapidly and reversibly hydrolyzes (hydrolysis is a chemical reaction whereby a substance is split or decomposed by water (Baes and Mesmer, 1976)) in dilute solutions (<0.001 m) with a low $\bar{n}$ value (<0.15), where $\bar{n}$ is the OH/Al molar ratio. The hydrolysis of Al forming monomeric (contains one metal ion or in this case, one Al^{3+}) species is shown below (Bertsch, 1989):

$$[Al(H_2O)_6]^{3+} + H_2O \rightleftharpoons [Al(OH)(H_2O)_5]^{2+} + H_3O^+ \quad (9.2)$$

$$[Al(OH)(H_2O)_5]^{2+} + H_2O \rightleftharpoons [Al(OH)_2 (H_2O)_4]^{1+} + H_3O^+ \quad (9.3)$$

$$[Al(OH)_2 (H_2O)_4]^{1+} + H_2O \rightleftharpoons [Al(OH)_3 (H_2O)_3]^0 + H_3O^+ \quad (9.4)$$

$$[Al(OH)_3 (H_2O)_3]^0 + H_2O \rightleftharpoons [Al(OH)_4 (H_2O)_2]^{1-} + H_3O^+ \quad (9.5)$$

The monomeric hydrolysis products shown above, i.e., $[Al(OH)(H_2O)_5]^{2+}$, $[Al(OH)_2 (H_2O)_4]^{1+}$, $[Al(OH)_3 (H_2O)_3]^0$, and $[Al(OH)_4 (H_2O)_2]^{1-}$, are produced as coordinated water is deprotonated. The formation quotients for the monomeric hydrolysis products are shown in Table 9.2. The formation quotients for the +2 and −1 products are best known. One of the important aspects of the reactions in Eqs. (9.2)–(9.5) is that H_3O^+ or H^+ is produced, resulting in a decrease in pH or increased acidity. The magnitude of the pH decrease depends on the Al concentration in the solution.

The form of monomeric Al in the soil solution depends on the pH. One can see the effect of pH on the solubilities of Al in water solutions in Fig. 9.2. At pH values below 4.7, Al^{3+} predominates. Between pH's of 4.7 and 6.5 the $Al(OH)_2^{1+}$ species predominates and from pH 6.5 to pH 8.0 $Al(OH)_3^0$ is the primary species. At a pH above 8.0, the aluminate species $Al(OH)_4^-$ predominates. From pH 4.7 to 7.5 the solubility of Al is low. This is the pH range where Al is precipitated and remains as $Al(OH)_3^0$. Below pH 4.7 and above 7.5 the concentration of Al in solution increases rapidly.

The structure of the free aqueous Al^{3+} ion is shown in Fig. 9.3. It is coordinated by 6 H_2O molecules in an octahedral coordination, $Al(H_2O)_6^{3+}$. Due to the high positive charge of the Al^{3+} ion, the water molecules form a tightly bound primary hydration shell (Nordstrom and May, 1996).

TABLE 9.2. *Monomeric Hydrolysis Products of Al at Infinite Dilution and 298 K[a]*

Species	Log Q_{ly}[c]
$Al(OH)^{2+}$ [b]	−4.95
$Al(OH)_2^+$	−10.01
$Al(OH)_3^0$	−16.8
$Al(OH)_4^-$	−22.87

[a] Reprinted from Bertsch and Parker (1996) with permission. Copyright 1996 CRC Press, Boca Raton, FL.
[b] Waters of hydration are omitted for simplicity.
[c] Q_{ly} is an equilibrium formation quotient of the hydrolysis of Al. The formation quotient for the reactions in Eqs. (9.2)–(9.5) can be expressed as the ratio of product concentrations to reactant concentrations. For example, Q_{ly} for Eq. (9.2) resulting in the hydrolysis product, $[Al(OH) (H_2O)_5]^{2+}$, would be $[Al(OH) (H_2O)_5{}^{2+}] [H_3O^+]/Al (H_2O)_6^{3+}]$ where brackets represent concentration.

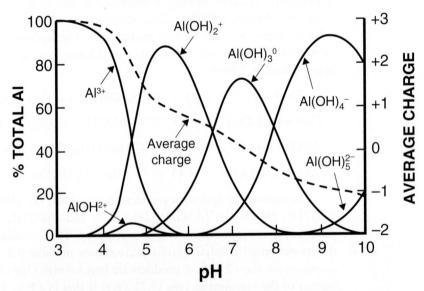

FIGURE 9.2. *Relationship between pH and the distribution and average charge of soluble aluminum species. From Marion et al. (1976), with permission.*

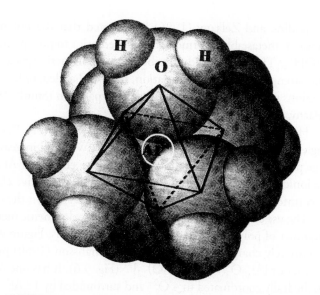

FIGURE 9.3.   *Structure of the free aqueous aluminum*
[Al(H$_2$O)$_6^{3+}$] ion. From Nordstrom and May (1996), with permission.
Copyright 1996 CRC Press, Boca Raton, FL.

It should be noted that free Al^{3+} may comprise a small fraction of the total soil solution Al. Much of the Al may be complexed with inorganic species such as F$^-$ and SO$_4^{2-}$ or with organic species such as humic substances and organic acids. For example, Wolt (1981) found that free Al^{3+} comprised 2–61% of the total Al in the soil solutions from acid soils where SO$_4^{2-}$ was a major complexing ligand. David and Driscoll (1984) found that 6–28% of the total Al in soil solutions occurred as free Al^{3+}. Most of the soil solution Al was complexed with organic species and with F$^-$.

Polymeric Al Species

In addition to monomeric Al species, polymeric Al species can also form by hydrolysis reactions in aqueous solutions. The presence of Al polymers in soil solutions has not been proven, and thermodynamic data necessary to calculate stability constants for Al polymeric species are also lacking. One of the reasons it has been difficult to determine the significance of Al polymers in the soil solution is that they are preferentially adsorbed on clay minerals and organic matter and are usually difficult to exchange.

A number of polymeric Al species have been proposed based on solution experiments in the laboratory. The hydrolysis of Al forming polymeric Al species can be represented as (Bertsch and Parker, 1996)

$$xAl^{3+} + yH_2O \rightleftharpoons Al_x(OH)_y^{(3x-y)^+} + yH^+ , \qquad (9.6)$$

where Al$_x$(OH)$_y^{(3x-y)^+}$ represents the polymeric Al species.

Jardine and Zelazny (1996) have noted that the polymeric species are transient, metastable intermediates formed prior to precipitation of crystalline $Al(OH)_3$. The nature and distribution of polymeric species depend on the ionic strength, total Al concentration, total OH added, pH, temperature, types of anions present, time, and method of preparation (Smith, 1971; Jardine and Zelazny, 1996).

Hsu (1977) proposed a polymerization scheme (Fig. 9.4) that consists of single or double gibbsite-like rings at $\bar{n} \leq 2.1$. With pH increases ($\bar{n} = 2.2–2.7$), large polymers that have a reduced net positive charge per Al atom form, with the ionic charge being constant until $\bar{n} = 3$. This positive charge is balanced by counteranions in solution or the negative charge on the clay minerals.

The use of ^{27}Al NMR spectroscopy and of colorimetric methods has verified a number of polymeric species in aqueous solutions. Figure 9.5 shows a series of positively charged OH–Al polymers. Johansson (1960) proposed the Al_{13} polymer or $[Al_{13}O_4(OH)_{24}(H_2O)_{12}]^{7+}$ (Fig. 9.6). It has one Al^{3+} at the center, tetrahedrally coordinated to 4 O^{2-} and surrounded by 12 Al^{3+}, each coordinated to 6 OH^-, H_2O, or the O^{2-} shared with the Al^{3+} at the center. Later studies using ^{27}Al NMR (Denney and Hsu, 1986) indicate that the Al_{13} polymer is present under only limited conditions, is transient, and does not represent all of the polymers that are present. Bertsch (1987) has shown that the quantity of Al_{13} polymers depends on the Al concentration, the OH/Al ratio, and the rate of base additions to the Al solutions. It is not known whether Al_{13} polymers form in soils, but alkalinization of the microenvironment at root apexes could cause their formation (Kinraide, 1991). However, Al_{13} polymers may not be stable in natural systems over long time periods (Bertsch and Parker, 1996).

In general, one can say that as polymerization increases, the number of Al atoms increases, the average charge/Al decreases, and the OH/Al ratio increases (Fig. 9.7).

Hsu (1989) has noted that some investigators have found that only polymeric or monomeric species appear in partially neutralized solutions. Whether monomeric or polymeric hydrolysis products result may depend on how the Al solutions were prepared (Hsu, 1989). If the Al solutions were prepared by dissolving an Al salt in water, monomeric species would predominate. Polymeric species would tend to predominate if the solutions were prepared by addition of base.

Exchangeable and Nonexchangeable Aluminum

Exchangeable Al in soils is primarily associated with the monomeric hexa-aqua ion, $[Al(H_2O)_6]^{3+}$. Exchangeable Al^{3+} is bound to the negatively charged surfaces of clay minerals and soil organic matter (SOM). It is readily displaced with a neutral, unbuffered salt such as 1 M KCl, $CaCl_2$, or $BaCl_2$. Unbuffered KCl is the most commonly used extractant. The extracting solution should be fairly concentrated to remove the Al^{3+} and at a low pH to maintain the Al in a soluble form.

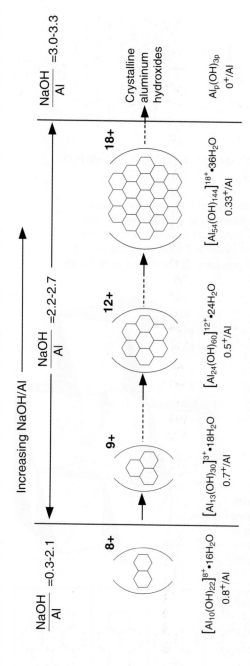

FIGURE 9.4. Polymerization scheme at several $\bar{n}$ (OH/Al molar ratio) values. From Hsu and Bates (1964), with permission of the Mineralogical Society.

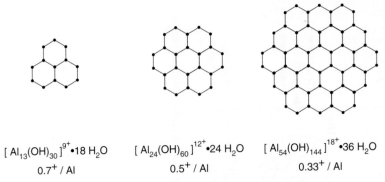

$[Al_{13}(OH)_{30}]^{9+}$•18 H_2O $[Al_{24}(OH)_{60}]^{12+}$•24 H_2O $[Al_{54}(OH)_{144}]^{18+}$•36 H_2O

0.7$^+$ / Al   0.5$^+$ / Al   0.33$^+$ / Al

FIGURE 9.5.   *Schematic representation of a series of positively charged OH–Al polymers of structures resembling fragments of gibbsite. Two OH$^-$ are shared between two adjacent Al^{3+} (black dots). Each edge Al^{3+} is coordinated by 4 OH$^-$ and 2 H$_2$O. OH$^-$ and H$_2$O are not shown in the sketch for the sake of clarity. From Hsu and Bates (1964), with permission of the Mineralogical Society.*

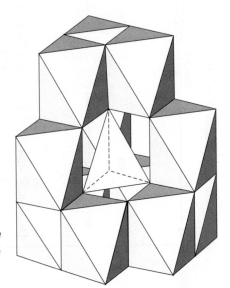

FIGURE 9.6.   *The $[Al_{13}O_4(OH)_{24}(H_2O)_{12}]^{7+}$ species. The drawing shows how the 12 AlO$_6$ octahedra are joined together by common edges. The tetrahedra of oxygen atoms in the center of the structure contain one 4-coordinate Al atom. From Johansson (1960), with permission.*

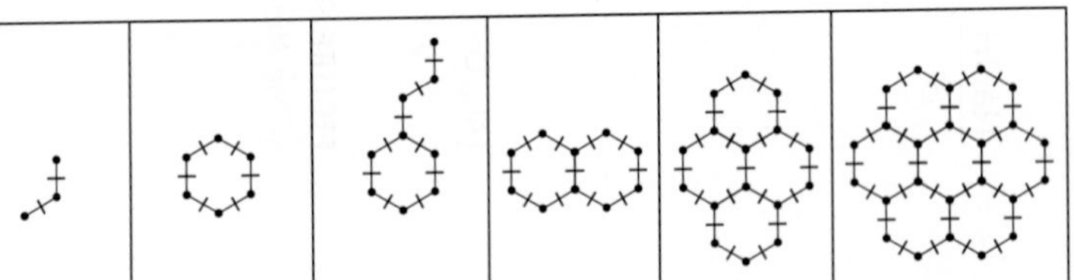

	Al^{3+}	$Al_2(OH)_2^{4+}$	$Al_3(OH)_4^{5+}$	$Al_6(OH)_{12}^{6+}$	$Al_9(OH)_{18}^{9+}$	$Al_{10}(OH)_{22}^{8+}$	$Al_{16}(OH)_{38}^{10+}$	$Al_{24}(OH)_{60}^{12+}$
Average charge / Al	3.0	2.0	1.67	1.0	1.0	0.80	0.63	0.50
OH / Al ratio	0	1.0	1.33	2.0	2.0	2.20	2.38	2.50

FIGURE 9.7.   *Summary of Al species with progressive polymerization, which can be deduced from the solid-state structure. The OH/Al ratio refers to the molar ratio in the complexes and not the OH/Al ratio of the system. From Stol et al. (1976), with permission.*

As pH increases, the amount of the monomeric Al^{3+} ion decreases and hydroxy-Al species, both as hydroxy-monomeric and hydroxy-polymeric species, predominate. These species are complicated and can be grouped collectively into the noncrystalline Al fraction (Bertsch and Bloom, 1996). This fraction consists of hydroxy-Al polymeric interlayers of smectite and vermiculites, organically complexed hydroxy-Al species, hydroxy-Al monomeric and -polymeric species of various sizes and basicities, and metastable forms present as Al and Fe oxyhydroxide coatings and as discrete amorphous to quasi-crystalline phases (Bertsch and Bloom, 1996). These species are considered nonexchangeable forms of Al since they are not readily replaced with an unbuffered salt like 1 M KCl. However, it has been shown that KCl can cause hydrolysis of nonexchangeable Al, which is then measured as exchangeable Al^{3+}. Nonexchangeable Al is titratable with a base such as NaOH and is a major source of variable charge in soils (Thomas and Hargrove, 1984).

Since the polymers are large and positive, they remain in the clay interlayer space and are nonexchangeable. In fact, the polymeric Al species are preferred over monomeric Al species in the interlayer. The polymeric Al species may exist as uniformly distributed islands (Fig. 9.8A) in the interlayer space (Grim and Johns, 1954) or be concentrated at the edges of the clay mineral surfaces (Chakravarti and Talibudeen, 1961; Dixon and Jackson, 1962) as "atolls" (Frink, 1965), as seen in Fig. 9.8B. If the polymers are at the edge, they can block exchange sites in the center of the interlayer space. Moreover, the clay mineral structure will collapse to 1 nm when K-saturated if the polymers are at the clay edges. However, if the polymers occur as islands in the interlayer space, they can serve as props to keep the interlayer space open and enhance the exchange of ions.

Soil Acidity

Forms of Soil Acidity

The main forms of acidity in mineral soils are associated with Al, which can be exchangeable (extractable is a preferred term since the technique, kind,

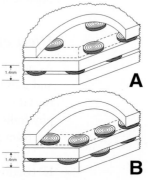

FIGURE 9.8. *Illustration of the distribution of hydroxy-aluminum polymers in the interlayer space of 2:1 clay minerals: (A) A uniform distribution and (B) an "atoll" arrangement. From Dixon and Jackson (1962), with permission.*

and concentration of extractant, and pH can all affect the removal; Jardine and Zelazny (1996)), nonexchangeable, or precipitated as an array of solid phases such as bayerite, gibbsite, or nordstrandite. Only in very acid soils with a pH <4 and in soils high in organic matter does one find major quantities of exchangeable H^+.

Exchangeable acidity is the amount of the total CEC due to H^+ and primarily Al^{3+}. As a proportion of the total acidity its quantity depends on the type of soil (e.g., type and quantity of soil components) and the percentage of the CEC composed of exchangeable bases such as Ca^{2+}, Mg^{2+}, K^+, and Na^+, or the percentage base saturation (Thomas and Hargrove, 1984). The ratio of exchangeable acidity to total acidity is greatest for montmorillonite, intermediate for dioctahedral vermiculites, and the least for kaolin clay minerals. This is related to the higher CEC of montmorillonite, which is predominantly composed of constant charge sites, as opposed to kaolinite, which has a low CEC and is variably charged. With dioctahedral vermiculites a significant portion of the total acidity is nonexchangeable—as hydroxy-Al polymeric material in the interlayer space that is pH-dependent. When Al and Fe oxides coat clay surfaces, the ratio of exchangeable acidity to total acidity is decreased since the oxide coatings reduce the quantity of exchangeable acidity that can be extracted (Coleman et al., 1964; Thomas and Hargrove, 1984). Aluminum oxides appear to be more effective in reducing the quantity of exchangeable acidity (Coleman et al., 1964) than Fe oxides.

Coleman et al. (1959) showed that soils high in montmorillonite had greater amounts of exchangeable acidity than soils high in kaolinite due to the higher CEC of the montmorillonitic soils and the predominance of constant charge. Evans and Kamprath (1970) found less exchangeable Al in organic soils than in mineral soils, even when the pH of the organic soil was lower. This can be ascribed to hydroxy-metal–organic complexes involving Al^{3+} and Fe^{3+}. Extraction of Al from these complexes is difficult (Mortensen, 1963; Schnitzer and Skinner, 1963; White and Thomas, 1981). Another reason that organic soils often have lower exchangeable Al content than mineral soils is that due to their lower pH, a large portion of the exchangeable acidity appears to be H^+ but there may be significant amounts of Al that are difficult to exchange from the organic matter. Moreover, much of the H^+ that appears to be exchangeable in organic soils may be due to hydrolysis of Al on the organic matter resulting in H^+.

Effect of Adsorbed Aluminum on Soil Chemical Properties

As noted earlier, nonexchangeable hydroxy-Al can reduce the CEC of clays and soils (Rich and Obenshain, 1955). Removal of the interlayer Al increases the CEC. The polymers may cause a lower CEC by occupying exchange sites, physically blocking the exchange reaction, and preventing the exchangeable ion from getting in contact with the exchange site.

Since the hydroxy-Al polymers can keep the clay interlayer space open, K^+ fixation can be reduced. The fact that the interlayer space is open may also enhance the exchange of ions like K^+ or NH_4^+ by ions of similar size, e.g., NH_4^+ or H_3O^+ exchanging for K^+, and K^+ and H_3O^+ exchanging for NH_4^+, from "wedge zones" that are partially collapsed regions in the interlayer spaces of weathered micas and vermiculites (Rich, 1964; Rich and Black, 1964). Larger hydrated cations such as Ca^{2+} and Mg^{2+} cannot fit into the wedge zones (Fig. 9.9). However, the position of the wedge zone in the interlayer space is important in affecting exchange and selectivity phenomena. If it is at the edge of the interlayer space, ion selectivity of K^+, for example, is less, but if it is deep within the interlayer space, the selectivity is greater (Rich, 1968). Monomeric Al^{3+} may also decrease fixation when KCl solutions are added since Al^{3+} is difficult to displace, particularly at low KCl concentrations.

Titration Analyses

One of the most useful ways to characterize soil acidity is to use potentiometric and conductometric titration analyses. With potentiometric titration analysis, one is investigating the relationship between pH and the amount of base added (Fig. 9.10). The first inflection point or break in the titration curve is ascribed to the end of the neutralization of exchangeable H^+ and the beginning of the titration of exchangeable Al^{3+}.

Another important tool for differentiating between H^+ and Al^{3+} is the use of conductometric titration curves. Their use in differentiating exchangeable Al^{3+} from nonexchangeable Al is limited. Figure 9.11 shows a conductometric titration curve for various acids where conductivity is plotted versus base added. With HCl, one notes a sharp drop in conductivity as base is added and then a sharp rise in conductivity after the acidity has been neutralized or after the equivalence point is reached. When a strong acid like HCl is titrated with a strong base the conductivity of the solution at first decreases, due to the replacement of H^+ ions that have a high mobility (36.30×10^{-8} m^2 sec^{-1} V^{-1}) by slower-moving cations, e.g., Na^+ (the mobility is 5.19×10^{-8} m^2 sec^{-1} V^{-1}). At the equivalence point, the decrease in conductance stops, and as further base is added the conductance increases since the OH^- ions, which have a high mobility (20.50×10^{-8} m^2 sec^{-1} V^{-1}), are no longer neutralized but remain free to carry electricity (Moore, 1972). One sees in Fig. 9.11 that with HCl the negative slope is sharp before the equivalence point is reached,

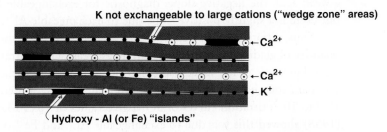

K not exchangeable to large cations ("wedge zone" areas)

← Ca^{2+}
← Ca^{2+}
← K^+

Hydroxy - Al (or Fe) "islands"

FIGURE 9.9. *Effect of hydroxy-Al polymers in an expansible clay mineral on cation fixation. From Rich (1968), with permission.*

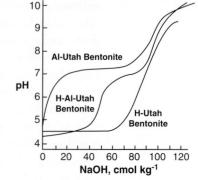

FIGURE 9.10.   *Potentiometric titration
curve of H- and Al-bentonite
(a smectitic clay). Reprinted with permission
from Coleman and Harward (1953).
Copyright 1953 American Chemical Society.*

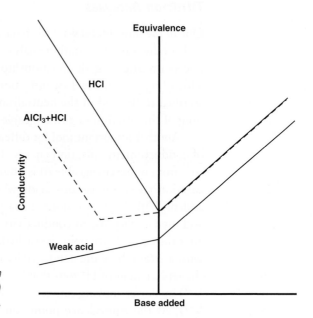

FIGURE 9.11.   *Conductimetric titration
curves of a strong (HCl) and a weak (AlCl₃)
acid and of a mixture of HCl and AlCl₃.*

indicating H^+, whereas with a combination of $AlCl_3$ and HCl there is a negative slope due to H^+ and a slightly positive slope due to exchangeable Al^{3+}. With a weak acid, the negative slope, diagnostic for exchangeable H^+, is missing, but there is a slightly positive slope due to exchangeable Al^{3+}.

Thus, titration curves can be used to measure both the type and the quantity of acidity. Soils and clays titrate as weak acids. Titration of SOM is a good example of the weak acid behavior. While carboxylic acids, which are the main source of acidity in SOM, give acid dissociation constants (K_a) of 10^{-4} to 10^{-5}, titration curves of SOM give values of 10^{-6}. Martin and Reeve (1958) showed this was due to exchangeable Al^{3+} and Fe^{3+} on the exchange sites. Removal of the Fe^{3+} and Al^{3+} from the organic matter resulted in a K_a of 10^{-4}. One can see the effect of Al on the titration of SOM in Fig. 9.12.

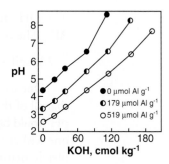

FIGURE 9.12. *Potentiometric titration curves of soil organic matter of varying Al content (from a muck soil). From Hargrove and Thomas (1982), with permission.*

With muck, Hargrove and Thomas (1982) used $Ca(OH)_2$ to differentiate between exchangeable H^+, Al^{3+}, and nonexchangeable Al (Fig. 9.13). The negative slope is strong acidity due to exchangeable H^+, the horizontal slope is weak acidity due to exchangeable Al^{3+}, and the third segment, which has a small positive slope, is due to very weak acidity. As Al was added, the amount of strong and weak acidity decreased, while the amount of very weak acidity increased or did not change. Aluminum must have reacted with the first two types of acidity and little reaction with the third type probably occurred. The total amount of acidity titrated decreased as the Al content increased. Thus, it appears that the Al that complexed with SOM is not exchangeable and apparently not titratable at pH values below 8 (Thomas and Hargrove, 1984). Another practical ramification of these results is that since the Al^{3+} blocks sites and limits dissociation of functional groups, the SOM in acid soils containing large quantities of SOM complexed with Al^{3+} may not contribute much to the soil's CEC, except at low pH.

Liming Soils

Soil acidity is neutralized by liming soils. Details on liming materials and liming practices can be found in Adams (1984). The general reaction that explains the interaction of a liming material such as $CaCO_3$ with water to form OH^- ions is (Thomas and Hargrove, 1984)

$$CaCO_3 + H_2O \rightarrow Ca^{2+} + HCO_3^- + OH^-. \tag{9.7}$$

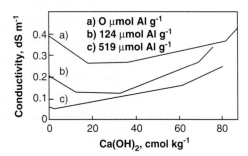

FIGURE 9.13. *Conductimetric titration curves of Michigan muck of various Al contents. From Hargrove and Thomas (1982), with permission.*

The OH^- reacts with indigenous H^+ or H^+ formed from the hydrolysis of Al^{3+}. The overall reaction of lime with an acid soil can be expressed as

$$2Al\text{-soil} + 3CaCO_3 + 3H_2O \rightarrow 3\ Ca\text{-soil} + 2Al(OH)_3 + 3CO_2. \qquad (9.8)$$

One sees that the products are exchangeable Ca^{2+} and $Al(OH)_3$. Assuming that all of the acidity was completely neutralized one would find that the soil pH would be 8.3 and the soil would be completely saturated. However, usually one wishes to lime to a lower pH so that only exchangeable acidity is completely neutralized. However, there is still titratable acidity (the total acidity in a soil at a particular pH such as 8.2) ascribable to weak forms of acidity such as hydroxy-Al and hydroxy-Fe species and functional groups of SOM and clay minerals that are not deprotonated.

Suggested Reading

Baes, C. F., Jr., and Mesmer, R. E. (1976). "The Hydrolysis of Cations." Wiley, New York.

Bertsch, P. M., and Bloom, P. R. (1996). Aluminum. In "Methods of Soil Analysis Part 3—Chemical Methods" (D. L. Sparks, Ed.), Soil Sci. Soc. Am. Book Ser. 5, pp 517–550. Soil Sci. Soc. Am., Madison, WI.

Bertsch, P. M., and Parker, D. R. (1996). Aqueous polynuclear aluminum species. In "The Environmental Chemistry of Aluminum" (G. Sposito, Ed.), 2nd ed., pp. 117–168. CRC Press, Boca Raton, FL.

Coleman, N. T., and Thomas, G. W. (1967). The basic chemistry of soil acidity. In "Soil Acidity and Liming" (R. W. Pearson and F. Adams, Eds.), Agron. Monogr. 12, pp. 1–41. Am. Soc. Agron., Madison, WI.

Hsu, P. H. (1989). Aluminum oxides and oxyhydroxides. In "Minerals in Soil Environments" (J. B. Dixon and S. B. Weed, Eds.), Soil Sci. Soc. Am. Book Ser. 1, pp. 331–378. Soil Sci. Soc. Am., Madison, WI.

Huang, P. M. (1988). Ionic factors affecting aluminum transformations and the impact on soil and environmental sciences. Adv. Soil Sci. 8, 1–78.

Jackson, M. L. (1963). Aluminum bonding in soils: A unifying principle in soil science. Soil Sci. Soc. Am. Proc. 27, 1–10.

Jardine, P. M., and Zelazny, L. W. (1996). Surface reactions of aqueous aluminum species. In "The Environmental Chemistry of Aluminum" (G. Sposito, Ed.), 2nd ed., pp. 221–270. CRC Press, Boca Raton, FL.

Jenny, H. (1961). Reflections on the soil acidity merry-go-round. Soil Sci. Soc. Am. Proc. 25, 428–432.

McLean, E. O. (1976). The chemistry of soil aluminum. Commun. Soil Sci. Plant Anal. 7, 619–636.

Sparks, D. L. (1984). Ion activities: An historical and theoretical overview. Soil Sci. Soc. Am. J. 48, 514–518.

Sposito, G., Ed. (1996). "The Environmental Chemistry of Aluminum," 2nd ed. CRC Press (Lewis), Boca Raton, FL.

Thomas, G. W. (1977). Historical developments in soil chemistry: Ion exchange. *Soil Sci. Soc. Am. J.* **41**, 230–237.

Thomas, G. W., and Hargrove, W. L. (1984). The chemistry of soil acidity. *In* "Soil Acidity and Liming" (F. Adams, Ed.), 2nd ed., Agron. Monogr. 12, pp. 3–56. Am. Soc. Agron., Madison, WI.

10 The Chemistry of Saline and Sodic Soils

Introduction

Oceans contain about 97.3% of the Earth's water, continents about 2.8%, and the atmosphere about 0.001% (Todd, 1970). About 77.2% of the water associated with continents occurs in ice caps and glaciers and about 22% is groundwater. The remaining 0.8% occurs as surface waters (lakes and rivers). The land surface of the Earth is 13.2×10^9 ha; of this area, 7×10^9 ha is arable and only 1.5×10^9 ha is cultivated (Massoud, 1981). Of the cultivated land, approximately 0.34×10^9 ha (23%) is saline and 0.56×10^9 ha (37%) is sodic, containing excessive levels of Na^+. Salinity can be defined as "the concentration of dissolved mineral salts present in waters and soils on a unit volume or weight basis" (Tanji, 1990b). Figure 10.1 and Table 10.1 show the global distribution of salt-affected soils.

Salt-affected soils can be classified as saline, sodic, and saline–sodic soils. Briefly, saline soils are plagued by high levels of soluble salts, sodic soils have high levels of exchangeable sodium, and saline–sodic soils have high contents of both soluble salts and exchangeable sodium. These soils will be described more completely later.

FIGURE 10.1.   *Global distribution of salt-affected soils. Reprinted with permission from Szabolcs, I. (1989). "Salt-Affected Soils." CRC Press, Boca Raton, FL.*

TABLE 10.1.   *Global Distribution of Salt-Affected Soils[a]*

	Area in millions of ha		
Continent	Saline	Sodic (alkali)	Total
North America	6.2	9.6	15.8
Central America	2.0	—	2.0
South America	69.4	59.6	129.0
Africa	53.5	27.0	80.5
South Asia	83.3	1.8	85.1
North and Central Asia	91.6	120.1	211.7
Southeast Asia	20.0	—	20.0
Australasia	17.4	340.0	357.4
Europe	7.8	22.9	30.7
Total	351.5	581.0	932.2

[a] From I. Szabolcs, "Review of Research on Salt-Affected Soils." Copyright © 1979 UNESCO, Paris. "Salt-Affected Soils." Copyright © 1989 CRC Press. Reprinted by permission of CRC Press.

   Salt-affected soils occur most often in arid and semiarid climates but they can also be found in areas where the climate and mobility of salts cause saline waters and soils for short periods of time (Tanji, 1990b). However, for the most part, in humid regions salt-affected soils are not a problem because rainfall is sufficient to leach excess salts out of the soil, into groundwater, and eventually into the ocean. Some salt-affected soils may occur along seacoasts or river delta regions where seawater has inundated the soil (Richards, 1954).

Causes of Soil Salinity

Soluble Salts

In arid and semiarid climates, there is not enough water to leach soluble salts from the soil. Consequently, the soluble salts accumulate, resulting in salt-affected soils. The major cations and anions of concern in saline soils and waters are Na^+, Ca^{2+}, Mg^{2+}, and K^+, and the primary anions are Cl^-, SO_4^{2-}, HCO_3^-, CO_3^{2-}, and NO_3^-. In hypersaline waters or brines, B, Sr, Li, SiO_2, Rb, F, Mo, Mn, Ba, and Al (since the pH is high Al would be in the $Al(OH)_4^-$ form) may also be present (Tanji, 1990b). Bicarbonate ions result from the reaction of carbon dioxide in water. The source of the carbon dioxide is either the atmosphere or respiration from plant roots or other soil organisms. Carbonate ions are normally found only at $pH \geq 9.5$. Boron results from weathering of boron-containing minerals such as tourmaline (Richards, 1954). When soluble salts accumulate, Na^+ often becomes the dominant counterion on the soil exchanger phase, causing the soil to become dispersed. This results in a number of physical problems such as poor drainage. The predominance of Na^+ on the exchanger phase may occur due to Ca^{2+} and Mg^{2+} precipitating as $CaSO_4$, $CaCO_3$, and $CaMg(CO_3)_2$. Sodium then replaces exchangeable Ca^{2+} and Mg^{2+} on the exchanger phase.

Evapotranspiration

An additional factor in causing salt-affected soils is the high potential evapotranspiration in these areas, which increases the concentration of salts in both soils and surface waters. It has been estimated that evaporation losses can range from 50 to 90% in arid regions, resulting in 2- to 20-fold increases in soluble salts (Cope, 1958; Yaalon, 1963).

Drainage

Poor drainage can also cause salinity and may be due to a high water table or to low soil permeability caused by sodicity (high sodium content) of water. Soil permeability is "the ease with which gases, liquids or plant roots penetrate or pass through a bulk mass of soil or a layer of soil" (Glossary of Soil Science Terms, 1997). As a result of the poor drainage, salt lakes can form like those in the western United States. Irrigation of nonsaline soils with saline water can also cause salinity problems. These soils may be level, well drained, and located near a stream. However, after they are irrigated with saline water drainage may become poor and the water table may rise.

Irrigation Water Quality

An important factor affecting soil salinity is the quality of irrigation water. If the irrigation water contains high levels of soluble salts, Na, B, and trace elements, serious effects on plants and animals can result (Ayers and Westcot, 1976).

Salinity problems are common in irrigated lands, with approximately one-third of the irrigated land in the United States seriously salt-affected (Rhoades, 1993). In some countries it may be as high as 50% (Postel, 1989). Areas affected include humid climate areas such as Holland, Sweden, Hungary, and Russia, and arid and semiarid regions such as the southwestern United States, Australia, India, and the Middle East. About 100,000 acres of irrigated land each year are no longer productive because of salinity (Yaron, 1981).

One of the major problems in these irrigated areas is that the irrigation waters contain dissolved salts, and when the soils are irrigated the salts accumulate unless they are leached out. Saline irrigation water, low soil permeability, inadequate drainage, low rainfall, and poor irrigation management all cause salts to accumulate in soils, which deleteriously affects crop growth and yields. The salts must be leached out for crop production. However, it is the leaching out of these salts, resulting in saline drainage waters, that causes pollution of waters, a major concern in saline environments.

The presence of selenium and other toxic elements (Cr, Hg) in subsurface drainage waters is also a problem in irrigated areas. Selenium (resulting from shale parent material) in drainage waters has caused massive death and deformity to fish and waterfowl in the Kesterson Reservoir of California.

Sources of Soluble Salts

The major sources of soluble salts in soils are weathering of primary minerals and native rocks, residual fossil salts, atmospheric deposition, saline irrigation and drainage waters, saline groundwater, seawater intrusion, additions of inorganic and organic fertilizers, sludges and sewage effluents, brines from natural salt deposits, and brines from oil and gas fields and mining (Jurinak and Suarez, 1990; Tanji, 1990b).

As primary minerals in soils and exposed rocks weather the processes of hydrolysis, hydration, oxidation, and carbonation occur and soluble salts are released. The primary source of soluble salts is fossil salts derived from prior salt deposits or from entrapped solutions found in earlier marine sediments.

Salts from atmospheric deposition, both as dry and wet deposition, can range from 100 to 200 kg $year^{-1}$ ha^{-1} along seacoasts and from 10 to 20 kg $year^{-1}$ ha^{-1} in interior areas of low rainfall. The composition of the salt varies with distance from the source. At the coast it is primarily NaCl. The salts become higher in Ca^{2+} and Mg^{2+} farther inland (Bresler *et al.*, 1982).

Important Salinity and Sodicity Parameters

The parameters determined to characterize salt-affected soils depend primarily on the concentrations of salts in the soil solution and the amount of exchangeable Na^+ on the soil. Exchangeable Na^+ is determined by exchanging the Na^+

from the soil with another ion such as Ca^{2+} and then measuring the Na^+ in solution by flame photometry or spectrometry (e.g., atomic absorption or inductively coupled plasma emission spectrometries). The concentration of salts in the solution phase can be characterized by several indices (Table 10.2) and can be measured by evaporation, or using electroconductometric or spectrometric techniques.

Total Dissolved Solids (TDS)

Total dissolved solids (TDS) can be measured by evaporating a known volume of water from the solid material to dryness and weighing the residue. However, this measurement is variable since in a particular sample various salts exist in varying hydration states, depending on the amount of drying. Thus, if different conditions are employed, different values for TDS will result (Bresler et al., 1982).

TDS is a useful parameter for measuring the osmotic potential, $-\tau_o$, an index of the salt tolerance of crops. For irrigation waters in the range of 5–1000 mg liter^{-1} TDS, the relationship between osmotic potential and TDS is (Bresler et al., 1982)

$$-\tau_o \approx -5.6 \times 10^{-4} \times \text{TDS (mg liter}^{-1}). \tag{10.1}$$

Without the minus sign for osmotic potential in Eq. (10.1), one could also use the same equation to determine osmotic pressure (τ_o) values. Further details on osmotic potential and osmotic pressure, as they affect plant growth, will be discussed later in this chapter.

The TDS (in mg liter^{-1}) can also be estimated by measuring an extremely important salinity index, electrical conductivity (EC), which is discussed below, to determine the effects of salts on plant growth. The TDS may be estimated by multiplying EC (dS m^{-1}) by 640 (for EC between 0.1 and 5.0 dS m^{-1}) for lesser saline soils and a factor of 800 (for EC > 5.0 dS m^{-1}) for hypersaline samples. The 640 and 800 are factors based on large data sets relating EC to TDS. To obtain the total concentration of soluble cations (TSC) or total concentration of soluble anions (TSA), EC (dS m^{-1}) is usually multiplied by a factor of 0.1 for mol liter^{-1} and a factor of 10 for mmol liter^{-1} (Tanji, 1990b).

TABLE 10.2. Salinity Parameters

Salinity index	Units of measurement
Total dissolved solids (TDS) or total soluble salt concentration (TSS)	mg liter^{-1}
Total concentration of soluble cations (TSC)	mol$_c$ liter^{-1}
Total concentration of soluble anions (TSA)	mol$_c$ liter^{-1}
Electrical conductivity (EC)	dS m^{-1} = mmhos cm^{-1} (higher saline soils); dS m^{-1} × 10^{-3} or μS cm^{-1} = μmhos cm^{-1} (lower saline soils)

Electrical Conductivity (EC)

The preferred index to assess soil salinity is electrical conductivity. Electrical conductivity measurements are reliable, inexpensive to do, and quick. Thus, EC is routinely measured in many soil testing laboratories. The EC is based on the concept that the electrical current carried by a salt solution under standard conditions increases as the salt concentration of the solution increases. A sample solution is placed between two electrodes of known geometry; an electrical potential is applied across the electrodes, and the resistance (R) of the solution between the electrodes is measured in ohms (Bresler *et al.*, 1982). The resistance of a conducting material (e.g., a salt solution) is inversely proportional to the cross-sectional area (A) and directly proportional to the length (L) of the conductivity cell that holds the sample and the electrodes. Specific resistance (R_s) is the resistance of a cube of a sample volume 1 cm on edge. Since most commercial conductivity cells are not this large, only a portion of R_s is measured. This fraction is the cell constant ($K = R/R_s$). The reciprocal of resistance is conductance (C). It is expressed in reciprocal ohms or mhos. When the cell constant is included, the conductance is converted, at the temperature of the measurement, to specific conductance or the reciprocal of the specific resistance (Rhoades, 1993). The specific conductance is the EC (Rhoades, 1993), expressed as

$$EC = 1/R_s = K/R. \tag{10.2}$$

Electrical conductivity is expressed in micromhos per centimeter (μmho cm^{-1}) or in millimhos per centimeter (mmho cm^{-1}). In SI units the reciprocal of the ohm is the siemen (S) and EC is given as S m^{-1} or as decisiemens per meter (dS m^{-1}). One dS m^{-1} is one mmho cm^{-1}. The EC at 298 K can be measured using the equation

$$EC_{298} = EC_t f_t, \tag{10.3}$$

where f_t is a temperature coefficient that can be determined from the relation $f_t = 1 + 0.019 \, (t\text{-}298 \text{ K})$ and t is the temperature at which the experimental measurement is made in degrees Kelvin (Richards, 1954).

A number of EC values can be expressed according to the method employed: EC_e, the EC of the extract of a saturated paste of a soil sample; EC_p, the EC of the soil paste itself; EC_w, the EC of a soil solution or water sample; and EC_a, the EC of the bulk field soil (Rhoades, 1990).

The electrical conductivity of the extract of a saturated paste of a soil sample (EC_e) is a very common way to measure soil salinity. In this method, a saturated soil paste is prepared by adding distilled water to a 200- to 400-g sample of air-dry soil and stirring. The mixture should then stand for several hours so that the water and soil react and the readily soluble salts dissolve. This is necessary so that a uniformly saturated and equilibrated soil paste results. The soil paste should shine as it reflects light, flow some when the beaker is tipped, slide easily off a spatula, and easily consolidate when the container is tapped after a trench is formed in the paste with the spatula. The extract of the saturation paste can be obtained by suction using a Büchner funnel and filter paper. The EC and temperature of the extract are measured using conductance meters/cells and thermometers and EC_{298} is calculated using Eq. (10.3).

The EC_w values for many waters used in irrigation in the western United States are in the range 0.15–1.50 dS m^{-1}. Soil solutions and drainage waters normally have higher EC_w values (Richards, 1954). The EC_w of irrigation water < 0.7 dS m^{-1} is not a problem, but an EC_w > 3 dS m^{-1} can affect the growth of many crops (Ayers and Westcot, 1976).

It is often desirable to estimate EC based on soil solution data. Marion and Babcock (1976) developed a relationship between EC_w (dS m^{-1}) to total soluble salt concentration (TSS in mmol liter^{-1}) and ionic concentration (C in mmol liter^{-1}), where C is corrected for ion pairs. If there is no ion complexation, TSS = C (Jurinak and Suarez, 1990). The equations of Marion and Babcock (1976) are

$$\log C = 0.955 + 1.039 \log EC_w \qquad (10.4)$$

$$\log TSS = 0.990 + 1.055 \log EC_w. \qquad (10.5)$$

These work well to 15 dS m^{-1}, which covers the range of EC_e and EC_w for slightly to moderately saline soils (Bresler et al., 1982).

Griffin and Jurinak (1973) also developed an empirical relationship between EC_w and ionic strength (I) at 298 K that corrects for ion pairs and complexes

$$I = 0.0127 \, EC_w \qquad (10.6)$$

where EC_w is in dS m^{-1} at 298 K. Figure 10.2 shows the straight line relationship between I and EC_w predicted by Eq. (10.6), as compared to actual values for river waters and soil extracts.

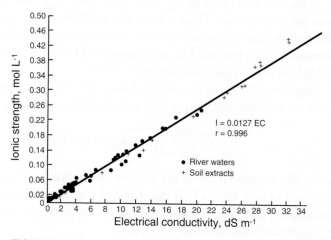

FIGURE 10.2. Relationship between ionic strength and electrical conductivity of natural aqueous solutions. •, River waters; +, soil extracts. From Griffin and Jurinak (1973), with permission.

In addition to measuring EC and other salinity indices in the laboratory, it is often important in the management of salt-affected soils, particularly those that are irrigated, to measure, monitor, and map soil salinity of large soil areas (Rhoades, 1993). This would assist in ascertaining the degree of salinity, in determining areas of under- and overirrigation, and in predicting trends in salinity. There are a number of rapid instrumental techniques for determining EC and computer-based mapping techniques that allow one to measure soil salinity over large areas. The use of geographic information systems (GIS) and remote sensing techniques will also augment these techniques.

There are three types of soil conductivity sensors that can measure bulk soil electrical conductivity (EC_a): a four-electrode sensor, an electromagnetic induction sensor, and a sensor based on time domain reflectometry technology. These are comprehensively discussed in Rhoades (1993).

Parameters for Measuring the Sodic Hazard

There are several important parameters commonly used to assess the status of Na^+ in the solution and on the exchanger phases. These are the sodium adsorption ratio (SAR), the exchangeable sodium ratio (ESR), and the exchangeable sodium percentage (ESP). The SAR is commonly measured using the equation

$$SAR = [Na^+]/[Ca^{2+} + Mg^{2+}]^{1/2}, \tag{10.7}$$

where brackets indicate the total concentration of the ions expressed in mmol liter^{-1} in the solution phase.

Total concentrations, not activities, are used in Eq. (10.7), and thus the SAR expression does not consider decreases in free ion concentrations and activities due to ion pair or complex formation (Sposito and Mattigod, 1977), which can be significant with Ca^{2+} and Mg^{2+}.

One also notes that in Eq. (10.7) Ca^{2+} and Mg^{2+} are treated as if they were the same species. There is not a theoretical basis for this other than the observation that ion valence is more important in predicting ion exchange phenomena than ion size. The concentration of Ca^{2+} is much higher than that of Mg^{2+} in many waters (Bresler et al., 1982).

Equation (10.7) can be simplified since Na^+, Ca^{2+}, and Mg^{2+} are the most common exchangeable ions in arid soils (Jurinak and Suarez, 1990) to

$$\frac{[Na\text{-soil}]}{CEC - [Na\text{-soil}]} = k'_G \, SAR = ESR, \tag{10.8}$$

where the concentration of the ions on the exchanger phase and CEC are expressed in mol$_c$ kg^{-1}, k'_G, is the modified Gapon selectivity coefficient (see Chapter 6), and ESR is the exchangeable Na^+ ratio (Richards, 1954). The k'_G, expressed in (mmol liter^{-1})$^{-1/2}$, is

$$\frac{[Na\text{-soil}][Ca^{2+} + Mg^{2+}]^{1/2}}{[Ca\text{-soil} + Mg\text{-soil}][Na^+]}, \tag{10.9}$$

where the concentrations of Ca^{2+} and Mg^{2+} on the exchanger phase are expressed in $cmol_c\ kg^{-1}$.

The U.S. Salinity Lab (Richards, 1954) reported a linear regression equation between ESR and SAR as ESR = −0.0126 + 0.014645 SAR with a correlation coefficient for 59 soils from the western United States of 0.923 (Fig. 10.3). Bower and Hatcher (1964) improved the relationship by adding ranges in the saturation extract salt concentration. The value of k'_G can be determined from the slope of the ESR–SAR linear relationship (Richards, 1954). The k'_G describes Na–Ca exchange well over the range of 0–40% exchangeable sodium percentage (ESP) where ESP = [Na-soil] × 100/CEC and has an average value of 0.015 for many irrigated soils from the western United States (Richards, 1954).

In terms of the ESP, Eq. (10.8) is (Richards, 1954; Jurinak and Suarez, 1990)

$$ESP/100 - ESP = k'_G\ SAR = ESR. \qquad (10.10)$$

Soils with an ESP >30 are very impermeable, which seriously affects plant growth. For many soils the numerical values of the ESP of the soil and the SAR of the soil solution are approximately equal up to ESP levels of 25 to 30.

While the ESP is used as a criterion for classification of sodic soils with an ESP of <15, indicating a nonsodic soil, and an ESP >15, indicating a sodic soil, the accuracy of the number is often a problem due to errors that may arise in measurement of CEC and exchangeable Na^+. Therefore, the more easily obtained SAR of the saturation extract should be used to diagnose the sodic hazard of soils. Although ESP and SAR are not precisely equal numerically, an SAR of 15 has also been used as the dividing line between sodic and nonsodic soils. However, the quantity and type of clay present in the soil are considerations in assessing how SAR and ESP values affect soil sodicity. For

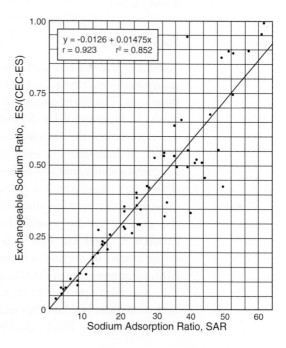

FIGURE 10.3. *Exchangeable sodium ratio (ESR) as related to the sodium adsorption ratio (SAR) of the saturation extract. ES, exchangeable sodium; CEC, cation exchange capacity. From Richards (1954).*

example, a higher SAR value may be of less concern if the soil has a low clay content or contains low quantities of smectite.

Classification and Reclamation of Saline and Sodic Soils

Saline Soils

Saline soils have traditionally been classified as those in which the EC_e of the saturation extract is >4 dS m^{-1} and ESP <15%. Some scientists have recommended that the EC_e limit for saline soils be lowered to 2 dS m^{-1} as many crops, particularly fruits and ornamentals, can be harmed by salinity in the range of 2–4 dS m^{-1}.

The major problem with saline soils is the presence of soluble salts, primarily Cl$^-$, SO$_4^{2-}$, and sometimes NO$_3^-$. Salts of low solubility, such as CaSO$_4$ and CaCO$_3$, may also be present. Because exchangeable Na$^+$ is not a problem, saline soils are usually flocculated and water permeability is good (Richards, 1954).

Saline soils can be reclaimed by leaching them with good-quality (low electrolyte concentration) water. The water causes dissolution of the salts and their removal from the root zone. For successful reclamation, salinity should be reduced in the top 45 to 60 cm of the soil to below the threshold values for the particular crop being grown (Keren and Miyamoto, 1990). Reclamation can be hampered by several factors (Bresler et al., 1982): restricted drainage caused by a high water table, low soil hydraulic conductivity due to restrictive soil layers, lack of good-quality water, and the high cost of good-quality water.

Sodic Soils

Sodic soils have an ESP >15, the EC_e is <4 dS m^{-1}, and the lower limit of the saturation extract SAR is 13. Consequently, Na$^+$ is the major problem in these soils. The high amount of Na$^+$ in these soils, along with the low EC_e, results in dispersion. Clay dispersion occurs when the electrolyte concentration decreases below the flocculation value of the clay (Keren and Miyamoto, 1990). Sodium-affected soils, which contain low levels of salt, have weak structural stability, and low hydraulic conductivities (HC) and infiltration rates (IR). These poor physical properties result in decreased crop productivity caused by poor aeration and reduced water supply. Low infiltration rates can also cause severe soil erosion (Sumner et al., 1998). Sodic soils have a pH between 8.5 and 10. The high pH is due to hydrolysis of Na$_2$CO$_3$. The major anions in the soil solution of sodic soils are Cl$^-$, SO$_4^-$, and HCO$_3^-$, with lesser amounts of CO$_3^{2-}$. Since the pH is high and CO$_3^{2-}$ is present, Ca^{2+} and Mg^{2+} are precipitated, and therefore soil solution Ca^{2+} and Mg^{2+} are low. Besides Na$^+$, another exchangeable and soluble cation that may occur in these soils is K$^+$ (Richards, 1954).

Historically, sodic soils were often called black alkali soils, which refers to the dispersion and dissolution of humic substances, resulting in a dark color. Sodic soils may be coarser-textured on the surface and have higher clay contents in the subsurface horizon due to leaching of clay material that is Na^+-saturated. Consequently, the subsoil is dispersed, permeability is low, and a prismatic soil structure may result.

In sodic soils reclamation is effected by applying gypsum $(CaSO_4 \cdot 2H_2O)$ or $CaCl_2$ to remove the exchangeable Na^+. The Ca^{2+} exchanges with the Na^+, which is then leached out as a soluble salt, Na_2SO_4 or $NaCl$. The $CaSO_4$ and $CaCl_2$ also increase permeability by increasing electrolyte concentration. Sulfur can also be applied to correct a sodium problem in calcareous soils (where $CaCO_3$ is present). Sulfuric acid can also be used to correct sodium problems in calcareous soils.

Saline–Sodic Soils

Saline–sodic soils have an EC_e >4 dS m^{-1} and an ESP >15. Thus, both soluble salts and exchangeable Na^+ are high in these soils. Since electrolyte concentration is high, the soil pH is usually <8.5 and the soil is flocculated. However, if the soluble salts are leached out, usually Na^+ becomes an even greater problem and the soil pH rises to >8.5 and the soil can become dispersed (Richards, 1954).

In saline–sodic soils reclamation involves the addition of good-quality water to remove excess soluble salts and the use of a Ca^{2+} source $(CaSO_4 \cdot 2H_2O$ or $CaCl_2)$ to exchange Na^+ from the soil as a soluble salt, Na_2SO_4. In saline–sodic soils a saltwater-dilution method is usually effective in reclamation. In this method the soil is rapidly leached with water that has a high electrolyte concentration with large quantities of Ca^{2+} and Mg^{2+}. After leaching, and the removal of Na^+ from the exchanger phase of the soil, the soil is leached with water of lower electrolyte concentration to remove the excess salts.

In both saline–sodic and sodic soils the cost and availability of a Ca^{2+} source are major factors in reclamation. It is also important that the Ca^{2+} source fully react with the soil. Thus, it is better to incorporate the Ca^{2+} source into the soil rather than just putting it on the surface so that Na^+ exchange from the soil exchanger phase is enhanced. Gypsum can also be added to irrigation water to increase the Ca/Na ratio of the water and improve reclamation (Keren and Miyamoto, 1990).

Effects of Soil Salinity and Sodicity on Soil Structural Properties

Soil salinity and sodicity can have a major effect on the structure of soils. Soil structure, or the arrangement of soil particles, is critical in affecting permeability and infiltration. Infiltration refers to the "downward entry of water into the soil through the soil surface" (Glossary of Soil Science Terms, 1997). If a soil has high quantities of Na^+ and the EC is low, soil permeability, hydraulic conductivity, and the infiltration rate are decreased due to swelling and dispersion of clays and slaking of aggregates (Shainberg, 1990). Infiltration rate can be defined as "the volume flux of water flowing into the soil profile per unit of surface area" (Shainberg, 1990). Typically, soil infiltration rates are initially high, if the soil is dry, and then they decrease until a steady state is reached. Swelling causes the soil pores to become more narrow (McNeal and Coleman, 1966), and slaking reduces the number of macropores through which water and solutes can flow, resulting in the plugging of pores by the dispersed clay. The swelling of clay has a pronounced effect on permeability and is affected by clay mineralogy, the kind of ions adsorbed on the clays, and the electrolyte concentration in solution (Shainberg *et al.*, 1971; Oster *et al.*, 1980; Goldberg and Glaubig, 1987). Swelling is greatest for smectite clays that are Na^+-saturated. As the electrolyte concentration decreases, clay swelling increases.

As ESP increases, particularly above 15, swelling clays like montmorillonite retain a greater volume of water (Fig. 10.4). Hydraulic conductivity and permeability decrease as ESP increases and salt concentration decreases (Quirk and Schofield, 1955; McNeal and Coleman, 1966). Permeability can be maintained if the EC of the percolating water is above a threshold level, which is the concentration of salt in the percolating solution, which causes a 10 to 15% decrease in soil permeability at a particular ESP (Shainberg, 1990).

Effects of Soil Salinity on Plant Growth

Salinity and sodicity have pronounced effects on the growth of plants (Fig. 10.5). Sodicity can cause toxicity to plants and create mineral nutrition

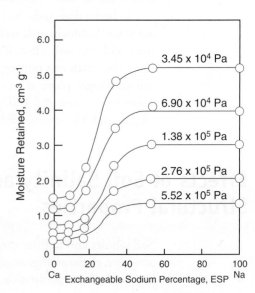

FIGURE 10.4. *Water retention as a function of ESP and pressure applied on montmorillonite. From Shainberg et al. (1971), with permission.*

problems such as Ca^{2+} deficiencies. In saline soils soluble ions such as Cl^-, SO_4^{2-}, HCO_3^-, Na^+, Ca^{2+}, Mg^{2+}, and sometimes NO_3^- and K^+ can harm plants by reducing the osmotic potential. However, plant species, and even different varieties within a particular species, will differ in their tolerance to a particular ion (Bresler *et al.*, 1982).

Soil water availability can be expressed as the sum of the matric and osmotic potentials. As the water content decreases, through evaporation and transpiration, both the matric and osmotic potentials decrease and are more negative (Läuchli and Epstein, 1990). The soluble ions cause an osmotic pressure effect. The osmotic pressure of the soil solution (τ_o) in kPa, which is a useful index for predicting the effect of salinity on plant growth, is calculated from (Jurinak and Suarez, 1990)

$$\tau_o = 2480 \, \Sigma_i m_i v_i \phi_i, \tag{10.11}$$

where m_i is the molal concentration of the ith ion, ϕ_i is the osmotic coefficient of the ith salt, and v_i is the stoichiometric number of ions yielded by the ith salt. The relationship between τ_o and EC at 298 K is (Jurinak and Suarez, 1990)

$$\tau_o \, (kPa) = 40EC. \tag{10.12}$$

At 273 K, the proportionality constant in Eq. (10.12) is 36 (Richards, 1954).

The tolerance of plants to salts can be expressed as (Maas, 1990; Rhoades, 1990)

$$Y_r = 100 - b(EC_e - a), \tag{10.13}$$

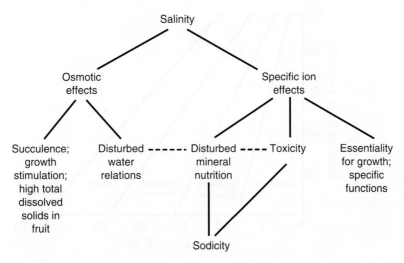

FIGURE 10.5. *Effects of salinity and sodicity on plants. From Läuchli, A., and Epstein, E. (1990a), Plant response to saline and sodic conditions, in "Agricultural Salinity Assessment and Management" (K. K. Tanji, Ed.), pp. 113–137. Am. Soc. Civ. Eng., New York. Reprinted by permission of the ASCE.*

where Y_r is the percentage of the yield of the crop grown under saline conditions compared to that obtained under nonsaline, but otherwise comparable
conditions, a is the threshold level of soil salinity at which yield decreases
begin, and b is the percentage yield loss per increase of salinity in excess of a.

The effect of salinity on plant growth is affected by climate, soil conditions, agronomic practices, irrigation management, crop type and variety, stage
of growth, and salt composition (Maas and Hoffman, 1977; Rhoades, 1990).
Salinity does not usually affect the yield of a crop until the EC_e exceeds a
certain value for each crop. This is known as the threshold salinity level or
the threshold EC_e value, which differs for various crops (Table 10.3). The
yields of many crops, for example, most food and fiber crops, will linearly
decrease as EC_e increases. Maas and Hoffman (1977) divided plants into five
different tolerance categories based on EC_e (Fig. 10.6).

Effects of Sodicity and Salinity on Environmental Quality

Degradation of soils by salinity and sodicity profoundly affects environmental
quality. In particular, the dispersive behavior of sodic soils, coupled with human
activities such as agriculture, forestry, urbanization, and soil contamination,
can have dire effects on the environment and humankind. The enhanced
dispersion promotes surface crusts or seals, which lead to waterlogging,

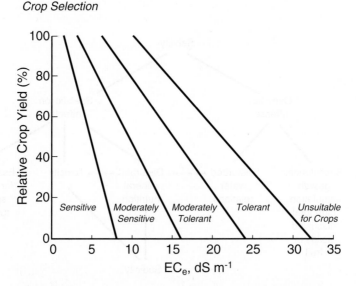

FIGURE 10.6.   *Divisions for classifying crop tolerance to salinity.
From Maas, E. V., and Hoffman, G. J. (1977), Crop salt tolerance—
Current assessment, J. Irrig. Drain Div., Am. Soc. Civ. Eng. **103**(IR 2),
114–134. Reprinted by permission of the ASCE.*

TABLE 10.3.   *Salt Tolerance of Agronomic Crops[a]*

Crop	Threshold EC$_e$ (dS m^{-1})	Tolerance to salinity[b]	Reference
		Fiber, grain, and special crops	
Barley	8.0	T	Maas and Hoffman (1977)
Corn	1.7	MS	Maas and Hoffman (1977)
Cotton	7.7	T	Maas and Hoffman (1977)
Peanut	3.2	MS	Maas and Hoffman (1977)
Rice, paddy	3.0	S	Maas and Hoffman (1977)
Rye	11.4	T	François *et al.* (1989a)
Sorghum	6.8	MT	François *et al.* (1984)
Soybean	5.0	MT	Maas and Hoffman (1977)
Wheat	6.0	MT	Maas and Hoffman (1977)
		Grasses and forage crops	
Alfalfa	2.0	MS	Maas and Hoffman (1977)
Clover, red	1.5	MS	Maas and Hoffman (1977)
Fescue, tall	3.9	MT	Maas and Hoffman (1977)
Orchardgrass	1.5	MS	Maas and Hoffman (1977)
Vetch	3.0	MS	Maas and Hoffman (1977)

[a] Adapted from Maas (1990).
[b] These data serve only as a guideline to relative tolerances among crops. Absolute tolerances vary, depending on climate, soil conditions, and cultural practices; S, sensitive; MS, moderately sensitive; MT, moderately tolerant; and T, tolerant.

surface runoff, and erosion. Consequently, high levels of inorganic and organic colloids can be mobilized, which can transport organic and inorganic contaminants such as pesticides, metals, and radionuclides in soils and waters (Sumner *et al.*, 1998).

The enhanced erosion potential of sodic soils also results in increased sediments that can contaminate waters. Suspended sediments in water increase turbidity. This causes less light to pass through, which negatively affects aquatic life. Additionally, increased levels of dissolved organic carbon (DOC) generated in sodic soils can discolor water (Sumner *et al.*, 1998).

Salinization of soils results in soluble salts that can be mobilized in soil profiles, causing land and water degradation. The salts can also effect release and solubilization of heavy metals into solution, with potential adverse effects on water quality and plant growth (Gambrell *et al.*, 1991; McLaughlin and Tiller, 1994).

Suggested Reading

Ayers, R. S., and Westcot, D. W. (1976). "Water Quality for Agricultural," Irrig. Drain. Pap. 29. Food and Agriculture Organization of the United Nations, Rome.

Bresler, E., McNeal, B. L., and Carter, D. L. (1982). "Saline and Sodic Soils. Principles-Dynamics-Modeling." Springer-Verlag, Berlin.

Rhoades, J. D. (1993). Electrical conductivity methods for measuring and mapping soil salinity. *Adv. Agron.* **49**, 201–251.

Richards, L. A., Ed. (1954). "Diagnosis and Improvement of Saline and Sodic Soils," USDA Agric. Handb. 60. USDA, Washington, DC.

Sumner, M. E., and Naidu, R. (1998). "Sodic Soils: Distribution, Properties, Management and Environmental Consequences." Oxford Univ. Press, New York.

Tanji, K. K., Ed. (1990). "Agricultural Salinity Assessment and Management," ASCE Manuals Rep. Eng. Pract. 71. Am. Soc. Civ. Eng., New York.

Yaron, D., Ed. (1981). "Salinity in Irrigation and Water Resources." Dekker, New York.

Appendix A: Periodic Table of the Elements

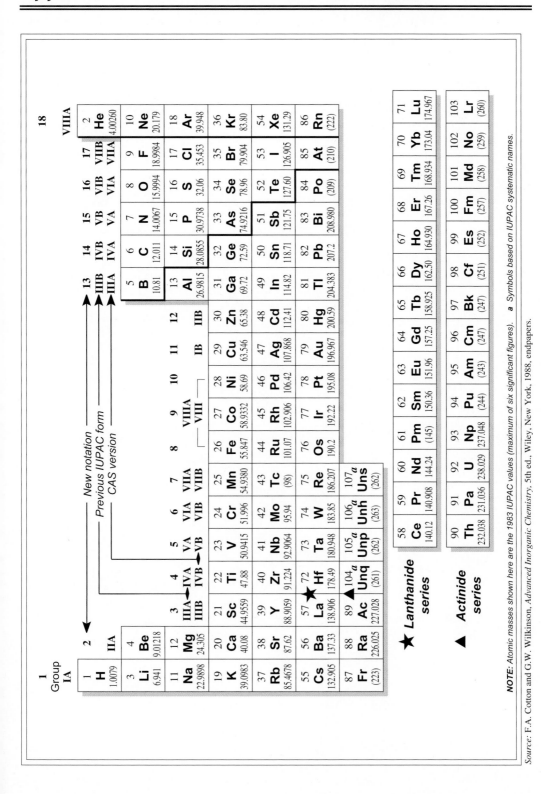

Source: F.A. Cotton and G.W. Wilkinson, *Advanced Inorganic Chemistry*, 5th ed., Wiley, New York, 1988, endpapers.

NOTE: *Atomic masses shown here are the 1983 IUPAC values (maximum of six significant figures).*   *a Symbols based on IUPAC systematic names.*

References

Abendroth, R.P. (1970). Behavior of a pyrogenic silica in simple electrolytes. *J. Colloid Interface Sci.* **34,** 591–596.

Acton, C.J., Paul, E.A., and Rennie, D.A. (1963). Measurements of the polysaccharide content of soils. *Can. J. Soil Sci.* **43,** 141–150.

Adams, F. (1971). Ionic concentrations and activities in the soil solution. *Soil Sci. Soc. Am. Proc.* **35,** 420–426.

Adams, F., ed. (1984). "Soil Acidity and Liming." Agronomy 12. Am. Soc. Agron., Madison, WI.

Adriano, D.C. (1986). "Trace Elements in the Terrestrial Environment." Springer-Verlag, New York.

Aharoni, C. (1984). Kinetics of adsorption: The S-shaped Z(t) plot. *Adsorpt. Sci. Technol.* **1,** 1–29.

Aharoni, C., and Sparks, D.L. (1991). Kinetics of soil chemical reactions: A theoretical treatment. *In* "Rates of Soil Chemical Processes" (D.L. Sparks and D.L. Suarez, eds.), SSSA Spec. Publ. No. 27, pp. 1–18, Soil Sci. Soc. Am., Madison, WI.

Aharoni, C., and Suzin, Y. (1982a). Application of the Elovich equation to the kinetics of occlusion: Part 1. Homogenous microporosity. *J. Chem. Soc., Faraday Trans. 1* **78,** 2313–2320.

Aharoni, C., and Suzin, Y. (1982b). Application of the Elovich equation to the kinetics of occlusion: Part 3. Heterogenous microporosity. *J. Chem. Soc., Faraday Trans. 1* **78,** 2329–2336.

Aharoni, C., and Ungarish, M. (1976). Kinetics of activated chemisorption. Part 1. The non-Elovichian part of the isotherm. *J. Chem. Soc., Faraday Trans.* **172,** 400–408.

Aiken, G.R., McKnight, D.M., and Wershaw, R.L., eds. (1985a). "Humic Substances in Soil, Sediments, and Water." John Wiley & Sons (Interscience), New York.

Aiken, G.R., McKnight, D.M., Wershaw, R.L., and MacCarthy, P. (1985b). An introduction to humic substances in soil, sediment, and water. *In* "Humic Substances in Soil, Sediments, and Water" (G.R. Aiken, D.M. McKnight, R.L. Wershaw, eds.), pp. 1–9. John Wiley & Sons (Interscience), New York.

Ainsworth, C. C., Pilou, J. L., Gassman, P. L., and Van Der Sluys, W. G. (1994). Cobalt, cadmium, and lead sorption to hydrous iron oxide: Residence time effect. *Soil Sci. Soc. Am. J.* **58,** 1615–1623.

Alexander, M. (1995). How toxic are toxic chemicals in soils? *Environ. Sci. Technol.* **29,** 2713–2717.

Alexander, M. (2000). Aging, bioavailability, and overestimation of risk from environmental pollutants. *Environ. Sci. Technol.* **34,** 4259–4265.

Allmann, R. (1970). Doppelschichtstrukturen mit brucitaehnlichen Schichtionen $[Me(II)_{1-x} Me(III)_x(OH)_2]^{x+}$. *Chimia* **24,** 99–108.

Altaner, S.P., Weiss, C.A. and Kirkpatrick, R.J. 1988. Evidence from ^{29}Si NMR for the structure of mixed-layer illite/smectite clay minerals. *Nature* **331:** 699–702.

Altug, I., and Hair, M.L. (1967). Cation exchange in porous glass. *J. Phys. Chem.* **71,** 4260–4263.

Amacher, M.C. (1991). Methods of obtaining and analyzing kinetic data. *In* "Rates of Soil Chemical Processes" (D.L. Sparks and D.L. Suarez, eds.) SSSA Spec. Publ. No. 27, pp. 19–59. Soil Sci. Soc. Am., Madison, WI.

Amacher, M.C., and Baker, D.E. (1982). "Redox Reactions Involving Chromium, Plutonium, and Manganese in Soils," DOE/DP/04515-1. Penn. State University, University Park, PA.

Anderson, M.A., and Rubin, A.J., eds. (1981). "Adsorption of Inorganics at Solid-Liquid Interfaces." Ann Arbor Sci., Ann Arbor, MI.

Arai, Y., Elzinga, E.J., and Sparks, D.L. (2001). X-ray absorption spectroscopic investigation of arsenite and arsenate adsorption at the aluminum oxide-water interface. *J. Colloid Interf. Sci.* **235,** 80–88.

Arai, Y., and Sparks, D.L. (2001). ATR-FTIR spectroscopic investigation on phosphate adsorption mechanisms at the ferrihydrite-water interface. *J. Colloid Interf. Sci.* **241,** 317–326.

Argersinger, W.J., Davidson, A.W., and Bonner, O.D. (1950). Thermodynamics and ion exchange phenomena. *Trans. Kans. Acad. Sci.* **53,** 404–410.

Arshad, M.A., Ripmeester, J.A., and Schnitzer, M. (1988). Attempts to improve solid-state ^{13}C NMR spectra of whole mineral soils. *Can J. Soil. Sci.* **68,** 593–602.

Atkinson, R.J., Hingston, F.J., Posner, A.M., and Quirk, J.P. (1970). Elovich equation for the kinetics of isotope exchange reactions at solid-liquid interfaces. *Nature (London)* **226,** 148–149.

Axe, L., Bunker, G.B., Anderson, P.R., and Tyson, T.A. (1997). An XAFS analysis of strontium at the hydrous ferric oxide surface. *J. Colloid Interf. Sci.* **199,** 44–52.

Ayers, R.S., and Westcot, D.W. (1976). "Water Quality for Agriculture," *Irrig. Drain. Pap. No. 29,* Food and Agriculture Organization of the United Nations, Rome.

Baas Becking, L.G.M., Kaplan, I.R., and Moore, D. (1960). Limits of the natural environment in terms of plant and oxidation-reduction potentials. *J. Geol.* **68,** 243–284.

Babcock, K.L. (1963). Theory of the chemical properties of soil colloidal systems at equilibrium. *Hilgardia* **34,** 417–542.

Babcock, K.L., and Doner, H.E. (1981). A macroscopic basis for soil chemistry. *Soil Sci.* **131,** 276–283.

Backes, C. A., McLaren , R. G., Rate, A. W., and Swift, R. S. (1995). Kinetics of cadmium and cobalt desorption from iron and manganese oxides. *Soil Sci. Soc. Am. J.* **59,** 778–785.

Baes, C.F., Jr., and Mesmer, R.E. (1976). "The Hydrolysis of Cations." John Wiley & Sons, New York.

Baham, J. (1984). Prediction of ion activities in soil solutions: Computer equilibrium modeling. *Soil Sci. Soc. Am. J.* **48,** 523–531.

Bailey, G.W., and White, J.L. (1970). Factors influencing the adsorption, desorption, and movement of pesticides in soil. *Residue Rev.* **32,** 29–92.

Baker, W.E. (1973). The role of humic acids from Tasmania podzolic soils in mineral degradation and metal mobilization. *Geochim. Cosmochim. Acta* **37**, 269–281.

Ball, J.W., Jenne, E.A., and Nordstrom, D.K. (1979). WATEQ2 – a computerized chemical model for trace and major element speciation and mineral equilibria of natural waters. *In* "Chemical Modeling in Aqueous Systems" (E.A. Jenne, ed.), Am. Chem. Soc. Symp. Series No. 93, pp. 815–835. Am. Chem. Soc., Washington, D.C.

Ball, J.W., Jenne, E.A., and Cantrell, M.W. (1981). "WATEQ3: A geochemical model with uranium added," *US Geol. Surv. Open-File Rep. 81–1183.*

Ball, W.P. (1989). Equilibrium sorption and diffusion rate studies with halogenated organic chemicals and sandy aquifer materials. Ph.D. Dissertation, Stanford University, Palo Alto, CA.

Ball, W.P., and Roberts, P.V. (1991). Long-term sorption of halogenated organic chemicals by aquifer material: 1. Equilibrium. *Environ. Sci. Technol.* **25**[7], 1223–1236.

Ballard, T.M. (1971). Role of humic carrier substances in DDT movement through forest soil. *Soil Sci. Soc. Am. Proc.* **35,** 145–147.

Bargar, J.R., Brown, G.E., Jr., and Parks, G.A. (1997a). Surface complexation of Pb(II) at oxide-water interfaces: I. XAFS and bond-valence determination of mononuclear and polynuclear Pb(II) sorption products on aluminum oxides. *Geochim. Cosmochim. Acta.* **61**(13), 2617–2637.

Bargar, J.R., Brown, G.E., Jr., and Parks, G.A. (1997b). Surface complexation of Pb(II) at oxide-water interfaces: II. XAFS and bond-valence determination of mononuclear Pb(II) sorption products and surface functional groups on iron oxides. *Geochim. Cosmochim. Acta* **61**(13), 2639–2652.

Bargar, J.R., Towle, S.N., Brown, G.E., Jr., and Parks, G.A. (1996). Outer-sphere Pb(II) at specific surface sites on single crystal α-alumina. *Geochim. Cosmochim. Acta.* **60**(18), 3541–3547.

Barnes, I., and Clarke, F.E. (1969). "Chemical properties of ground water and their corrosion and encrustation effects on wells," *US Geol. Surv. Prof. Pap. 498-D.*

Barrer, R.M., and Klinowski, J. (1974). Ion-exchange selectivity and electrolyte concentration. *J. Chem. Soc., Faraday Trans. 1* **70**, 2080–2091.

Barrow, N. J. (1986). Testing a mechanistic model: II. The effects of time and temperature on the reaction of zinc with a soil. *J. Soil Sci.* **37,** 277–286.

Barrow, N.J. (1985). Reactions of anions and cations with variable-charged soils. *Adv. Agron.* **38,** 183–230.

Barrow, N.J., Bowden, J.W., Posner, A.M., and Quirk, J.P. (1980). An objective method for fitting models of ion adsorption on variable charge surfaces. *Aust. J. Soil Res.* **18,** 37–47.

Barrow, N.J., Bowden, J.W., Posner, A.M., and Quirk, J.P. (1981). Describing the adsorption of copper, zinc and lead on a variable charge mineral surface. *Aust. J. Soil Res.* **19,** 309–321.

Bar-Tal, A., Sparks, D.L., Pesek, J.D., and Feigenbaum, S. (1990). Analyses of adsorption kinetics using a stirred-flow chamber: I. Theory and critical tests. *Soil Sci. Soc. Am. J.* **54,** 1273–1278.

Bartell, F.E., and Fu, Y. (1929). Adsorption from aqueous solutions by silica. *J. Phys. Chem.* **33,** 676–687.

Bartlett, R.J. (1981). Nonmicrobial nitrite-to-nitrate transformation in soils. *Soil Sci. Soc. Am. J.* **45,** 1054–1058.

Bartlett, R.J. (1986). Soil redox behavior. *In* "Soil Physical Chemistry" (D.L. Sparks, ed.), pp. 179–207. CRC Press, Boca Raton, FL.

Bartlett, R.J., and James, B.R. (1979). Behavior of chromium in soils. III. Oxidation. *J. Environ. Qual.* **8,** 31–35.

Bartlett, R.J., and James, B.R. (1993). Redox chemistry of soils. *Adv. Agron.* **50,** 151–208.

Batjes, N.H. (1996). Total C and N in the soils of the world. *Eur. J. Soil. Sci.* **47,** 151–163.

Bell, L.C., Posner, A.M., and Quirk, J.P. (1973). The point of zero charge of hydroxyapatite and fluorapatite in aqueous solutions. J. *Colloid Interf. Sci.* **42,** 250–261.

Bell, A.T. (1980). Applications of fourier transform infrared spectroscopy to studies of adsorbed species. *In* "Vibrational Spectroscopies for Adsorbed Species" (A.T. Bell and M.L. Hair, eds.), ASC Symp. Ser. 37:13–35. Washington, DC.

Belot, Y., Gailledreau, C., and Rzekiecki, R. (1966). Retention of strontium-90, calcium-45, and barium-140 by aluminum oxide of large area. *Health Phys.* **12,** 811–823.

Berggren, B., and Oden, S. (1972). "Analysresultat Rorande Fungmetaller Och Kloerade Kolvaten I Rotslam Fran Svenska Reningsverk 1968-1971." Institutione fur Markvetenskap Lantbrukshogskolan, Uppsala, Sweden.

Bernasconi, E.C. (1976). "Relaxation Kinetics." Academic Press, New York.

Berrow, M.L., and Webber, J. (1972). Trace elements in sewage sludges. *J. Sci. Food Agric.* **23**[1], 93–100.

Berry, L.G., and Mason, B. (1959). "Mineralogy-Concepts, Descriptions, Determinations." Freeman, San Francisco.

Bertsch, P.M. (1987). Conditions for Al_{13} polymer formation in partially neutralized Al solutions. *Soil Sci. Soc. Am. J.* **51,** 825–828.

Bertsch, P.M. (1989). Aqueous polynuclear aluminum species. *In* "The Environmental Chemistry of Aluminum" (G. Sposito, ed.), pp. 87–115. CRC Press, Boca Raton, FL.

Bertsch, P.M., and Bloom, P.R. (1996). Aluminum. *In* "Methods of Soil Analysis, Part 3 – Chemical Methods" (D.L. Sparks, ed.), pp. 517–550. Soil Sci. Soc. Am. Book Series 5, SSSA, Madison, WI.

Bertsch, P.M., and Hunter, D. B., (1998). Elucidating fundamental mechanisms in soil and environmental chemistry. The role of advanced analytical, spectroscopic and microscopic methods. *In* "Future Prospects for Soil Chemistry" (P.M. Huang, D. L. Sparks and S.A. Boyd, eds.), Soil Sci. Soc. Amer. Spec. Publ. No. 55, pp. 103–122, SSSA, Madison, WI.

Bertsch, P.M. and Parker, D.R. (1996). Aqueous polynuclear aluminum species. *In* "The Environmental Chemistry of Aluminum" 2nd ed. (G. Sposito, ed.), pp. 117–168. CRC Press, Boca Raton, FL.

Bérubé, Y.G., and de Bruyn, P.L. (1968). Adsorption at the rutile solution interface. I. Thermodynamic and experimental study. *J. Colloid Interf. Sci.* **27,** 305–318.

Bleam, W.F., and McBride, M.B. (1986). The chemistry of adsorbed Cu (II) and Mn (II) in aqueous titanium dioxide suspensions. *J. Colloid Interf. Sci.* **110,** 335–346.

Bleam, W.F., Pfeffer, P.E., Goldberg, S., Taylor, R.W., and Dudley, R. (1991). A ^{31}P solid-state nuclear magnetic resonance study of phosphate adsorption at the boehmite/aqueous solution. *Langmuir* **7,** 1702–1712.

Bloom, P.R., and Erich, M.S. (1987). Effect of solution composition on the rate and mechanism of gibbsite dissolution in acid solutions. *Soil Sci. Soc. Am. J.* **51,** 1131–1136.

Bochatay, L., Persson, P., and Sjöberg, S. (2000a). Metal ion coordination at the water-manganite (γ-MnOOH) interface. I. An EXAFS study of cadmium(II). *J. Colloid Interf. Sci.* **229,** 584–592.

Bochatay, L., and Persson, P. (2000b). Metal ion coordination at the water-manganite (γ-MnOOH) interface. II. An EXAFS study of zinc(II). *J. Colloid Interf. Sci.* **229,** 593–599.

Bohn, H.L. (1968). Electromotive force of inert electrodes in soil suspensions. *Soil Sci. Soc. Am. Proc.* **32,** 211–215.

Bohn, H.L., McNeal, B.L., and O'Conner, G.A. (1985). "Soil Chemistry." 2nd ed. John Wiley & Sons, New York.

Bolt, G.H., ed. (1979). "Soil Chemistry. B: Physico-Chemical Models." Elsevier, Amsterdam.

Bolt, G.H., de Boodt, M.E., Hayes, M.H.B., and McBride, M.B., eds. (1991). "Interactions at the Soil Colloid-Soil Solution Interface." NATO ASI Ser. E. Vol. 190. Kluwer Academic Publishers, Dordrecht, The Netherlands.

Bolt, G.H., and van Riemsdijk, W.H. (1982). Ion adsorption on inorganic variable charge constituents. *In* "Soil Chemistry. Part B. Physico-Chemical Methods" (G.H. Bolt, ed.), pp. 459–503. Elsevier, Amsterdam.

Bowden, J.W., Posner, A.M., and Quirk, J.P. (1977). Ionic adsorption on variable charge mineral surfaces. Theoretical charge development and titration curves. *Aust. J. Soil Res.* **15,** 121–136.

Bowden, J.W., Nagarajah, S., Barrow, N.J., Posner, A.M., and Quirk, J.P. (1980). Describing the adsorption of phosphate, citrate and selenite on a variable charge mineral surface. *Aust. J. Soil Res.* **18,** 49–60.

Bowen, H.J.M. (1979). "Environmental Chemistry of the Elements." Academic Press, London.

Bower, C.A., and Hatcher, J.T. (1964). Estimation of the exchangeable-sodium percentage of arid zone soils from solution cation composition. *In* West. Soc. Soil Sci. Abstr., Vancouver, British Columbia, Canada.

Bradfield, R. (1923). The nature of the acidity of the colloidal clay of acid soils. *J. Am. Chem. Soc.* **45,** 2669–2678.

Bradfield, R. (1925). The chemical nature of colloidal clay. *J. Am. Soc. Agron.* **17,** 253–370.

Brady, N.C. (1984). "The Nature and Properties of Soils," 9th ed. Macmillan, New York.

Breeuwsma, A., and Lyklema, J. (1971). Interfacial electrochemistry of hematite (α-Fe$_2$O$_3$). *Disc. Faraday Soc.* **522,** 3224–3233.

Bresler, E., McNeal, B.L., and Carter, D.L. (1982). "Saline and Sodic Soils. Principles-Dynamics-Modeling." Springer-Verlag, Berlin.

Bricker, O.P. (1965). Some stability relationships in the system MnO_2-H_2O at 25°C and 1 atm total pressure. *Am. Mineral.* **50**, 1296–1354.

Brindley, G.W., and Brown, G., eds. (1980). "Crystal Structures of Clay Minerals and Their X-ray Identification." Monogr. No. 5. Mineralogical Society, London.

Britz, D., and Nancollas, G.H. (1969). Thermodynamics of cation exchange of hydrous zirconia. *J. Inorg. Nucl. Chem.* **31**, 3861–3868.

Broadbent, F.E., and Bradford, G.R. (1952). Cation exchange groupings in soil organic fraction. *Soil Sci.* **74**, 447–457.

Brooks, R.R., ed. (1988). "Plants that Hyperaccumulate Heavy Metals." Cambridge Univ. Press, Cambridge.

Brown, G. (1980). Associate minerals. *In* "Clay Structures of Clay Minerals and Their X-ray Identification" (G.W. Brindley and G. Brown, eds.), Monograph No. 5, pp. 361–410. Mineralogical Society, London.

Brown, G.E., Jr. (1990). Spectroscopic studies of chemisorption reaction mechanisms at oxide-water interfaces. *In* "Mineral-Water Interface Geochemistry" (M.F. Hochella, Jr. and A.F. White, eds.), Rev. in Mineralogy 23, pp. 309–353. Mineral Soc. Am., Washington, D.C.

Brown, G.E., Jr., Calas, G., Waychunas, G.A. and Petiau, J. (1988). X-ray absorption spectroscopy and its applications in mineralogy and geochemistry. In "Spectroscopic Methods in Mineralogy and Geology" (F. Hawthorne, ed.), Rev. in Mineralogy 18, pp. 431–512. Mineral Soc. Am., Washington, D.C.

Brown, G.E., Jr., Parks, G.A., and O'Day, P.A. (1995). Sorption at mineral-water interfaces: Macroscopic and microscopic perspectives. *In* "Mineral Surfaces" (D.J. Vaughan and R.A.D. Pattrick, eds.), pp. 129–183. Chapman and Hall, London.

Bruemmer, G.W., Gerth, J., and Tiller, K.G. (1988). Reaction kinetics of the adsorption and desorption of nickel, zinc and cadmium by goethite: I. Adsorption and diffusion of metals. *J. Soil Sci.* **39**, 37–52.

Brunauer, S., Emmett, P.H., and Teller, E. (1938). Adsorption of gases in multi-molecular layers. *J. Am. Chem. Soc.* **60**, 309–319.

Brunnix, E. (1975). The coprecipitation of Zn, Cd and Hg with ferric hydroxide. *Phillips Res. Repts.* **30**, 177–191.

Brusseau, M. L. and Rao, P. S. C. (1989). Sorption nonideality during organic contaminant transport in porous media. *CRC Critical Reviews in Environ. Control.* **19**, 33–99.

Buffle, J. (1984). Natural organic matter and metal-organic interactions in aquatic systems. *In* "Metal Ions in Biological Systems" (H. Sigel, ed.), pp. 165–221. Dekker, New York.

Buffle, J., and Stumm, W. (1994). General chemistry of aquatic systems. *In* "Chemical and Biological Regulation of Aquatic Systems" (J. Buffle and R.R. DeVitre, eds.), pp. 1–42. CRC Press, Boca Raton, FL.

Bunnett, J.F. (1986). Kinetics in solution. *In* "Investigations of Rates and Mechanisms of Reactions" (C.F. Bernasconi, ed.), 4th ed., pp. 171–250. John Wiley & Sons, New York.

Bunzl, K., Schmidt, W., and Sansoni, B. (1976). Kinetics of ion exchange in soil organic matter. IV. Adsorption and desorption of Pb^{2+}, Cu^{2+}, Zn^{2+}, and Ca^{2+} by peat. *J. Soil Sci.* **27,** 32–41.

Caccavo, F., Jr., Blakemore, R.P., and Lovely, D.R. (1992). A hydrogen-oxidizing, Fe(III)-reducing micro-organism from the Great Bay Estuary. *Appl. Environ. Microbiol.* **58,** 3211–3216.

Calas, G., 1988. Electron paramagnetic resonance. *In* "Spectroscopic Methods in Mineralogy and Geology" (F.C. Hawthorne, Ed.), Reviews in Mineralogy 18, 513–563. Mineralogical Society of America, Washington, DC.

Carroll, K.M., Harkness, M.R., Bracco, A.A., and Balcarcel, R.B. (1994). Application of a permeant/polymer diffusional model to the desorption of polychlorinated biphenyls from Hudson River sediments. *Environ. Sci. Technol.* **28,** 253–258.

Carter, D.L., Mortland, M.M., and Kemper, W.D. (1986). Specific surface. *In* "Methods of Soil Analysis, Part 1 – Physical and Mineralogical Methods" (A. Klute, ed.), 2nd ed., pp. 413–423. Am. Soc. Agron., Madison, WI.

Casey, W.H., Westrich, H.R., Arnold, G.W., and Banfield, J.F. (1989). The surface chemistry of dissolving labradorite feldspar. *Geochim. Cosmochim. Acta* **53,** 821–832.

Chakravarti, S.N., and Talibudeen, O. (1961). Phosphate interaction with clay minerals. *Soil Sci.* **92**[4], 232–242.

Chapman, D.L. (1913). A contribution to the theory of electrocapillarity. *Philos. Mag.* **25**[6], 475–481.

Charberek, S., and Martell, A.E. (1959). "Organic Sequestering Agents." John Wiley & Sons, New York.

Charlet, L., and Manceau, A.A. (1992). X-ray absorption spectroscopic study of the sorption of Cr(III) at the oxide-water interface. II. Adsorption, coprecipitation, and surface precipitation on hydrous ferric oxide. *J. Colloid Interf. Sci.* **148**(2), 442–458.

Charlet, L., and Manceau, A. (1993). Structure, formation, and reactivity of hydrous oxide particles: Insights from x-ray absorption spectroscopy. *In* "Environmental Particles" (J. Buffle and A.P. vanLeeuwen, eds.), pp. 117–164. Lewis Publishers, Boca Raton, FL.

Cheah, S.F., Brown, G.E., Jr., and Parks, G.A. (1998). XAFS spectroscopy study of Cu(II) sorption on amorphous SiO_2 and γ-Al_2O_3: Effect of substrate and time on sorption complexes. *J. Colloid Interf. Sci.* **208,** 110–128.

Chen, Y. and Schnitzer, M. (1976), Scanning electron microscopy of a humic acid and a fulvic acid and its metal and clay complexes. *Soil Sci. Soc. Am. J.* **40,** 682–686.

Cheng, H.H., ed. (1990). "Pesticides in the Soil Environment: Processes, Impacts, and Modeling." Soil Sci. Soc. Am. Book Ser. No. 2. Soil Sci. Soc. Am., Madison, WI.

Chernov, V.A. (1947). "The Nature of Soil Acidity" (English translation furnished by Hans Jenny-translator unknown), Press of Academy of Sciences, Moscow. Soil Sci. Soc. Am. (1964).

Chien, S.H., and Clayton, W.R. (1980). Application of Elovich equation to the kinetics of phosphate release and sorption in soils. *Soil Sci. Soc. Am. J.* **44,** 265–268.

Chiou, C.T., Freed, V.H., Schmedding, D.W., and Kohnert, R.L. (1977). Partition coefficient and bioaccumulation of selected organic compounds. *Environ. Sci. Technol.* **11,** 475–478.

Chiou, C.T., Peters, L.J., and Freed, V.H. (1979). A physical concept of soil-water equilibria for nonionic organic compounds. *Science* **206,** 831–832.

Chiou, C.T., Porter, P.E., and Schmedding, D.W. (1983). Partition equilibria of nonionic organic compounds between soil organic matter and water. *Environ. Sci. Technol.* **17,** 227–231.

Chisholm-Brause, C.J., Hayes, K.F., Roe, A.L., Brown, G.E., Jr., Parks, G.A., and Leckie, J.O. (1990a). Spectroscopic investigation of Pb(II) complexes at the γ-Al$_2$O$_3$/water interface. *Geochim. Cosmochim. Acta.* **54,** 1897–1909.

Chisholm-Brause, C. J., O'Day , P. A., Brown, G. E., Jr., and Parks, G. A. (1990b). Evidence for multinuclear metal-ion complexes at solid/water interfaces from X-ray absorption spectroscopy. *Nature* **348,** 528–530.

Christman, R.F., and Gjessing, E.T., eds. (1983). "Aquatic and Terrestrial Humic Materials." Ann Arbor Sci., Ann Arbor, MI.

Churms, S.C. (1966). The effect of pH on the ion-exchange properties of hydrated alumina. Part 1. Capacity and selectivity. *J. S. Afr. Chem. Inst.* **19,** 98–107.

Chute, J.H., and Quirk, J.P. (1967). Diffusion of potassium from mica-like materials. *Nature (London)* **213,** 1156–1157.

Coates, J.T., and Elzerman, A.W. (1986). Desorption kinetics for selected PCB congeners from river sediments. *J. Contam. Hydrol.* **1,** 191–210.

Coleman, N.T., and Craig, D. (1961). The spontaneous alteration of hydrogen clay. *Soil Sci.* **91,** 14–18.

Coleman, N.T., and Harward, M.E. (1953). The heats of neutralization of acid clays and cation exchange resins. *J. Am. Chem. Soc.* **75,** 6045–6046.

Coleman, N.T., Kamprath, E.J., and Weed, S.B. (1959). Liming. *Adv. Agron.* **10,** 475–522.

Coleman, N.T., Thomas, G.W., LeRoux, F.H., and Bredell, G. (1964). Salt-exchangeable and titratable acidity in bentonite-sesquioxide mixtures. *Soil Sci. Soc. Am. Proc.* **28,** 35–37.

Coleman, N.T. and Thomas, G.W. (1967). The basic chemistry of soil acidity. *In* "Soil Acidity and Liming" (R. W. Pearson and F. Adams, eds.), pp. 1–41, Agron. Monogr. 12. Am. Soc. Agron., Madison, WI.

Comans, R. N. J. and Hockley, D. E. (1992). Kinetics of cesium sorption on illite. *Geochim. Cosmochim. Acta.* **56,** 1157–1164.

Connaughton, D.F., Stedinger, J.R., Lion, L.W., and Shuler, M.L. (1993). Description of time-varying desorption kinetics: Release of naphthalene from contaminated soils. *Environ. Sci. Technol.* **27,** 2397–2403.

Cope, F. (1958). Catchment salting in Victoria. *Soil Conserv. Auth. Victoria Bull.* **1,** 1–88.

Cornelissen, G., van Noort, P.C.M., Parsons, J.R., and Govers, H.A.J. (1997). The temperature dependence of slow adsorption and desorption kinetics of organic compounds in sediments. *Environ. Sci. Technol.* **31,** 454–460.

Crank, J. (1976). "The Mathematics of Diffusion." 2nd ed., Oxford University Press (Clarendon), London.

David, M.B., and Driscoll, C.T. (1984). Aluminum speciation and equilibria in soil solutions of a Haplorthod in the Adirondack Mountains (New York, U.S.A.). *Geoderma* **33,** 297–318.

Davies, C.W. (1962). "Ion Association." Butterworth, London.

Davis, J.A., and Leckie, J.O. (1978). Surface ionization and complexation at the oxide/water interface. 2. Surface properties of amorphous iron oxyhydroxide and adsorption of metal ions. *J. Colloid Interf. Sci.* **67**, 90–107.

Davis, J.A., and Leckie, J.O. (1980). Surface ionization and complexation at the oxide/water interface. 3. Adsorption of anions. *J. Colloid Interf. Sci.* **74**, 32–43.

Davis, J.A., James, R.O., and Leckie, J.O. (1978). Surface ionization and complexation at the oxide/water interface. 1. Computation of electrical double layer properties in simple electrolytes. *J. Colloid Interf. Sci.* **63**, 480–499.

Deist, J., and Talibudeen, O. (1967a). Ion exchange in soils from the ion pairs K-Ca, K-Rb, and K-Na. *J. Soil Sci.* **18**[1], 1225–1237.

Deist, J., and Talibudeen, O. (1967b). Thermodynamics of K-Ca exchange in soils. *J. Soil Sci.* **18**(1), 1238–1248.

Dennen, W.H. (1960). "Principles of Mineralogy." The Ronald Press, New York.

Denney, D.Z., and Hsu, P.H. (1986). ^{27}Al nuclear magnetic resonance and ferron kinetic studies of partially neutralized $AlCl_3$ solutions. *Clays Clay Miner.* **34**, 604–607.

DiToro, D.M., and Horzempa, L.M. (1982). Reversible and resistant components of PCB adsorption-desorption: Isotherms. *Environ. Sci. Technol.* **16**, 594–602.

Dixon, J.B., and Jackson, M.L. (1962). Properties of intergradient chlorite-expansible layer silicates of soils. *Soil Sci. Soc. Am. Proc.* **26**, 358–362.

Dixon, J.B., and Weed, S.B., eds. (1989). "Minerals in Soil Environments." SSSA Book Ser. 1. Soil Sci. Am., Madison, WI.

Donaldson, J.D., and Fuller, M.J. (1968). Ion exchange properties of tin (IV) materials. I. Hydrous tin (IV) oxide and its cation exchange properties. *J. Inorg. Nucl. Chem.* **30**, 1083–1092.

Doner, H.E., and Lynn, W.C. (1989). Carbonate, halide, sulfate, and sulfide minerals. *In* "Minerals in Soil Environments" (J.B. Dixon and S.B. Weed, eds.), SSSA Book Ser. 1, pp. 279–330. Soil Sci. Soc. Am., Madison, WI.

Dove, P.M., and Hochella, M.F., Jr. (1993). Calcite precipitation mechanisms and inhibition by orthophosphate: *In situ* observations by scanning force microscopy. *Geochim. Cosmochim. Acta.* **57**, 705–714.

Dragun, J. (1988). "The Soil Chemistry of Hazardous Materials." Hazardous Materials Control Research Institute, Silver Spring, MD.

Dragun, J., and Chiasson. (1991). Elements in North American Soils. Hazardous Materials Control Resources Institute, Greenbelt, MD.

Drever, J.I., ed. (1985). "The Chemistry of Weathering." NATO ASI Ser. C. 149. Reidel, Dordrecht, The Netherlands.

Driscoll, C.T., Lawrence, G.B., Bulger, A.J., Butler, T.J., Cronan, C.S., Eagar, C., Lambert, K.F., Likens, G.E., Stoddard, J.L., and Weathers, K.C. (2001). Acidic deposition in the northeastern United States: Sources and inputs, ecosystem effects, and management strategies. *Bioscience* **51**, 180–198.

Driscoll, C.T., Yan, C., Schofield, C.L., Munson, R., and Holsapple, J. (1994). The mercury cycle and fish in the Adirondack lakes. *Environ. Sci. Technol.* **28**, 136A–143A.

Dubach, P., and Mehta, N.C. (1963). The chemistry of soil humic substances. *Soils Fert.* **26,** 293–300.

Ducker, W.A., Senden, T.J. and Pashley, R.M. (1992). Measurement of forces in liquids using a force microscope. *Langmuir* **8,** 1831–1836.

Dumont, F., and Watillon, A. (1971). Stability of ferric oxide hydrosols. *Discuss. Faraday Soc.* **52,** 352–360.

Durfer, C.N., and Becker, E. (1964). "Public water supplies of the 100 largest cities in the United States." *U.S. Geol. Surv., Water-Supply Pap. 1812.*

Dzombak, D.A., and Morel, F.M.M. (1990). "Surface Complexation Modeling. Hydrous Ferric Oxide." John Wiley & Sons, New York.

Eary, L.E., and Rai, D. (1987). Kinetics of chromium(VI) by reaction with manganese dioxide. *Environ. Sci. Technol.* **21,** 1187–1193.

Eaton, F.M., Harding, R.B., and Ganje, T.J. (1960). Soil solution extractions at tenth-bar moisture percentages. *Soil Sci.* **90,** 253–258.

Ebens, R.J., and Schaklette, H.T. (1982). "Geochemistry of some rocks, mine spoils, stream sediments, soils, plants, and waters in the western energy region of the conterminous United States." *U.S. Geol. Surv. Prof. Pap. 1237.*

Eggleston, C.M. and Hochella, M.F., (1990). Scanning tunneling microscopy of sulfide surfaces. *Geochim. Cosmochim. Acta* **54,** 1511–1517.

Eigen, M. (1954). Ionic reactions in aqueous solutions with half-times as short as 10^{-9} second. Applications to neutralization and hydrolysis reactions. *Discuss. Faraday Soc.* **17,** 194–205.

Ekedahl, E., Hogfeldt, E., and Sillén, L.G. (1950). Activities of the components in ion exchangers. *Acta Chem. Scand.* **4,** 556–558.

Elrashidi, M.A., and O'Connor, G.A. (1982). Influence of solution composition on sorption of zinc. *Soil Sci. Soc. Am. J.* **46,** 1153–1158.

El-Sayed, M.H., Burau, R.G., and Babcock, K.L. (1970). Thermodynamics of copper (II)-calcium exchange on bentonite clay. *Soil Sci. Soc. Am. Proc.* **34,** 397–400.

Elzinga, E.J., Peak, D., and Sparks, D.L. (2001). Spectroscopic studies of Pb(II)-sulfate interactions at the goethite-water interface. *Geochim. Cosmochim. Acta* **65,** 2219–2230.

Elzinga, E.J., and Sparks, D.L. (1999). Nickel sorption mechanisms in a pyrophyllite-montmorillonite mixture. *J. Colloid Interf. Sci.* **213,** 506–512.

Elzinga, E.J., and Sparks, D.L. (2001). Reaction condition effects on nickel sorption mechanisms in illite-water suspensions. *Soil Sci. Soc. Am. J.* **65,** 94–101.

Environmental Protection Agency. (1987). "Agricultural Chemicals in Ground Water: Proposed Strategy," Office of Pesticides and Toxic Substances, U.S. Environ. Prot. Agency, Washington, D.C.

Evans, C.E., and Kamprath, E.J. (1970). Lime response as related to percent Al saturation, solution Al and organic matter response. *Soil Sci. Soc. Am. Proc.* **34,** 893–896.

Evans, J.C., Bryce, R.W., Bates, D.J., and Kemner, M.L. (1990). "Hanford Site Ground-Water Surveillance for 1989," PNL-7396. Pacific Northwest Laboratory, Richland, WA.

Farley, K.J., Dzombak, D.A., and Morel, F.M.M. (1985). A surface precipitation model for the sorption of cations on metal oxides. *J. Colloid Interf. Sci.* **106,** 226–242.

Faulkner, S.P., Patrick, W.H., Jr., and Gambrell, R.P. (1989). Field techniques for measuring wetland soil parameters. *Soil Sci. Soc. Am. J.* **53**, 883–890.

Felmy, A.R., Girvin, D.C., and Jenne, E.A. (1984). MINTEQ. A computer program for calculating aqueous geochemical equilibria. NTIS PB84-157418. EPA-600/3-84-032. Natl. Tech. Inform. Serv., Springfield, VA.

Fendorf, S.E. (1992). Oxidation and sorption mechanisms of hydrolysable metal ions on oxide surfaces. Ph.D. Dissertation, University of Delaware, Newark, DE.

Fendorf, S.E. (1999). Fundamental aspects and applications of x-ray absorption spectroscopy in clay and soil science. *In* "Synchrotron X-ray Methods in Clay Science" (D.G. Schulze, J.W. Stucki, and P.M. Bertsch, eds.), pp.19–68. The Clay Minerals Society, Boulder, CO.

Fendorf, S.E., Eick, M.J., Grossl, P., and Sparks, D.L. (1997). Arsenate and chromate retention mechanisms on goethite. 1. Surface structure. *Environ. Sci. Technol.* **31**, 315–320.

Fendorf, S.E., Fendorf, M., Sparks, D.L., and Gronsky, R. (1992). Inhibitory mechanisms of Cr (III) oxidation by γ-MnO_2. *J. Colloid Interf. Sci.* **153**, 37–54.

Fendorf, S.E., Lamble, G.M., Stapleton, M.G., Kelley, M.J., and Sparks, D.L. (1994a). Mechanisms of chromium (III) sorption on silica: 1. Cr(III) surface structure derived by extended x-ray absorption fine structure spectroscopy. *Environ. Sci. Technol.* **28**, 284–289.

Fendorf, S.E., Li, G., and Gunter, M.E. (1996). Micromorphologies and stabilities of Cr (III) surface precipitates elucidated by scanning force microscopy. *Soil Sci. Soc. Am, J.* **60**, 99–106.

Fendorf, S.E., and Sparks, D.L. (1996). X-ray absorption fine structure. *In* "Methods of Soil Analysis: Part 3—Chemical Methods (D.L. Sparks, ed.), pp. 377–416. SSSA Book Ser. Soil Sci. Soc. Am., Madison, WI.

Fendorf, S.E., and Sparks, D.L. (1994b). Mechanisms of chromium (III) sorption on silica: 2. Effect of reaction conditions. *Environ. Sci. Technol.* **28**, 290–297.

Fendorf, S.E., Sparks, D.L., Franz, J.A., and Camaioni, D.M. (1993). Electron paramagnetic resonance stopped-flow kinetic study of manganese (II) sorption-desorption on birnessite. *Soil Sci. Soc. Am. J.* **57**, 57–62.

Fendorf, S.E., Sparks, D.L., Lamble, G.M., and Kelly, M.J. (1994c). Applications of x-ray absorption fine structure spectroscopy to soils. *Soil. Sci. Soc. Am. J.* **58**, 1585–1595.

Fendorf, S.E., and Zasoski, R.J. (1992). Chromium (III) oxidation by γ-MnO_2: I. Characterization. *Environ. Sci. Technol.* **26**, 79–85.

Ferris, A.P., and Jepson, W.R. (1975). The exchange capacities of kaolinite and the preparation of homoionic clays. *J. Colloid Interf. Sci.* **51**, 245–259.

Ford, R.G., Scheinost, A.C., and Sparks, D.L. (2001). Frontiers in metal sorption/precipitation mechanisms on soil mineral surfaces. *Adv. Agron.* **74**, 41–62.

Ford, R. G., and Sparks D. L. (2000). The nature of Zn precipitates formed in the presence of pyrophyllite. *Environ. Sci. Technol.* **34**, 2479–2483.

Forbes, E.A., Posner, A.M., and Quirk, J.P. (1976). The specific adsorption of divalent Cd, Co, Cu, Pb, and Zn on goethite. *J. Soil Sci.* **27**, 154–166.

Foy, C.D. (1984). Physiological effects of hydrogen, aluminum, and manganese toxicities in acid soil. *In* "Soil Acidity and Liming" (F. Adams, ed.), 2nd ed., Agronomy 12, pp. 57–97. Am. Soc. Agron., Madison, WI.

François, L.E., Donovan, T.J., and Maas, E.V. (1984). Salinity effects on seed yield, growth and germination of grain sorghum. *Agron. J.* **76,** 741–744.

François, L.E., Donovan, T.J., Lorenz, K., and Maas, E.V. (1989). Salinity effects on rye grain yield, quality, vegetative growth, and emergence. *Agron. J.* **81,** 707–712.

Fried, M., and Shapiro, G. (1956). Phosphate supply pattern of various soils. *Soil Sci. Soc. Am. Proc.* **20,** 471–475.

Frink, C.R. (1965). Characterization of aluminum interlayers in soil clays. *Soil Sci. Soc. Am. Proc.* **29,** 379–382.

Fuller, C. C., Davis, J. A. and Waychunas, G. A. (1993). Surface chemistry of ferrihydride: Part 2. Kinetics of arsenate adsorption and coprecipitation. *Geochim. Cosmochim. Acta.* **57,** 2271–2282.

Furr, K.A., Lawrence, A.W., Tong, S.S.C., Grandolfo, M.G., Hofstader, R.A., Bache, C.A., Gutemann, W.H., and Lisk, D.J. (1976). Multielement and chlorinated hydrocarbon analysis of municipal sewage sludge of American cities. *Environ. Sci. Technol.* **10,** 683–687.

Furrer, G., and Stumm, W. (1986). The coordination chemistry of weathering. I. Dissolution kinetics of γ-Al_2O_3 and BeO. *Geochim. Cosmochim. Acta.* **50,** 1847–1860.

Gabano, J.P., Etienne, P., and Laurent, J.F. (1965). Étude des proprietes de surface du bioxyde de manganese. *Electrochim. Acta* **10,** 947–963.

Gadde, R.R., and Laitinen, H.A. (1974). Studies of heavy metal adsorption by hydrous iron and manganese oxides. *Anal. Chem.* **46,** 2022–2026.

Gaines, G.L., and Thomas, H.C. (1953). Adsorption studies on clay minerals. II. A formation of the thermodynamics of exchange adsorption. *J. Chem. Phys.* **21,** 714–718.

Gambrell, R.P., Khalid, R.A., Verlow, M.G., and Patrick, W.H., Jr. (1977). "Transformation of Heavy Metals and Plant Nutrients in Dredged Sediments as Affected by Oxidation-Reduction Potential and pH. Part II. Materials and Methods, Results and Discussion," DACW-39-74-C-0076. U.S. Army Engineer Waterways Exp. Stn., Vicksburg, MS.

Gambrell, R.P., Wiesepape, J.B., Patrick, W.H., Jr., and Duff, M.C. (1991). The effects of pH, redox, and salinity on metal release from a contaminated sediment. *Water Air Soil Pollut.* **57–58,** 359–367.

Gapon, Y.N. (1933). On the theory of exchange adsorption in soils. *J. Gen. Chem. USSR (Engl. Transl.)* **3,** 144–160.

Gardiner, W.C., Jr. (1969). "Rates and Mechanisms of Chemical Reactions." Benjamin, New York.

Garrels, R.M., and Christ, C.L. (1965). "Solutions, Minerals, and Equilibria." Freeman, Cooper, San Francisco.

Gast, R.G. (1972). Alkali metal cation exchange on Chambers montmorillonite. *Soil Sci. Soc. Am. Proc.* **36,** 14–19.

Gast, R.G., and Klobe, W.D. (1971). Sodium-lithium exchange equilibria on vermiculite at 25° and 50°C. *Clays Clay Miner.* **19,** 311–319.

Ghosh, K. and Schnitzer, M. (1980). Macromolecular structure of humic substances. Soil Sci. **129,** 266–276.

Gieseking, J.E., ed. (1975). "Soil Components. Inorganic Components," Vol. 2, Springer-Verlag, New York.

Gillman, G.P., and Bell, L.C. (1978). Soil solution studies on weathered soils from tropical North Queensland. *Aust. J. Soil Res.* **16,** 67–77.

Gillman, G.P., and Sumner, M.E. (1987). Surface charge characterization and soil solution composition of four soils from the southern piedmont of Georgia. *Soil Sci. Soc. Am. J.* **51,** 589–594.

Glossary of Soil Science Terms. (1987). Soil Sci. Soc. Am., Madison, WI.

Glossary of Soil Science Terms. (1997). Soil Sci. Soc. Am., Madison, WI.

Goldberg, S. (1992). Use of surface complexation models in soil chemical systems. *Adv. Agron.* **47,** 233–329.

Goldberg, S., and Glaubig, R.A. (1987). Effect of saturating cation, pH and aluminum and iron oxides on the flocculation of kaolinite and montmorillonite. *Clays Clay Miner.* **35,** 220–227.

Goldberg, S., and Johnston, C.T. (2001). Mechanisms of arsenic adsorption on amorphous oxides evaluated using macroscopic measurements, vibrational spectroscopy, and surface complexation modeling. *J. Colloid Interf. Sci.* **234,** 204–216.

Goulding, K.W.T. (1983a). Thermodynamics and potassium exchange in soils and clay minerals. *Adv. Agron.* **36,** 215–261.

Goulding, K.W.T. (1983b). Adsorbed ion activities and other thermodynamic parameters of ion exchange defined by mole or equivalent fractions. *J. Soil Sci.* **34,** 69–74.

Goulding, K.W.T., and Talibudeen, O. (1984). Thermodynamics of K-Ca exchange in soils. II. Effects of mineralogy, residual K and pH in soils from long-term ADAS experiments. *J. Soil Sci.* **35,** 409–420.

Gouy, G. (1910). Sur la constitution de la charge électrique à la surface d'un electrolyte. *Ann. Phys. (Paris), [IV]* **9,** 457–468.

Grahame, D.C. (1947). The electrical double layer and the theory of electrocapillarity. *Chem. Rev.* **41,** 441–501.

Grantham, H.C. and Dove, P.M. (1996). Investigation of bacterial-mineral interactions using fluid tapping mode atomic force microscopy. *Geochim. Cosmochim. Acta,* **60,** 2473–2480.

Greenland, D.J. (1971). Interactions between humic and fulvic acids and clays. *Soil Sci.* **111,** 34–41.

Griffin, R.A., and Jurinak, J.J. (1973). Estimation of acidity coefficients from the electrical conductivity of natural aquatic systems and soil extracts. *Soil Sci.* **116,** 26–30.

Grim, R.E. (1968). "Clay Mineralogy." 2nd ed. McGraw-Hill, New York.

Grim, R.E., and Johns, W.D. (1954). Clay mineral investigation of sediments in the northern Gulf of Mexico. *Clays Clay Miner.* **2,** 81–103.

Grim, R.E., Bray, R.H., and Bradley, W.F. (1937). The mica in argillaceous sediments. *Amer. Minerl.* **22,** 813–829.

Grimme, H. (1968). Die adsorption von Mn, Co, Cu and Zn durch goethite aus verdünnten lösungen. *Z. Pflanzenernähr. Düng. Bodenkunde* **121,** 58–65.

Grossl, P.R., Sparks, D.L., and Ainsworth, C.C. (1994). Rapid kinetics of Cu (II) adsorption/desorption on goethite. *Environ. Sci. Technol.* **28**, 1422–1429.

Grossl, P.R., Eick, M.J., Sparks, D.L., Goldberg, S., and Ainsworth, C.C. (1997). Arsenate and chromate retention mechanisms on goethite. 2. Kinetic evaluation using a pressure-jump relaxation technique. *Environ. Sci. Technol.* **31,** 321–326.

Guggenheim, E.A. (1967). "Thermodynamics." North-Holland, Amsterdam.

Gupta, R.K., Bhumbla, D.K., and Abrol, I.P. (1984). Sodium-calcium exchange equilibria in soils as affected by calcium carbonate and organic matter. *Soil Sci.* **138**[2], 109–114.

Haderlein, S.B., and Pecher, K. (1998). Pollutant reduction in heterogeneous Fe(II)-Fe(III) systems. *In* "Mineral-Water Interfacial Reactions" (D.L. Sparks and T.J. Grundl, eds.), pp. 323–341, ACS Symp. Ser. 715, Am. Chem. Soc., Washington, DC.

Hair, M.I., (1967). "Infrared Spectroscopy in Surface Chemistry." Marcel Dekker, New York.

Hamaker, J. W. and Thompson, J. M. (1972). Adsorption. *In* "Organic Chemicals in the Environment" (C. A. I. Goring and J. W. Hamaker, eds.), pp. 39–151, Dekker, New York.

Haque, R., Lindstrom, F.T., Freed, V.H., and Sexton, R. (1968). Kinetic study of the sorption of 2,4-D on some clays. *Environ. Sci. Technol.* **2,** 207–211.

Hargrove, W.L., and Thomas, G.W. (1982). Titration properties of Al-organic matter. *Soil Sci.* **134,** 216–225.

Harris, D.C. (1987). "Quantitative Chemical Analysis." 2nd ed. Freeman, New York.

Harris, D.K., Ingle, S.E., Magnunson, V.R., and Taylor, D.K. (1981). "REDEQL.UMD," Dept. of Chemistry, University of Minnesota, Duluth.

Harter, R.D., and Smith, G. (1981). Langmuir equation and alternate methods of studying "adsorption" reactions in soils. *In* "Chemistry in the Soil Environment" (R.H. Dowdy, J.A. Ryan, V.V. Volk, D.E. Baker, eds.), Am. Soc. Agron. Spec. Publ. No. 40, pp. 167–182. Am. Soc. Agron./Soil Sci. Soc. Am., Madison, WI.

Hartman, H., Sposito, G., Yang, A., Manne, S., Gould, A.A., and Hansina, P.K. (1990). Molecular-scale imaging of clay mineral surfaces with atomic force microscope. *Clays Clay Miner.* **38,** 337–342.

Hassett, J.J., and Banwart, W.L. (1992). "Soils and their Environment." Prentice-Hall, Englewood Cliffs, NJ.

Hatcher, P.G., Bortiatynski, J.M. and Knicker, H., (1994). NMR techniques (C, H, and N) in soil chemistry, pp. 23–44 in 15th World Congress of Soil Science, Transactions Vol. 3a, Commission II, Acapulco, Mexico.

Havlin, J.L., and Westfall, D.G. (1985). Potassium release kinetics and plant response in calcareous soils. *Soil Sci. Soc. Am. J.* **49,** 366–370.

Hayes, K.F., and Katz, L.E. (1996). Application of x-ray absorption spectroscopy for surface complexation modeling of metal ion sorption. *In* "Physics and Chemistry of Mineral Surfaces" (P.V. Brady, ed.), pp. 147–223. CRC Press, Boca Raton, FL.

Hayes, K.F., and Leckie, J.O. (1986). Mechanism of lead ion adsorption at the goethite-water interface. *In* "Geochemical Processes at Mineral Surfaces" (J.A. Davis and K.F. Hayes, eds.). *ACS Symp. Ser.* **323,** 114–141.

Hayes, K.F., and Leckie, J.O. (1987). Modeling ionic strength effects on cation adsorption at hydrous oxide/solution interfaces. *J. Colloid Interf. Sci.* **115,** 564–572.

Hayes, K.F., Roe, A.L., Brown, G.E., Jr., Hodgson, K.O., Leckie, J.O., and Parks, G.A. (1987). In situ x-ray absorption study of surface complexes: Selenium oxyanions on α-FeOOH. *Science* **238,** 783–786.

Hayes, M.H.B., MacCarthy, P., Malcolm, R.L., and Swift, R.S., eds. (1989). "Humic Substances. II: In Search of Structure." John Wiley & Sons, New York.

Hayes, M.H.B., and Swift, R.S. (1978). The chemistry of soil organic colloids. *In* "The Chemistry of Soil Constituents" (D.J. Greenland and M.H.B. Hayes, eds.), pp. 179–230. John Wiley & Sons (Interscience), New York.

Helfferich, F. (1962a). Ion exchange kinetics. III. Experimental test of the theory of particle-diffusion controlled ion exchange. *J. Phys. Chem.* **66,** 39–44.

Helfferich, F. (1962b). "Ion Exchange." McGraw-Hill, New York.

Helling, C.S., Chester, G., and Corey, R.B. (1964). Contribution of organic matter and clay to soil cation-exchange capacity as affected by the pH of the saturation solution. *Soil Sci. Soc. Am. Proc.* **28,** 517–520.

Hem, J.D. (1970). "Study and interpretation of the chemical characteristics of natural water," *U.S. Geol. Surv. Water-Supply Pap. 2254.*

Hem, J.D. (1978). Redox processes at the surface of manganese oxide and their effects on aqueous metal ions. *Chem. Geol.* **21,** 199-218.

Hem, J.D., Roberson, C.E., and Lind, C.J. (1985). Thermodynamic stability of CoOOH and its coprecipitation with manganese. *Geochim. Cosmochim. Acta* **49,** 801–810.

Hendricks, S.B., and Fry, W.H. (1930). The results of X-ray and mineralogical examination of soil colloids. *Soil Sci.* **29,** 457–476.

Herrero, C.P., Sanz, J. and Serratosa, J.M., (1989). Dispersion of charge deficits in the tetrahedral sheet of phyllosilicates. Analysis from ^{29}Si NMR spectra. *J. Phys. Chem.* **93,** 4311–4315.

Hiemenz, P.C. (1986). "Principles of Colloid and Surface Chemistry." 2nd ed. Dekker, New York.

Hiemstra, T., van Riemsdijk, W.H., and Bruggenwert, M.G.M. (1987). Proton adsorption mechanism at the gibbsite and aluminum oxide solid/solution interface. *Neth. J. Agric. Sci.* **35, 281–293.**

Hileman, B. (1999). Energy department supports CO_2 sequestration research. Chem. and Eng. News, May 3rd.

Hingston, F.J. (1981). A review of anion adsorption. *In* "Adsorption of Inorganics at Solid-Liquid Interfaces" (M.A. Anderson and A.J. Rubin, eds.), pp. 51–90. Ann Arbor Sci., Ann Arbor, MI.

Hingston, F.J., Posner, A.M., and Quirk, J.P. (1972). Anion adsorption by goethite and gibbsite. I. The role of the proton in determining adsorption envelopes. *J. Soil Sci.* **23,** 177–192.

Hochella, M.F., Jr. (1990). Atomic structure, microtopography, composition, and reactivity of mineral surfaces. *In* "Mineral Water-Interface Geochemistry" (M.F. Hochella, Jr. and A.F. White, eds.), Rev. in Mineralogy 23, pp.87–132. Mineral. Soc. Am., Washington, D.C.

Hochella, M.F., Jr, Eggleston, C. M., Elings, V. B., Parks, G. A., Brown, G. E., Jr., Kjoller, K. K. and Wu, C. M. (1989). Mineralogy in two dimensions: Scanning tunneling microscopy of semiconducting minerals with implications for geochemical reactivity. *Amer. Miner.* **74,** 1233–1246.

Hochella, M.F., Jr, Rakovan, J.F., Rosso, K.M., Bickmore, B.R. and Rufe, E. (1998). New directions in mineral surface geochemical research using scanning probe microscopies. *In* "Mineral-Water Interfacial Reactions" (D.L. Sparks and T.J. Grundl, eds.) ACS Symp. Ser., 715, pp. 37–56. Am. Chem Soc., Washington, D.C.

Hogfeldt, E., Ekedahl, E., and Sillén, L.G. (1950). Activities of the components in ion exchangers with multivalent ions. *Acta Chem. Scand.* **4,** 828–829.

Hohl, H., and Stumm, W. (1976). Interaction of Pb^{2+} with hydrous γ-Al_2O_3. *J. Colloid Interf. Sci.* **55,** 281–288.

Hohl, H., Sigg, L., and Stumm, W. (1980). Characterization of surface chemical properties of oxides in natural waters. *Adv. Chem. Ser.* **189,** 1–31.

Hopkins, C.G., Knox, W.H., and Pettit, J.H. (1903). A quantitative method for determining the acidity of soils. *USDA Bur. Chem. Bull.* **73,** 114–119.

Hsu, P.H. (1977). Aluminum hydroxides and oxyhydroxides. *In* "Minerals in Soil Environments" (J.B. Dixon and S.B. Weed, eds.) pp. 99–143. Soil Sci. Soc. Am., Madison, WI.

Hsu, P.H. (1989). Aluminum oxides and oxyhydroxides. *In* "Minerals in Soil Environments" (J.B. Dixon and S.B. Weed, eds.), SSSA Book Ser. 1, pp. 331–378. Soil Sci. Soc. Am., Madison, WI.

Hsu, P.H., and Bates, T.E. (1964). Formation of x-ray amorphous and crystalline aluminum hydroxides. *Mineral. Mag.* **33,** 749–768.

Hsu, P.H., and Rich, C.I. (1960). Aluminum fixation in a synthetic cation exchanger. *Soil Sci. Soc. Am. Proc.* **24,** 21–25.

Huang, C.P., and Stumm, W. (1973). Specific adsorption of cations on hydrous γ-Al_2O_3. *J. Colloid Interf. Sci.* **43,** 409–420.

Huang, P.M. (1988). Ionic factors affecting aluminum transformations and the impact on soil and environmental sciences. *Adv. Soil Sci.* **8,** 1–78.

Huang, P.M. (1989). Feldspars, olivines, pyroxenes, and amphiboles. *In* "Minerals in Soil Environments" (J.B. Dixon and S.B. Weed, eds.), SSSA Book Ser. No. 1, pp. 975–1050. Soil Sci. Soc. Am., Madison, WI.

Huang, P.M. (1991). Kinetics of redox reactions on manganese oxides and its impact on environmental quality. *In* "Rates of Soil Chemical Processes" (D.L. Sparks and D.L. Suarez, eds.), SSSA Spec. Pub. No. 27, pp. 191–230. Soil Sci. Soc. Am., Madison, WI.

Huang, P.M., and Schnitzer, M., eds. (1986). "Interactions of Soil Minerals with Natural Organics and Microbes." SSSA Spec. Pub. No. 17. Soil Sci. Soc. Am., Madison, WI.

Hug, S.J. (1997). *In situ* Fourier transform infrared measurements of sulfate adsorption on hematite in aqueous solutions. *J. Colloid Interf. Sci.* **188,** 415–422.

Hurlbut, C.S., Jr. (1966). "Dana's Manual of Mineralogy." 17th ed. John Wiley & Sons, New York.

Hurlbut, C.S., Jr., and Klein, C. (1977). "Manual of Mineralogy" (after James D. Dana). 19th ed. John Wiley & Sons, New York.

Hutcheon, A.T. (1966). Thermodynamics of cation exchange on clay: Ca-Montmorillonite. *J. Soil Sci.* **17,** 339–355.

Ingle, S.E., Schultt, M.D., and Shults, D.W. (1978). "A User's Guide for REDEQL.EPA," EPA-600/3-78-024. U.S. Environ. Prot. Agency, Corvallis, OR.

Ishiwatari, R. (1975). Chemical nature of sedimentary humic acid. *In* "Humic Substances, Their Structure and Function in the Biosphere" (D. Povoledo and H.L. Golterman, eds.), pp. 87–107. Centre for Agricultural Publications and Documentation. Wageningen.

Jackson, M.L. (1963). Aluminum bonding in soils: A unifying principle in soil science. *Soil Sci. Soc. Am. Proc.* **27,** 1–10.

Jackson, M.L. (1964). Chemical composition of soils. *In* "Chemistry of the Soil" (F.E. Bear, ed.), pp. 87–112. Reinhold, New York.

James, B.R., and Bartlett, R.J. (1983). Behavior of chromium in soils. IV. Interactions between oxidation-reduction and organic complexation. *J. Environ. Qual.* **12,** 173–176.

James, B.R., and Bartlett, R.J. (2000). Redox phenomena. *In* "Handbook of Soil Science" (M. E. Sumner, ed.), pp. B169-B194, CRC Press, Boca Raton, FL.

James, R.O., and Healy, T.W. (1972). Adsorption of hydrolyzable metal ions at the oxide-water interface. III. A thermodynamic model of adsorption. *J. Colloid Interf. Sci.* **41,** 65–80.

Jardine, P. M., Dunnivant, F. M., Selim, H. M. and McCarthy, J. F. . (1992). Comparison of models for describing the transport of dissolved organic carbon in aquifer columns. *Soil Sci. Soc. Am. J.* **56,** 393–401.

Jardine, P.M., Jacobs, G.K., and O'Dell, J.D. (1993). Unsaturated transport processes in undisturbed heterogeneous porous media. II. Co-contaminants. *Soil Sci. Soc. Am. J.* **57,** 954–962.

Jardine, P.M., and Sparks, D.L. (1984a). Potassium-calcium exchange in a multireactive soil system. I. Kinetics. *Soil Sci. Soc. Am. J.* **48,** 39–45.

Jardine, P.M., and Sparks, D.L. (1984b). Potassium-calcium exchange in a multireactive soil system. II. Thermodynamics. *Soil Sci. Soc. Am. J.* **48,** 45–50.

Jardine, P.M., and Zelazny, L.W. (1996). Surface reactions of aqueous aluminum species. *In* "The Environmental Chemistry of Aluminum", 2nd ed. (G. Sposito, ed.), pp. 221–270. CRC Press (Lewis Publishers), Boca Raton, FL.

Jaworski, N.A., Howarth, R.W., and Hetling, L.J. (1997). Atmospheric deposition of nitrogen oxides onto the landscape contributes to coastal eutrophication in the northeastern United States. *Environ. Sci. Technol.* **31,** 1995–2004.

Jenne, E.A. (1979). Chemical modeling-goals, problems, approaches, and priorities. *In* "Chemical Modeling in Aqueous Systems" (E.A. Jenne, ed.), Am. Chem. Soc. Symp. Series No. 93, pp. 3–21. Am. Chem. Soc., Washington, DC.

Jenny, H. (1941). "Factors of Soil Formation." McGraw-Hill, New York.

Jenny, H. (1961). Reflections on the soil acidity merry-go-round. *Soil Sci. Soc. Am. Proc.* **25,** 428–432.

Jenny, H., Bingham, F., and Padilla-Saravia, B. (1948). Nitrogen and organic matter contents of equatorial soils of Columbia, South America. *Soil Sci.* **66,** 173–186.

Jensen, H.E. (1972). Potassium-calcium exchange on a montmorillonite and a kaolinite clay. I. A test on the Argersinger thermodynamic approach. *Agrochimica* **17,** 181–190.

Jensen, H.E., and Babcock, K.L. (1973). Cation exchange equilibria on a Yolo loam. *Hilgardia* **41,** 475–487.

Johansson, G. (1960). On the crystal structure of some basic aluminum salts. *Acta Chem. Scand.* **14,** 771–773.

Johnsson, P.A., Eggleston, C.M. and Hochella, M.F., Jr. (1991). Imaging molecular-scale structure and microtopography of hematite with atomic force microscope. *Am. Mineral.* **76,** 1442–1445.

Johnston, C.T., Sposito, G. and Earl, W.L. (1993). Surface spectroscopy of environmental particles by Fourier-transform infrared and nuclear magnetic resonance spectroscopy. *In* "Environmental Particles" (J. Buffle and H.P.v. Leeuwen, Eds.), pp. 1–36. Lewis Publ., Boca Raton, FL.

Josselyn, M.N., Faulkner, S.P., and Patrick, W.H., Jr. (1990). Relationships between seasonally wet soils and occurrence of wetland plants in California. *Wetlands* **10,** 7–26.

Junta, J.L., and Hochella, M.F., Jr., (1994). Manganese (II) oxidation at mineral surfaces: A microscopic and spectroscopic study. *Geochim. Cosmochim. Acta* **58,** 4985–4999.

Jurinak, J.J., and Suarez, D.L. (1990). The chemistry of salt-affected soils and water. In "Agricultural Salinity Assessment and Management" (K.K. Tanji, ed.), ASCE Manuals Prac. No. 71, pp. 42–63. Am. Soc. Civ. Eng., New York.

Kämpf, N., Scheinost, A.L., and Schulze, D.G. (2000). Oxide minerals. *In* "Handbook of Soil Science" (M. E. Sumner, ed.), pp. F–125–168. CRC Press, Boca Raton, FL.

Karickhoff, S.W. (1980). Sorption kinetics of hydrophobic pollutants in natural sediments. *In* "Contaminants and Sediments" (R.A. Baker, ed.), pp. 193–205. Ann Arbor Sci., Ann Arbor, MI.

Karickhoff, S.W., and Morris, K.R. (1985). Sorption dynamics of hydrophobic pollutants in sediment suspensions. *Environ. Toxicol. Chem.* **4,** 469–479.

Karickhoff, S.W., Brown, D.S., and Scott, T.A. (1979). Sorption of hydrophobic pollutants on natural sediments. *Water Res.* **13,** 241–248.

Katz, L.E., and Boyle-Wight, E.J. (2001). Application of spectroscopic methods to sorption model parameter estimation. *In* "Physical and Chemical Processes of Water and Solute Transport/Retention in Soils" (H. M. Selim and D. L. Sparks, eds.) SSSA Spec. Publ. 56, Soil Sci. Soc. Am., Madison, WI.

Katz, L. E., and Hayes, K. F. (1995a). Surface complexation modeling. I. Strategy for modeling monomer complex formation at moderate surface coverage. *J. Colloid Interf. Sci.* **170,** 477–490.

Katz, L. E., and Hayes, K. F. (1995b). Surface complexation modeling. II. Strategy for modeling polymer and precipitation reactions at high surface coverage. *J. Colloid Interf. Sci.* **170,** 491–501.

Kelley, W.P., and Brown, S.M. (1926). Ion exchange in relation to soil acidity. *Soil Sci.* **21,** 289–302.

Kelley, W.P., Dore, W.H., and Brown, S.M. (1931). The nature of the base exchange material of bentonite, soils and zeolites, as revealed by chemical investigation and X-ray analysis. *Soil Sci.* **31,** 25–55.

Keren, R., and Miyamoto, S. (1990). Reclamation of saline, sodic, and boron-affected soils. *In* "Agricultural Salinity Assessment and Management" (K.K. Tanji, ed.), ASCE Manuals Prac. No. 71, pp. 410–431. Am. Soc. of Civ. Eng., New York.

Keren, R., and Sparks, D.L. (1994). Effect of pH and ionic strength on boron adsorption by pyrophyllite. *Soil Sci. Soc. Am. J.* **58**, 1095–1100.

Kern, J.S. and Johnson, M.G. (1993). Conservation tillage impacts on national soil and atmospheric carbon levels. *Soil. Sci. Soc. Am. J.* **57**, 200–210.

Kerr, H.W. (1928). The nature of base exchange and soil acidity. *J. Am. Soc. Agron.* **20**, 309–355.

Khan, S.U. (1973). Equilibrium and kinetic studies on the adsorption of 2,4-D and picloram on humic acid. *Can. J. Soil Sci.* **53**, 429–434.

Kharaka, Y.K., and Barnes, J. (1973). "SOLMNEQ: Solution-Mineral Equilibrium Computations." 73-002, NTIS PB-215899. U.S. Geol. Surv. Water Res. Invest., Washington, DC.

Kielland, J. (1937). Individual activity coefficients of ions in aqueous solutions. *J. Am. Chem. Soc.* **59**, 1675–1678.

Kinniburgh, D.G., and Jackson, M.L. (1981). Cation adsorption by hydrous metal oxides and clay. *In* "Adsorption of Inorganics at Solid-Liquid Interfaces" (M.A. Anderson and A.J. Rubin, eds.), pp. 91–160. Ann Arbor Sci., Ann Arbor, MI.

Kinniburgh, D.G., Jackson, M.L., and Syers, J.K. (1976). Adsorption of alkaline earth, transition, and heavy metal cations by hydrous oxide gels of iron and aluminum. *Soil Sci. Soc. Am. J.* **40**, 796–799.

Kinniburgh, D.G., and Miles, D.L. (1983). Extraction and chemical analysis of interstitial water from soils and rocks. *Environ. Sci. Technol.* **17**, 372–368.

Kinraide, T.B. (1991). Identity of the rhizotoxic aluminum species. *Plant Soil* **134**, 167–178.

Klausen, J. (1995). Abiotic redox transformations of aromatic nitro and amino compounds in suspensions of soil minerals: Laboratory studies in batch and flow-through reactors (batch reactors, iron oxides, manganese dioxide), Ph.D. Thesis. ETH, Switzerland.

Klein, C., and Hurlbut, C.S., Jr. (1985). "Manual of Mineralogy." 20th ed. John Wiley & Sons, New York.

Kononova, M.M. (1966). "Soil Organic Matter." Pergamon, New York.

Kopp, J.F., and Kroner, R.C. "Trace Metals in Waters of the United States," U.S. Environ. Prot. Agency, Washington, DC.

Koskinen, W.C., and Harper, S.S. (1990). The retention process: Mechanisms. *In* "Pesticides in the Soil Environment: Processes, Impacts, and Modeling" (H.H. Cheng, ed.), SSSA Book Ser. 2, pp. 51–77. Soil Sci. Soc. Am., Madison, WI.

Kostka, J.E., and Nealson, K.H. (1995). Dissolution and reduction of magnetite by bacteria. *Environ. Sci. Technol.* **29**, 2535–2540.

Kozawa, A. (1959). On an ion exchange property. *J. Electrochem. Soc.* **106**, 552–556.

Krishnamoorthy, C., and Overstreet, R. (1949). Theory of ion exchange relationships. *Soil Sci.* **68**, 307–315.

Kunze, G.W., and Dixon, J.B. (1986). Pretreatment for mineralogical analysis. *In* "Methods of Soil Analysis. Part 1 Physical and Mineralogical Methods"

(A. Klute, ed.), 2nd ed. pp. 91–100. Am. Soc. Agron., Madison, WI.

Kuo, S., and Lotse, E.G. (1974). Kinetics of phosphate adsorption and desorption by lake sediments. *Soil Sci. Soc. Am. Proc.* **38,** 50–54.

Kurbatov, J.D., Kulp, J.L., and Mack, E. (1945). Adsorption of strontium and barium ions and their exchange on hydrous ferric oxide. *J. Am. Chem. Soc.* **67,** 1923–1929.

Ladeira, A.C.Q., Ciminelli, V.S.T., Duarte, H.A., Alves, M.C.M., and Ramos, A.Y. (2001). Mechanism of anion retention from EXAFS and density functional calculations: Arsenic(V) adsorbed on gibbsite. *Geochim. Cosmochim. Acta.* **65**(8), 1211–1217.

Lal, R. (2001). World cropland soils as a source of sink for atmospheric carbon. *Adv. Agron.* **71,** 145–191.

Langmuir, D. (1997). "Aqueous Environmental Geochemistry." Prentice-Hall, Englewood Cliffs, NJ.

Langmuir, I. (1918). The adsorption of gases on plane surface of glass, mica, and platinum. *J. Am. Chem. Soc.* **40,** 1361–1382.

Lasaga, A.C., and Kirkpatrick, R.J., eds. (1981). "Kinetics of Geochemical Processes." Mineralogical Society of America, Washington, DC.

Lasaga, A. C. (1998). "Kinectic Theory in the Earth Sciences." Princeton Univ. Press, Princeton, NJ.

Latimer, W.M. (1952). "Oxidation Potentials." 2nd ed. Prentice-Hall, Englewood Cliffs, NJ.

Läuchli, A., and Epstein, E. (1990). Plant response to saline and sodic conditions. *In* "Agricultural Salinity Assessment and Management" (K.K. Tanji, ed.), ASCE Manuals Prac. No. 71, pp. 113–137. Am. Soc. Civ. Eng., New York.

Laudelout, H., and Thomas, H.C. (1965). The effect of water activity on ion-exchange selectivity. *J. Phys. Chem.* **69,** 339–341.

Laudelout, H., van Bladel, R., Bolt, G.H., and Page, A.L. (1967). Thermodynamics of heterovalent cation exchange reactions in a montmorillonite clay. *Trans. Faraday Soc.* **64,** 1477–1488.

Lee, L. S., Rao, P. S. C., Brusseau, M. L., and Ogwada, R. A. (1988). Nonequilibrium sorption of organic contaminants during flow through columns of aquifer materials. *Environ. Toxicol. Chem.* **7,** 779–793.

Leenheer, J.A. (1980). Origin and nature of humic substances in the water of the Amazon River basin. *Acta Amazonica* **10,** 513–526.

Leenheer, J.A., and Ahlrichs, J.L. (1971). A kinetic and equilibrium study of the adsorption of carbaryl and parathion upon soil organic matter surfaces. *Soil Sci. Soc. Am. Proc.* **35,** 700–704.

Lewis, G.N., and Randall, M. (1961). "Thermodynamics," (rev. by K.S. Pitzer and L. Bremer) 2nd ed. McGraw-Hill, New York.

Lewis-Russ, A. (1991). Measurement of surface charge of inorganic geologic materials: Techniques and their consequences. *Adv. Agron.* **46,** 199–243.

Liberti, K., and Helfferich, F.G., eds. (1983). "Mass Transfer and Kinetics of Ion Exchange." NATO ASI Ser. E., Vol. 71. Martinus Nijhoff, The Hague, The Netherlands.

Lindsay, W.L. (1979). "Chemical Equilibria in Soils." John Wiley & Sons, New York.

Lindsay, W.L. (1981). Solid-phase solution equilibria in soils. *In* "Chemistry in the Soil Environment" (R.H. Dowdy, J.A. Ryan, V.V. Volk, D.E. Baker, eds.), Am. Soc.

Agron. Spec. Publ. 40, pp. 183–202. Am. Soc. of Agron./Soil Sci. Soc. Am., Madison, WI.

Liu, C.W., and Narasimhan, T.N. (1989). Redox-controlled multiple-species reactive chemical transport. 1. Model development. *Water Resour. Res.* **25**: 869–882.

Loach, P.A. (1976). Oxidation-reduction potentials, absorbance bonds, and molar absorbance of compounds used in biochemical studies. *In* "Handbook of Biochemistry and Molecular Biology: Physical and Chemical Data" (G.D. Fasman, ed.), Vol. 1, 3rd ed. pp. 122–130. Chemical Rubber, Cleveland, OH.

Loganathan, P., and Burau, R.G. (1973). Sorption of heavy metal ions by a hydrous manganese oxide. *Geochim. Cosmochim. Acta* **37**, 1277–1293.

Lovely, D.R. (1991). Dissimilatory Fe(III) and Mn(IV) reduction. *Microbiol. Rev.* **55**, 259–287.

Lövgren, L., Sjöberg, S. and Schindler, P. W. (1990). Acid/base reactions and Al(III) complexation at the surface of goethite. *Geochim. Cosmochim. Acta.* **54**, 1301–1306.

Low, M.J.D. (1960). Kinetics of chemisorption of gases on solids. *Chem. Rev.* **60**, 267–312.

Low, P.F. (1955). The role of aluminum in the titration of bentonite. *Soil Sci. Soc. Am. Proc.* **19**, 135–139.

Lyklema, J. (1991). "Fundamentals of Interface and Colloid Science, Vol. 1. Fundamentals." Academic Press, London.

Ma, L.Q., Komar, K.M., Tu, C., Zhang, W., Cai, Y., and Kennelley, E.D. (2001). A fern that hyperaccumulates arsenic. *Nature.* **409**, 579.

Maas, E.V., and Hoffman, G.J. (1977). Crop salt tolerance-current assessment. *J. Irrig. Drain. Div., Am. Soc. Civ. Eng.* **103(IR 2)**, 115–134.

Maas, E. V. (1990). Crop salt tolerance. *In* "Agricultural Salinity Assessment and Management" (K. K. Tanji, ed.), pp. 262–304. Am. Soc. of Civ. Eng. Manuals and Reports on Engineering Practice No. 71. Am. Soc. of Civ. Eng., New York.

MacCarthy, P. (2001). Principles of humic substances. Proc. Humic Substances Seminar, Boston, MA.

Malati, M.A., and Estefan, S.F. (1966). The role of hydration in the adsorption of alkaline earth ions onto quartz. *J. Colloid Interf. Sci.* 24, 306–307.

Malcolm, R.L., Wershaw, R.L., Thurman, E.M., Aiken, G.R., Pickney, D.J., and Kaakinen, J. (1981). Reconnaisance sampling and characterization of aquatic humic substances at the Yuma Desalting Test Facility, Arizona. *U.S. Geol. Surv. Water Resour. Invest. 81–42.*

Malm, W.C., Sisler, J.F., Huffman, D., Eldred, R.A., and Cahill, T.A. (1994). Spatial and seasonal trends in particle concentration and optical extinction in the United States. *J. Geophys. Res.* **99**, 1347–1370.

Manahan, S.E. (1991). "Environmental Chemistry." 5th ed. Lewis Publishers, Chelsea, MI.

Manceau, A., and Charlet, L. (1994). The mechanism of selenate adsorption on goethite and hydrous ferric oxide. *J. Colloid Interf. Sci.* **168**, 87–93.

Manceau, A., Lanson, B., Schlegel, M.L., Harge, J.C., Musso, M., Eybert-Berard, L., Hazemann, J.-L., Chateigner, D., and Lamble, G.M. (2000). Quantitative Zn speciation in smelter-contaminated soils by EXAFS spectroscopy. *Am. Jour. of Sci.* **300**, 289–343.

Mann, L.K. (1986). Changes in soil carbon storage after cultivation. *Soil Sci.* **142**, 279–288.

Manne, S., Cleveland, J.P., Gaub, H.E., Stucky, G.D. and Hansma, P.K. (1994). Direct visualization of surfactant hemimicelles by force microscopy of the electrical double layer. *Langmuir* **10**, 4409–4413.

Manning, B.A., Fendorf, S.E., and Goldberg, S. (1998). Surface structures and stability of arsenic(III) on goethite: Spectroscopic evidence for inner-sphere complexes. *Environ. Sci. Technol.* **32**, 2383–2388.

Marion, G.M., and Babcock, K.L. (1976). Predicting specific conductance and salt concentration of dilute aqueous solution. *Soil Sci.* **122,** 181–187.

Marion, G.M., Hendricks, D.M., Dutt, G.R., and Fuller, W.H. (1976). Aluminum and silica solution in soils. *Soil Sci.,* **127,** 76–85.

Marshall, C.E. (1964). "The Physical Chemistry and Mineralogy of Soils, Vol.1." John Wiley & Sons, New York.

Martell, A.E., and Smith, R.N. (1976). "Critical Stability Constants." Plenum, New York.

Martin, A.E., and Reeve, R. (1958). Chemical studies of podzolic illuvial horizons: III. Titration curves of organic matter suspensions. *J. Soil Sci.* **9,** 89–100.

Massoud, F.J. (1981). "Salt Affected Soils at a Global Scale and Concepts for Control," Tech. Pap. FAO Land and Water Development Div., Rome.

Mattigod, S.V., and Zachara, J.M. (1996). Equilibrium modeling in soil chemistry. *In* "Methods of Soil Analysis: Part 3—Chemical Methods" (D.L. Sparks, ed.), pp. 1309–1358. SSSA Book Ser. 5, Soil Sci. Soc. Am., Madison, WI.

Mattson, S. (1928). Anionic and cationic adsorption by soil colloidal materials of varying SiO_2/Al_2O_3 +Fe_2O_3 ratio. *Trans. Int. Congr. Soil Sci., 1st, Vol. II.* pp. 199–211, Washington, DC.

Maurice, P.A. (1996). Scanning probe microscopy of mineral surfaces. *In* "Environmental Particles, Structure and Surface Reactions of Soil Particles, Vol. 4" (P.M. Huang, N. Senesi, and J. Buffle, eds.), pp. 109–153. John Wiley & Sons, N.Y.

Maurice, P.A. (1998). Atomic force microscopy as a tool for studying the reactivities of environmental particles. *In* "Mineral-Water Interfacial Reactions" (D.L. Sparks and T.J. Grundl, eds.), ACS Symp. Ser., 715, pp. 57–66. Am. Chem Soc., Washington, D.C.

Maurice, P.A., Hochella, M.F., Jr., Parks, G.A., Sposito, G. and Schwertmann, U. (1995). Evolution of hematite surface microtopography upon dissolution by simple organic acids. *Clays Clay Miner.* **43**(1), 29–38.

McBride, M.B. (1991). Processes of heavy and transition metal sorption by soil minerals. *In* "Interactions at the Soil Colloid-Soil Solution Interface", Vol. 190. (G.H. Bolt, M.F.D. Boodt, M.H.B. Hayes, and M.B. McBride, Eds.), pp. 149–176. Kluwer Academic Publishers, Dordrecht.

McBride, M. B. (1994). "Environmental Chemistry of Soils." Oxford University Press, New York.

McBride, M. B., Fraser, A.R., and McHardy, W.J. (1984). Cu^{2+} interaction with microcrystalline gibbsite. Evidence for oriented chemisorbed copper ions. *Clays Clay Miner.* **32**, 12–18.

McCall, P.J., and Agin, G.L. (1985). Desorption kinetics of picloram as affected by residence time in the soil. *Environ. Toxicol. Chem.* **4**, 37–44.

McDuff, R.E., Morel, F.M.M., and Morgan, J.J. (1973). "Description and use of the chemical equilibrium program REDEQL2," California Institute of Technology, Pasadena, CA.

McFee, W.W., Byrnes, W.R., and Stockton, J.G. (1981). Characteristics of coal mine overburden important to plant growth. *J. Environ. Qual.* **10**, 300–308.

McKenzie, R.M. (1972). The sorption of some heavy metals by the lower oxides of manganese. *Geoderma* **8**, 29–35.

McKenzie, R.M. (1980). The adsorption of lead and other heavy metals on oxides of manganese and iron. *Aust. J. Soil Res.* **18**, 61–73.

McKenzie, R.M. (1989). Manganese oxides and hydroxides. *In* "Minerals in Soil Environments" (J.B. Dixon and S.B. Weed, eds.), SSSA Book Ser. 1, pp. 439–465. Soil Sci. Soc. Am., Madison, WI.

McLaren, R.G., Backes, C.A., Rate, A.W., and Swift, R.S. (1998). Cadmium and cobalt desorption kinetics from soil clays: Effect of sorption period. *Soil Sci. Soc. Am. J.* **62**, 332–337.

McLaughlin, M.J., and Tiller, K.G. (1994). Chloro-complexation of cadmium in soil solutions of saline/sodic soils increases phytoavailability of cadmium. 15th World Congress of Soil Science, Transactions Vol. 3b, pp. 195–196, Acapulco, Mexico.

McLean, E.O. (1976). The chemistry of soil aluminum. *Commun. Soil Sci. Plant Anal.* **7**, 619–636.

McMillan, P.F. and Hofmeister, A.M., (1988). Infrared and raman spectroscopy. *In* "Spectroscopic Methods in Mineralogy and Geology" (F.C. Hawthorne, ed.), pp. 573–630. Mineralogical Society of America, Washington D.C.

McNeal, B.L., and Coleman, N.T. (1966). Effect of solution composition on soil hydraulic conductivity. *Soil Sci. Soc. Am. Proc.* **30**, 308–312.

Mehlich, A. (1952). Effect of iron and aluminum oxides on the release of calcium and on the cation-anion exchange properties of soils. *Soil Sci.* **73**, 361–374.

Melchior, D.C., and Bassett, R.L., eds. (1990). "Chemical Modeling of Aqueous Systems II." ACS Symp. Ser. No. 416, Am. Chem. Soc., Washington, DC.

Mendelssohn, I.A., McKee, K.L., and Patrick, W.H., Jr. (1981). Oxygen deficiency in Spartina alterniflora roots: Metabolic adaptation to anions. *Science* **214**, 439–441.

Mesarzos, E. (1992). Occurrence of atmospheric acidity. *In* "Atmospheric Acidity: Sources, Consequences and Abatement" (M. Radojevic and R.M. Harrison, eds.), pp. 1–37. Elsevier Applied Science, London.

Miller, C.T., and Pedit, J. (1992). Use of a reactive surface-diffusion model to describe apparent sorption-desorption hysteresis and abiotic degradation of lindane in a subsurface material. *Environ. Sci. Technol.* **26**[7], 1417–1427.

Moore, D.M., and Reynolds, R.C., Jr. (1989). "X-ray Diffraction and the Identification and Analysis of Clay Minerals." Oxford Univ. Press, New York.

Moore, J.N., Walker, J.R., and Hayes, T.H. (1990). Reaction scheme for the oxidation of As (III) to As (V) by birnessite. *Clays Clay Miner.* **38**, 549–555.

Moore, M.V., Pace, M.L., Mather, J.R., Murdoch, P.S., Howarth, R.W., Folt, C.L., Chen, C.Y., Hemond, H.F., Flebbe, P.A., and Driscoll, C.T. (1997). Potential effects of climate change on freshwater ecosystems of the New England/Mid-Atlantic region. *Hydrological Processes* **11,** 925–947.

Moore, W.J. (1972). "Physical Chemistry." 4th ed. Prentice-Hall, Englewood Cliffs, NJ.

Morel, F.M.M. (1983). "Principles of Aquatic Chemistry." John Wiley & Sons, New York.

Morel, F.M.M., and Morgan, J.J. (1972). A numerical method for computing equilibrium in aqueous chemical systems. *Environ. Sci. Technol.* **6,** 58–67.

Mortensen, J.L. (1963). Complexing of metals by soil organic matter. *Soil Sci. Soc. Am. Proc.* **27,** 179–186.

Mortland, M.M. (1970). Clay-organic complexes and interactions. *Adv. Agron.* **22,** 75–117.

Mortland, M.M., Shaobai, S., and Boyd, S.A. (1986). Clay-organic complexes as adsorbents for phenol and chlorophenols. *Clays Clay Miner.* **34,** 581–585.

Murray, D.J., Healy, T.W., and Fuerstenau, D.W. (1968). The adsorption of aqueous metals on colloidal hydrous manganese oxide. *In* "Adsorption from Aqueous Solution," No. 79, Adv. Chem. Ser., pp. 74–81. Am. Chem. Soc., Washington, DC.

Murray, J.W. (1975). The interaction of cobalt with hydrous manganese dioxide. *Geochim. Cosmochim. Acta* **39,** 635–647.

Murray, J.W., and Dillard, J.G. (1979). The oxidation of cobalt (II) adsorbed on manganese dioxide. *Geochim. Cosmochim. Acta* **43,** 781–787.

Myneni, S.C.B., Brown, J.T., Martinez, G.A. and Meyer-Ilse, W. (1999). Imaging of humic substance macromolecular structures in water. *Science* **286,** 1335–1337.

NRC Report. (2001). "Arsenic in Drinking Water – 2001 Update." National Academy Press, Washington, DC.

Namjesnik-Dejanovic, K., and Maurice, P.A. (2000). Conformations and aggregate structure of sorbed natural organic matter on muscovite and hematite. *Geochem. Cosmochim. Acta.* **65,** 1047–1057.

Newman, A.C.D., ed. (1987). "Chemistry of Clays and Clay Minerals." Mineral. Soc. Eng. Monogr. No. 6. Longman Group Ltd., Harlow, Essex, England.

Newman, A.C.D., and Brown, G. (1987). The chemical constitution of clays. *In* "Chemistry of Clays and Clay Minerals" (A.C.D. Newman, ed.), Mineral. Soc. Eng. Monogr. No. 6, pp. 1–128. Longman Group Ltd., Harlow, Essex, England.

New York Times (1991). Military Has New Strategic Goal in Cleanup of Vast Toxic Waste. (K. Schneider) August 5, p. D3. New York.

New York Times. (1993). What price cleanup: New view calls environmental policy misguided. March 21. Sea-dumping ban, good politics, but not necessarily good policy. March 22. Animal tests as risk clues, the best data may fall short. March 23. How a rebellion over environmental rules grow from a patch of weeds. March 24. Second chance on the environment. March 25. New York.

New York Times. (2001). New pollution tool: Toxic avengers with leaves. March 6, pp. D1 and D7, (Andrew C. Revkin). New York.

Nkedi-Kizza, P., Biggar, J. W., Selim, H. M., van Genuchten, M. Th., Wierenga, P. J. Davidson, J. M. and Nielsen, D. R. (1984). On the equivalence of two conceptual models for describing ion exchange during transport through an aggregated Oxisol. *Water Resour. Res.* **20,** 1123–1130.

Nordstrom, D.K. and May, H. M. (1996). Aqueous equilibrium data for mononuclear aluminum species. *In* "The Environmental Chemistry of Aluminum," 2nd ed., (G. Sposito, ed.), pp. 39–80. CRC Press (Lewis Publishers), Boca Raton, FL.

O'Day, P.A. (1992). Structure, bonding, and site preference of cobalt (II) sorption complexes on kaolinite and quartz from solution and spectroscopic studies. Ph.D. Dissertation, Stanford University, CA, 208 pp.

O'Day, P.A. (1999). Molecular environmental geochemistry. *Rev. Geophys.* **37,** 249–274.

O'Day, P.A., Brown, G.E., Jr., and Parks, G.A. (1994a). X-ray absorption spectroscopy of cobalt (II) multinuclear surface complexes and surface precipitates on kaolinite. *J. Colloid Interf. Sci.* **165,** 269–289.

O'Day, P.A., Chisholm-Brause, C.J., Towle, S.N., Parks, G.A., and Brown, G.E., Jr. (1996). X-ray absorption spectroscopy of Co(II) sorption complexes on quartz (α-SiO_2) and rutile (TiO_2). *Geochim. Cosmochim. Acta.* **60**(14), 2515-2532.

O'Day, P.A., Parks, G.A., and Brown, G.E., Jr. (1994b). Molecular structure and binding sites of cobalt(II) surface complexes on kaolinite from X-ray absorption spectroscopy. *Clays Clay Miner.* **42,** 337–355.

Ogwada, R.A., and Sparks, D.L. (1986a). Use of mole or equivalent fractions in determining thermodynamic parameters for potassium exchange in soils. *Soil Sci.* **141,** 268–273.

Ogwada, R.A., and Sparks, D.L. (1986b). A critical evaluation on the use of kinetics for determining thermodynamics of ion exchange in soils. *Soil Sci. Soc. Am. J.* **50,** 300–305.

Oliver, B.G., and Cosgrove, E.G. (1975). Metal concentrations in the sewage, effluents, and sludges of some southern Ontario wastewater treatment plants. *Environ. Lett.* **9,** 75–90.

Olsen, S.R., and Watanabe, F.S. (1957). A method to determine a phosphorus adsorption maximum of soils as measured by the Langmuir isotherm. *Soil Sci. Soc. Am. Proc.* **21,** 144–149.

Onken, A.B., and Matheson, R.L. (1982). Dissolution rate of EDTA-extractable phosphate from soils. *Soil Sci. Soc. Am. J.* **46,** 276–279.

O'Reilly, S.E., Strawn, D.G., and Sparks, D.L. (2001). Residence time effects on arsenate adsorption/desorption mechanisms on goethite. *Soil Sci. Soc. Am. J.* **65,** 67–77.

Oscarson, D.W., Huang, P.M., and Liaw, W.K. (1980). The oxidation of arsenite by aquatic sediments. *J. Environ. Qual.* **9,** 700–703.

Oscarson, D.W., Huang, P.M., Defosse, C., and Herbillon, A. (1981). The oxidative power of Mn (IV) and Fe (III) oxides with respect to As (III) in terrestrial and aquatic environments. *Nature (London)* **291,** 50–51.

Oscarson, D.W., Huang, P.M., and Hammer, U.T. (1983). Oxidation and sorption of arsenite by manganese dioxide as influenced by surface coatings of iron and aluminum oxides and calcium carbonate. *Water, Air, Soil Pollut.* **20,** 233–244.

Oster, J.D., Shainberg, I., and Wood, J.D. (1980). Flocculation value and gel structure of Na/Ca montmorillonite and illite suspensions. *Soil Sci. Soc. Am. J.* **44,** 955–959.

Ostergren, J.D., Brown, G.E., Jr., Parks, G.A., and Persson, P. (2000a). Inorganic ligand effects on Pb(II) sorption to goethite (α-FeOOH). II. Sulfate. *J. Colloid. Interf. Sci.* **225,** 482–493.

Ostergren, J.D., Trainor, T.P., Bargar, J.R., Brown, G.E., Jr., and Parks, G.A. (2000b). Inorganic ligand effects on Pb(II) sorption to goethite (α-FeOOH). I. Carbonate. *J. Colloid Interf. Sci.* **225,** 466–482.

Overbeek, J.T. (1952). Electrochemistry of the double-layer. *Colloid Sci.* **1,** 115–193.

Page, H.J. (1926). The nature of soil acidity. *Int. Soc. Soil Sci. Trans. Comm.* **2A,** 232–244.

Papelis, C., and Hayes, K.F. (1996). Distinguishing between interlayer and external sorption sites using clay minerals and x-ray absorption spectroscopy. *Coll. Surfaces,* **107,** 89–96.

Parfitt, R.L. (1978). Anion adsorption by soils and soil materials. *Adv. Agron.* **30,** 1–50.

Parker, D.R., Kinraide, T.B., and Zelazny, L.W. (1988). Aluminum speciation and phytotoxicity in dilute hydroxy-aluminum solutions. *Soil Sci. Soc. Am. J.* **52,** 438–444.

Parker, D.R., Kinraide, T. B., and Zelazny, L.W. (1989). On the phytotoxicity of polynuclear hydroxy-aluminum complexes. *Soil Sci. Soc. Am. J.* **53,** 789–796.

Parker, D.R., Chaney, R.L., and Norvell, W.A. (1995). Chemical equilibrium models: Applications to plant nutrition research. *In* "Chemical Equilibrium and Reaction Models" (R.H. Loeppert, A.P. Schwab, and S. Goldberg, eds.), pp. 163–200. Soil Sci. Soc. Am., Madison, WI.

Parker, J.C., Zelazny, J.W., Sampath, S., and Harris, W.G. (1979). A critical evaluation of the extension of zero point of charge (ZPC) theory to soil systems. *Soil Sci. Soc. Am. J.* **43,** 668–674.

Parks, G.A. (1967). Aqueous surface chemistry of oxides and complex oxide minerals – Isoelectric point and zero point of charge. *In* "Equilibrium Concepts in Natural Water Systems" (W. Stumm, ed.), Adv. Chem. Ser. No. 67, pp. 121–160. Am. Chem. Soc., Washington, DC.

Patrick, W.H., Jr., Gambrell, R.P., and Faulkner, S.P. (1996). Redox measurements of soils. *In* "Methods of Soil Analysis: Part 3—Chemical Methods" (D.L. Sparks, ed.), pp. 1255–1273. SSSA Book Ser. 5, Soil Sci. Soc. Am., Madison, WI.

Pauling, L. (1929). The principles determining the structure of complex ionic crystals. *J. Am. Chem. Soc.* **51,** 1010–1026.

Pauling, L. (1930). The structure of micas and related minerals. *Proc. Natl. Acad. Sci. U.S.A.* **16,** 123–129.

Pauling, L. (1948). "The Nature of the Chemical Bond." 3rd ed. Cornell Univ. Press, Ithaca, NY.

Paver, H., and Marshall, C.E. (1934). The role of aluminum in the reactions of the clays. *Chem. Ind. (London)* 750–760.

Pavlostathis, S.G., and Mathavan, G.N. (1992). Desorption kinetics of selected volatile organic compounds from field contaminated soils. *Environ. Sci. Technol.* **26,** 532–538.

Pearson, R.G. (1963). Hard and soft acids and bases. *J. Am. Chem. Soc.* **85,** 3533–3539.

Pearson, R.G. (1968). Hard and soft acids and bases, HSAB. Part 1. Fundamental principles. *J. Chem. Educ.* **45,** 581–587.

Peak, J.D., Sparks, D.L., and Ford, R.G. (1999). An in situ ATR-FTIR investigation of sulfate bonding mechanisms on goethite. *J. Colloid Interf. Sci.* **218,** 289–299.

Pedit, J. A., and Miller, C. T. (1995). Heterogenous sorption processes in subsurface systems. 2. Diffusion modeling approaches. *Environ. Sci. Technol.* **29**, 1766–1772.

Perdue, E.M., and Lytle, C.R. (1983). Distribution model for binding of protons and metal ions by humic substances. *Environ. Sci. Technol.* **17**, 654–660.

Persson, P., Nilsson, N., and Sjoberg, S. (1996). Structure and bonding of orthophosphate ions at the iron oxide-aqueous interface. *J. Colloid Interf. Sci.* **177**, 263–275.

Petrovic, R., Berner, R.A., and Goldhaber, M.B. (1976). Rate control in dissolution of alkali feldspars. I. Study of residual feldspar grains by x-ray photoelectron spectroscopy. *Geochim. Cosmochim. Acta* **40**, 537–548.

Piccolo, A. (1994). Advanced infrared techniques (FT-IR, DRIFT, and ATR) applied to organic and inorganic soil materials, 15th World Congress of Soil Science, Acapulco, Mexico, Transactions Vol. 3a, Commission II, pp. 3–22.

Pignatello, J. J. (2000). The measurement and interpretation of sorption and desorption rates for organic compounds in soil media. *Adv. Agron.,* **69**, 1–73.

Pignatello, J.J., and Huang, L.Q. (1991). Sorptive reversibility of atrazine and metolachlor residues in field soil samples. *J. Environ. Qual.* **20**, 222–228.

Polyzopoulos, N.A., Keramidas, V.Z., and Pavlatou, A. (1986). On the limitations of the simplified Elovich equation in describing the kinetics of phosphate sorption and release from soils. *J. Soil Sci.* **37**, 81–87.

Ponnamperuma, F.N. (1955). The chemistry of submerged soils in relation to the yield of rice. Ph.D. Dissertation, Cornell University, Ithaca, NY.

Ponnamperuma, F.N. (1965). Dynamic aspects of flooded soils and the nutrition of the rice plant. *In* "The Mineral Nutrition of the Rice Plant" (R.F. Chandler, ed.), pp. 295–328. Johns Hopkins Press, Baltimore.

Ponnamperuma, F.N. (1972). The chemistry of submerged soils. *Adv. Agron.* **24**, 29–96.

Ponnamperuma, F.N., and Castro, R. U. (1964). *Trans. Int. Congr. Soil Sci.,* 8th, **3**, 379–386.

Postel, (1989). "Water for Agriculture: Facing the Limits," *Worldwatch Pap. 93.* Worldwatch Institute, Washington, DC.

Preslo, L.M., Robertson, J.B., Dworkin, D., Fleischer, E.J., Kostecki, P.T., and Calabrese, E.J. (1988). "Remedial Technologies for Leaking Underground Storage Tanks." Lewis Publishers, Chelsea, MI.

Puckett, L.J. (1995). Identifying the major sources of nutrient water pollution. *Environ. Sci. Technol.* **29**, 408–414.

Quirk, J.P., and Schofield, R.K. (1955). The effect of electrolyte concentration on soil permeability. *J. Soil Sci.* **6**, 163–178.

Rai, E., and Serne, R.J. (1977). Plutonium activities in soil solutions and the stability and formation of selected plutonium minerals. *J. Environ. Qual.* **6**, 89–95.

Randall, S.R., Sherman, D.M., and Ragnarsdottir, K.V. (2001). Sorption of As(V) on green rust ($Fe_4(II)Fe_2(III)(OH)_{12}SO_4 \cdot 3H_2O$) and lepidocrocite ($\gamma$-FeOOH): Surface complexes from EXAFS spectroscopy. *Geochim. Cosmochim. Acta.* **65**(7), 1015–1023.

Reddy, C.N., and Patrick, W.H., Jr. (1977). Effect of redox potential and pH on the uptake of Cd and Pb by rice plants. *J. Environ. Qual.* **6**, 259–262.

Reuss, J.O., and Walthall, P.M. (1989). Soil reaction and acidic deposition. *In* "Acidic

Precipitation" (S.A. Norton, S.E. Lindberg, A.L. Page, eds.), Vol. 4, pp. 1–33. Springer-Verlag, Berlin.

Rhoades, J.D. (1990). Overview: Diagnosis of salinity problems and selection of control practices. *In* "Agricultural Salinity Assessment and Management" (K.K. Tanji, ed.), ASCE Manuals Prac. No. 71, pp. 18–41. Am. Soc. of Civ. Engr., New York.

Rhoades, J.D. (1993). Electrical conductivity methods for measuring and mapping soil salinity. *Adv. Agron.* **49,** 201–251.

Rich, C.I. (1964). Effect of cation size and pH on potassium exchange in Nason soil. *Soil Sci.* **98,** 100–106.

Rich, C.I. (1968). Mineralogy of soil potassium. *In* "The Role of Potassium in Agriculture" (V.J. Kilmer, S.E. Younts, N.C. Brady, eds.), pp. 79–96. Am. Soc. Agron., Madison, WI.

Rich, C.I., and Black, W.R. (1964). Potassium exchange as affected by cation size, pH, and mineral structure. *Soil Sci.* **97,** 384–390.

Rich, C.I., and Obenshain, S.S. (1955). Chemical and clay mineral properties of a red-yellow podzolic soil derived from muscovite schist. *Soil Sci. Soc. Am. Proc.* **19,** 334–339.

Rich, C.I., and Thomas, G.W. (1960). The clay fraction of soils. *Adv. Agron.* **12,** 1–39.

Richards, L.A., ed. (1954). "Diagnosis and Improvement of Saline and Sodic Soils," USDA Agric. Handb. No. 60. Washington, DC.

Riley, R.G., Zachara, J.M., and Wobber, F.J. (1992). "Chemical Contaminants on DOE Lands and Selection of Contaminant Mixtures for Subsurface Science Research," DOE/ER-0547T. U.S. Dept. of Energy, Office of Energy Research, Subsurface Science Program, Washington, DC.

Ritsema, C.J., van Meusvoot, M.E.F., Dent, D.L., Tau, Y., vanden Bosch, H., and van Wijk, A.L.M. (2000). Acid sulfate soils. *In* "Handbook of Soil Science" (M.E. Sumner, ed.), pp. G-121–154. CRC Press, Boca Raton, FL.

Robarge, W.P., and Johnson, D.W. (1992). The effects of acidic deposition on forested soils. *Adv. Agron.* **47,** 1–83.

Roberts, D.R. (2001). Speciation and sorption mechanisms of metals in soils using bulk and micro-focused spectroscopic and microscopic techniques. Ph.D. Dissertation, University of Delaware, Newark, DE, 171 pp.

Roberts, D.R., A.M. Scheidegger, and Sparks, D.L. (1999). Kinetics of mixed Ni-Al precipitate formation on a soil clay fraction. *Environ. Sci. Technol.* **33,** 3749–3754.

Roberts, D.R., Scheinost, A.C., and Sparks, D.L. (2002). Zinc speciation in a smelter-contaminated soil profile using bulk and microspectroscopic techniques. *Environ. Sci. Technol.* **36,** 1742–1750.

Robertson, G.P., Paul, E.A., and Harwood, R.R. (2000). Greenhouse gases in intensive agriculture: Contributions of individual gases to the radioactive forcing of the atmosphere. *Science* **289,** 1922–1925.

Robie, R.A., Hemingway, B.S., and Fisher, J.R. (1978). "Thermodynamic Properties of Minerals and Related Substances at 298.15K and 1 Bar (10^5 Pascals) Pressure and at Higher Temperatures," *U.S. Geol. Surv. Bull. 1452.*

Roe, A.L., Hayes, K.F., Chisholm-Brause, C., Brown, G.E., Jr., Parks, G.A., Hodgson, K.O., and Leckie, J.O. (1991). In situ x-ray absorption study of lead ion surface complexes at the goethite-water interface. *Langmuir* **7,** 367–373.

Rogers, J.C., Daniels, K.L., Goodpasture, S.T., and Kimbrough, C.W. (1989). "Oak Ridge Reservation Environmental Report for 1988," ES/ESH-8/Vol. 1. Martin Marietta Energy Systems, Inc., Oak Ridge, TN.

Ross, C.S., and Hendricks, S.B. (1945). Minerals of the montmorillonite group, their origin and relation to soils and clays. *U.S. Geol. Surv. Tech. Pap. 205B.*

Sahai, N., Carroll, S.A., Roberts, S., and O'Day, P.A. (2000). X-ray absorption spectroscopy of strontium(II) coordination. II. Sorption and precipitation at kaolinite, amorphous silica, and goethite surfaces. *J. Colloid Interf. Sci.* **222,** 198–212.

Sawhney, B.L., and Brown, K., eds. (1989). "Reactions and Movement of Organic Chemicals in Soils." SSSA Spec. Publ. 22. Soil Sci. Soc. Am., Madison, WI.

Scheckel, K.G., Scheinost, A.C., Ford, R.G., and Sparks, D.L. (2000). Stability of layered Ni hydroxide surface precipitates – A dissolution kinetics study. *Geochim. Cosmochim. Acta* **64,** 2727–2735.

Scheckel, K.G., and Sparks, D.L. (2000). Kinetics of the formation and dissolution of Ni precipitates in a gibbsite/amorphous silica mixture. *J. Colloid Interf. Sci.* **229,** 222–229.

Scheckel, K. G. and Sparks, D. L. (2001). Dissolution kinetics of nickel suface precipitates on clay mineral and oxide surfaces. *Soil Sci. Soc. Am. J.* **65,** 685–694.

Scheidegger, A.M., Lamble, G.M., and Sparks, D.L. (1996). Investigation of Ni sorption on pyrophyllite: An XAFS study. *Environ. Sci. Tech.* **30,** 548–554.

Scheidegger, A.M., and Sparks, D.L. (1996). Kinetics of the formation and the dissolution of nickel surface precipitates on pyrophyllite. *Chem. Geol.* **132,** 157–164.

Scheidegger, A.M., Lamble, G.M., and Sparks, D.L. (1997). Spectroscopic evidence for the formation of mixed-cation, hydroxide phases upon metal sorption on clays and aluminum oxides. *J. Colloid Interf. Sci.* **186,** 118–128.

Scheidegger, A.M., Strawn, D.G., Lamble,G.M. and Sparks, D.L. (1998). The kinetics of mixed Ni-Al hydroxide formation on clay and aluminum oxide minerals: A time-resolved XAFS study. *Geochim. Cosmochim. Acta* **62,** 2233–2245.

Scheinost, A.C., Ford, R.G., and Sparks, D.L. (1999). The role of Al in the formation of secondary Ni precipitates on pyrophyllite, gibbsite, talc, and amorphous silica: A DRS study. *Geochim. Cosmochim. Acta* **63,** 3193–3203.

Schindler, P.W. (1981). Surface complexes at oxide-water interfaces. *In* "Adsorption of Inorganics at Solid-Liquid Interfaces" (M.A. Anderson and A.J. Rubin, eds.), pp. 1–49. Ann Arbor Sci., Ann Arbor, MI.

Schindler, P.W., Fürst, B., Dick, B., and Wolf, P.U. (1976). Ligand properties of surface silanol groups. I. Surface complex formation with Fe^{3+}, Cu^{2+}, Cd^{2+}, and Pb^{2+}. *J. Colloid Interf. Sci.* **55,** 469–475.

Schindler, P.W., and Gamsjager, H. (1972). Acid-base reactions of the TiO_2 (anatase)-water interface and the point of zero charge of TiO_2 suspensions. *Kolloid-Z. Z. Polym.* **250,** 759–763.

Schlesinger, W.H. (1984). Soil organic matter: A source of atmospheric CO_2. *In* "The Role of Terrestrial Vegetation in the Global Carbon Cycle," (G.M. Woodwell ed.), pp. 111–127. John Wiley & Sons, New York, NY.

Schlesinger, W.H. (1997). "Biogeochemistry: An Analysis of Global Change." 2nd ed. Academic Press, San Diego, CA.

Schlesinger, W.H. (1999). Carbon sequestration in soils. *Science* **284,** 2095.

Schnitzer, M. (1985). Nature of nitrogen in humic substances. *In* "Humic Substances in Soil, Sediments, and Water" (G.R. Aiken, D.M. McKnight, R.L. Wershaw, eds.), pp. 303–325. John Wiley & Sons, New York.

Schnitzer, M. (1986). Binding of humic substances by soil mineral colloids. *In* "Interactions of Soil Minerals with Natural Organics and Microbes" (P.M. Huang and M. Schnitzer, eds.), SSSA Spec. Publ. No. 17, pp. 77–101. Soil Sci. Soc. Am., Madison, WI.

Schnitzer, M. (1991). Soil organic matter – the next 75 years. *Soil Sci.* **151,** 41–58.

Schnitzer, M. (2000). A lifetime perspective on the chemistry of soil organic matter. *Adv. Agron.* **68,** 1–58.

Schnitzer, M., and Hansen, E.H. (1970). Organo-metallic interactions in soils: 8. An evaluation of methods for the determination of stability constants of metal-fulvic acid complexes. *Soil Sci.* **109,** 333–340.

Schnitzer, M., and Khan, S.U., eds. (1972). "Humic Substances in the Environment." Dekker, New York.

Schnitzer, M., and Khan, S.U., eds. (1978). "Soil Organic Matter." Elsevier, New York.

Schnitzer, M., and Preston, C.M., 1986. Analysis of humic acids by solution–and solid-state carbon-13 nuclear magnetic resonance. *Soil Sci. Soc. Am. J.* **50,** 326–331.

Schnitzer, M. and Schulten, H.-R. (1992). The analysis of soil organic matter by pyrolysis-field ionization mass spectrometry. *Soil Sci. Soc. Am. J.* **56,** 1811–1817.

Schnitzer, M. and Schulten, H.-R. (1995). Analysis of organic matter in soil extracts and whole soils by pyrolysis-mass spectrometry. *Adv. Agron.* **55,** 168–218.

Schnitzer, M., and Skinner, S.J.M. (1963). Organo-metallic interactions in soils: 1. Reactions between a number of metal ions and the organic matter of a podzol Bh horizon. *Soil Sci.* **96,** 86–93.

Schofield, R.K. (1947). Calculation of surface areas from measurements of negative adsorption. *Nature* **160,** 408–412.

Schofield, R.K. (1949). Effect of pH on electric charges carried by clay particles. *J. Soil Sci.* **1,** 1–8.

Schofield, R.K., and Samson, H.R. (1953). The deflocculation of kaolinite suspensions and the accompanying change-over from positive to negative chloride adsorption. *Clay Miner. Bull.* **2,** 45–51.

Schott, J., and Petit, J.C. (1987). New evidence for the mechanisms of dissolution of silicate minerals. *In* "Aquatic Surface Chemistry" (W. Stumm, ed.), pp. 293–312. John Wiley & Sons (Interscience), New York.

Schulten, H.-R., and Schnitzer, M. (1993). A state of the art structural concept for humic substances. *Naturwissenschaften* **80,** 29–30.

Schulten, H.-R., and Schnitzer, M. (1997). Chemical model structures for soil organic matter and soils. *Soil Sci.* **162,** 115–130.

Schulten, H.-R., Leinmeher, P., and Schnitzer, M. (1998). Analytical pyrolysis and computer modeling of humic and soil particles. *In* "Structure and Surface Reactions of Soil Particles" (P.M. Huang, N. Senesi and J. Buffle, eds.), pp. 282–324. John Wiley & Sons, New York, NY.

Schulthess, C.P., and Sparks, D.L. (1991). Equilibrium-based modeling of chemical sorption on soils and soil constituents. *Adv. Soil Sci.* **16,** 121–163.

Schulze, D.G. (1989). An introduction to soil mineralogy. *In* "Minerals in the Soil Environment" (J.B. Dixon and S.B. Weed, eds.), SSSA Book Ser. No. 1, pp. 1–34. Soil Sci. Soc. Am., Madison, WI.

Schulze, D.G and Bertsch, P.M. (1999). Overview of synchrotron x-ray sources and synchrotron x-rays. *In* "Synchrotron X-ray Methods in Clay Science" (D.G Schulze, J.W. Stucki and P.M. Bertsch, eds.), pp. 1–18. Clay Minerals Society, Boulder, CO.

Schwertmann, U., and Cornell, R.M. (1991). "Iron Oxides in the Laboratory." VCH, Weinheim.

Schwertmann, U., Susser, P., and Natscher, L. (1987). Protonenpuffer-substanzen in Boden. *Z. Pflanzernähr. Düng. Bodenkunde* **150,** 174–178.

Scott, M.J., and Morgan, J.J. (1995). Reactions at oxide surfaces. 1. Oxidation of As(III) by synthetic birnessite. *Environ. Sci. Technol.* **29,** 1898–1905.

Scott, M.J., and Morgan, J.J. (1996). Reactions at oxide surfaces. 2. Oxidation of Se(IV) by synthetic birnessite. *Environ. Sci. Technol.* **30,** 1990–1996.

Scribner, S.L., Benzing, T.R., Sun, S., and Boyd, S.A. (1992). Desorption and bioavailability of aged simazine residues in soil from a continuous corn field. *J. Environ. Qual.* **21,** 115–120.

Seinfeld, J.H. (1986). "Atmospheric Chemistry and Physics of Air Pollution." John Wiley and Sons, New York.

Senesi, N. and Sposito, G., (1984). Residual copper(II) complexes in purified soil and sewage sludge fulvic acids: An electron spin resonance study. *Soil Sci. Soc. Am. J.* **48,** 1247–1253.

Senesi, N., Bocian, D.F. and Sposito, G., (1985). Electron spin resonance investigation of copper(II) complexation by soil fulvic acid. *Soil Sci. Soc. Am. J.* **49,** 114–119.

Shainberg, I. (1990). Soil response to saline and sodic conditions. *In* "Agricultural Salinity Assessment and Management" (K.K. Tanji, ed.), ASCE Manuals Prac. No. 71, pp. 91–112. Am. Soc. of Civ. Eng., New York.

Shainberg, I., Bresler, E., and Klausner, Y. (1971). Studies on Na/Ca montmorillonite systems. I: The swelling pressure. *Soil Sci.* **111,** 214–219.

Sharpley, A.N. (1983). Effect of soil properties on the kinetics of phosphorus desorption. *Soil Sci. Soc. Am. J.* **47,** 462–467.

Shuman, L.M. (1975). The effect of soil properties on zinc adsorption by soils. *Soil Sci. Soc. Am. Proc.* **39,** 455–458.

Singh, U., and Uehara, G. (1986). Electrochemistry of the double-layer: Principles and applica-tions to soils. *In* "Soil Physical Chemistry" (D.L. Sparks, ed.), pp. 1–38. CRC Press, Boca Raton, FL.

Singh, U., and Uehara, G. (1999). Electrochemistry of the double layer: Principles and applications to soils. *In* "Soil Physical Chemistry," 2nd ed. (D.L. Sparks, ed.), pp. 1–46. CRC Press, Boca Raton, FL.

Skopp, J. (1986). Analysis of time dependent chemical processes in soils. *J. Environ. Qual.* **15,** 205–213.

Smith, R.W. (1971). Relations among equilibrium and nonequilibrium aqueous species of aluminum hydroxy complexes. *Adv. Chem. Ser.* **106,** 250–279.

Soundararajan, R., Barth, E.F., and Gibbons, J.J. (1990). Using an organophilic clay to chemically stabilize waste containing organic compounds. *Hazard. Mater. Control* **3**[1], 42–45.

Sowden, F.J., Griffith, S.M., and Schnitzer, M. (1976). The distribution of nitrogen in some highly organic tropical volcanic soils. *Soil Biol. Biochem.* **8,** 55–60.

Sparks, D.L. (1984). Ion activities: An historical and theoretical overview. *Soil Sci. Soc. Am. J.* **48,** 514–518.

Sparks, D.L. (1985). Kinetics of ionic reactions in clay minerals and soils. *Adv. Agron.* **38,** 231–266.

Sparks, D.L. (1987). Dynamics of soil potassium. *Adv. Soil Sci.* **6,** 1–63.

Sparks, D.L. (1989). "Kinetics of Soil Chemical Processes." Academic Press, San Diego, CA.

Sparks, D.L. (1992). Soil kinetics. *In* "Encyclopedia of Earth Systems Science" (W.A. Nirenberg, ed.), Vol. 4, pp. 219–229. Academic Press, San Diego, CA.

Sparks, D.L. (1993). Soil decontamination. *In* "Handbook of Hazardous Materials" (M. Corn, ed.), pp. 671–680. Academic Press, San Diego, CA.

Sparks, D.L. (1994). Soil chemistry. *In* "Encyclopedia of Agricultural Science" (C.J. Arntzen, ed.), pp. 75–81. Academic Press, San Diego, CA.

Sparks, D.L. (1995). "Environmental Soil Chemistry." Academic Press, San Diego.

Sparks, D.L. (1998). Kinetics of sorption/release reactions on natural particles. *In* "Structure and Surface Reactions of Soil Particles" (P.M. Huang, N. Senesi and J. Buffle, eds.), pp. 413–448. John Wiley & Sons, New York.

Sparks, D.L. (1998). Kinetics of soil chemical phenomena: Future directions. *In* "Future Prospects for Soil Chemistry" (P.M. Huang, D. L. Sparks, and S. A. Boyd, ed.), pp. 81–102. Soil Sci. Soc. Am. Spec. Pub. 55, Soil Science Society of America, Madison, WI.

Sparks, D.L. (1999). Kinetics and mechanisms of chemical reactions at the soil mineral/water interface. *In* "Soil Physical Chemistry," 2nd ed. (D.L. Sparks, ed.), pp. 135–191. CRC Press, Boca Raton, FL.

Sparks, D.L. (2000). Kinetics and mechanisms of soil chemical reactions. *In* "Handbook of Soil Science". (M.E. Sumner, ed.), pp. B-123–167. CRC Press, Boca Raton, FL.

Sparks, D.L., Fendorf, S.E., Toner, C.V., IV, and Carski, T.H. (1996). Kinetic methods and measurements. *In* "Methods of Soil Analysis: Part 3—Chemical Methods," (D.L. Sparks, ed.). SSSA Book Ser. 5, pp. 1275–1307. Soil Sci. Soc. Am., Madison, WI.

Sparks, D.L., and Huang, P.M. (1985). Physical chemistry of soil potassium. *In* "Potassium in Agriculture" (R.E. Munson, ed.), pp. 201–276. Am. Soc. Agron., Madison, WI.

Sparks, D.L., and Jardine, P.M. (1984). Comparison of kinetic equations to describe K-Ca exchange in pure and in mixed systems. *Soil Sci.* **138,** 115–122.

Sparks, D.L., and Suarez, D.L., eds. (1991). "Rates of Soil Chemical Processes." SSSA Spec. Publ. No. 27. Soil Sci. Soc. Am., Madison, WI.

Sparks, D.L., and Zhang, P.C. (1991). Relaxation methods for studying kinetics of soil chemical phenomena. *In* "Rates of Soil Chemical Processes" (D.L. Sparks and

D.L. Suarez, eds.), SSSA Spec. Publ. No. 27, pp. 61–94. Soil Sci. Soc. Am., Madison, WI.

Sposito, G. (1981a). Cation exchange in soils: An historical and theoretical perspective. *In* "Chemistry in the Soil Environment" (R.H. Dowdy, J.A. Ryan, V.V. Volk, and D.E. Baker, eds.), Am. Soc. Agron. Spec. Publ. 40, pp. 13–30. Am. Soc. Agron./Soil Sci. Soc. Am., Madison, WI.

Sposito, G. (1981b). "The Thermodynamics of the Soil Solution." Oxford Univ. Press (Clarendon), Oxford.

Sposito, G. (1984). "The Surface Chemistry of Soils." Oxford Univ. Press, New York.

Sposito, G. (1986). Distinguishing adsorption from surface precipitation. *In* "Geochemical Processes at Mineral Surfaces", (J.A. Davis and K.F. Hayes, eds.). ACS Symp. Ser. 323, 217–229. Am. Chem. Soc., Washington, DC.

Sposito, G. (1989). "The Chemistry of Soils." Oxford Univ. Press, New York

Sposito, G. (1994). "Chemical Equilibria and Kinetics in Soils." John Wiley & Sons, New York.

Sposito, G., ed. (1996). "The Environmental Chemistry of Aluminum." CRC Press (Lewis Publishers), Boca Raton, FL.

Sposito, G. (2000). Ion exchange phenomena. *In* "Handbook of Soil Science" (M.E. Sumner, ed.), pp. B-241–264, CRC Press, Boca Raton, FL.

Sposito, G., and Coves, J. (1988). "SOILCHEM: A Computer Program for the Calculation of Chemical Equilibria in Soil Solutions and Other Natural Water Systems," Kearney Foundation of Soil Science, University of California, Riverside.

Sposito, G., and Mattigod, S.V. (1977). On the chemical foundation of the sodium adsorption ratio. *Soil Sci. Soc. Am. J.* **41,** 323–329.

Sposito, G., and Mattigod, S.V. (1979). Ideal behavior in Na-trace metal cation exchange on Camp Berteau montmorillonite. *Clays Clay Miner.* **27,** 125–128.

Sposito, G., and Mattigod, S.V. (1980). "GEOCHEM: A Computer Program for the Calculation of Chemical Equilibria in Soil Solutions and Other Natural Water Systems," Kearney Foundation of Soil Science, University of California, Riverside.

Steelink, C. (1985). Implications of elemental characteristics of humic substances. *In* "Humic Substances in Soil, Sediments, and Water" (G.R. Aiken, D.M. McKnight, R.L. Wershaw, eds.), pp. 457–476. John Wiley & Sons (Interscience), New York.

Steelink, C., Wershaw, R.L., Thorn, K.A., and Wilson, M.A. (1989). Application of liquid-state NMR spectroscopy to humic substances. *In* "Humic Substances II. In Search of Structure" (M.H.B. Hayes, P. MacCarthy, R.L. Malcom, R.S. Swift, eds.), pp. 281–308. John Wiley & Sons, New York.

Steinberg, S.M., Pignatello, J.J., and Sawhney, B.L. (1987). Persistence of 1,2-dibromoethane in soils: Entrapment in intraparticle micropores. *Environ. Sci. Technol.* **21,** 1201–1208.

Stenner, R.D., Cramer, K.H., Higley, K.A., Jette, S.J., Lamar, D.A., McLaughlin, T.J., Sherwood, D.R., and Houten, N.C.V. (1988). "Hazard Ranking System Evaluation of CERCLA Inactive Waste Sites at Hanford: Evaluation Methods and Results," PNL-6456, Vol. 1. Pacific Northwest Laboratory, Richland, WA.

Stern, O. (1924). Zur Theorie der Elektrolytischen Doppelschicht. *Z. Elektrochem.* **30,** 508–516.

Stevenson, F.J. (1982). "Humus Chemistry." John Wiley & Sons, New York.

Stol, R.J., van Helden, A.K., and de Bruyn, P.L. (1976). Hydrolysis precipitation studies of aluminum (III) solutions. 2. A kinetic study and model. *J. Colloid Interf. Sci.* **57,** 115–131.

Stone, A.T. (1986). Adsorption of organic reductants and subsequent electron transfer on metal oxide surfaces. In "Geochemical Processes at Mineral Surfaces" (J.A. Davis and K.F. Hayes, eds.). ACS Symp. Ser. 323, pp. 446–461, Am. Chem. Soc., Washington, DC.

Stone, A.T. (1987a). Microbial metabolites and the reductive dissolution of manganese oxides: Oxalate and pyruvate. *Geochim. Cosmochim. Acta* **51,** 919–925.

Stone, A.T. (1987b). Reductive dissolution of manganese (III/IV) oxides by substituted phenols. *Environ. Sci. Technol.* **21,** 979–988.

Stone, A.T. (1991). Oxidation and hydrolysis of ionizable organic pollutants at hydrous metal oxide surfaces. *In* "Rates of Soil Chemical Processes" (D.L. Sparks and D.L. Suarez, eds.), SSSA Spec. Publ. No. 27, pp. 231–254. Soil Sci. Soc. Am., Madison, WI.

Stone, A.T., and Morgan, J.J. (1984a). Reduction and dissolution of manganese (III) and manganese (IV) oxides by organics. 1. Reaction with hydroquinone. *Environ. Sci. Technol.* **18,** 450–456.

Stone, A.T., and Morgan, J.J. (1984b). Reduction and dissolution of manganese (III) and manganese (IV) oxides by organics. 2. Survey of the reactivity of organics. *Environ. Sci. Technol.* **18,** 617–624.

Strawn, D.G., Scheidegger, A.M., and Sparks, D.L. (1998). Kinetics and mechanisms of Pb(II) sorption and desorption at the aluminum oxide-water interface. *Environ. Sci. Technol.* **32,** 2596–2601.

Strawn, D.G., and Sparks, D.L. (1999). The use of XAFS to distinguish between inner- and outer-sphere lead adsorption complexes on montmorillonite. *J. Colloid Interf. Sci.* **216,** 257–269.

Strawn, D.G., and Sparks, D.L. (2000). Effects of soil organic matter on the kinetics and mechanisms of Pb(II) sorption and desorption in soil. *Soil Sci. Soc. Am. J.* **64,** 145–156.

Stumm, W., ed. (1987). "Aquatic Surface Chemistry." John Wiley & Sons (Interscience), New York.

Stumm, W., ed. (1990). "Aquatic Chemical Kinetics." John Wiley & Sons, New York.

Stumm, W., ed. (1992). "Chemistry of the Solid-Water Interface." John Wiley & Sons, New York.

Stumm, W., and Furrer, G. (1987). The dissolution of oxides and aluminum silicates: Examples of surface-coordination-controlled kinetics. *In* "Aquatic Surface Chemistry" (W. Stumm, ed.), pp. 197–219. John Wiley & Sons (Interscience), New York.

Stumm, W., Kummert, K., and Sigg, L. (1980). A ligand exchange model for the adsorption of inorganic and organic ligands at hydrous oxide interfaces. *Croat. Chem. Acta* **53,** 291–312.

Stumm, W., and Morgan, J.J. (1981). "Aquatic Chemistry." John Wiley & Sons, New York.

Stumm, W., and Morgan, J.J. (1996). "Aquatic Chemistry." 3rd ed. John Wiley & Sons, New York.

Stumm, W., and Wollast, R. (1990). Coordination chemistry of weathering. Kinetics of the surface-controlled dissolution of oxide minerals. *Rev. Geophys.* **28,** 53–69.

Su, C., and Suarez, D.L. (1995). Coordination of adsorbed boron: A FTIR spectroscopic study. *Environ. Sci. Technol.* **29,** 302–311.

Su, C., and Suarez, D.L. (1997). *In situ* infrared speciation of adsorbed carbonate on aluminum and iron oxides. *Clays Clay Min.* **45,** 814–825.

Suarez, D.L., Goldberg, S., and Su,C. (1998). Evaluation of oxyanion adsorption mechanisms on oxides using FTIR spectroscopy and electrophoretic mobility. *In* "Mineral-Water Interfacial Reactions-Kinetics and Mechanisms", (D.L. Sparks, and T.J.Grundl, eds.), pp. 136–178. ACS Symp. Ser. 715, Am. Chem. Soc., Washington, DC.

Suffett, I.H., and MacCarthy, P., eds. (1989). "Aquatic Humic Substances. Influence of Fate and Treatment of Pollutants." Adv. Chem. Ser. 219. Am. Chem. Soc., Washington, DC.

Sumner, M.E. (1991). Soil acidity control under the impact of industrial society. In "Interactions at the Soil Colloid-Soil Solution Interface" (G.H. Bolt, M.F.D. Boodt, M.H.B. Hayes, M.B. McBride, eds.), NATO ASI, Ser. E, 190, pp. 517–541. Kluwer Academic Publ., Dordrecht, The Netherlands.

Sumner, M.E., and Miller, W.P. (1996). Cation exchange capacity and exchange coefficients. *In* "Methods of Soil Analysis, Part 3—Chemical Methods" (D.L. Sparks, ed.), pp.1201–1229. Soil Sci. Soc. Am. Book Series 5, SSSA, Madison, WI.

Sumner, M.E., Miller, W.P., Kookana, R.S., and Hazelton, P. (1998). Sodicity, dispersion, and environmental quality. *In* "Sodic Soils: Distribution, Properties, Management, and Environmental Consequences," (M.E. Sumner and R. Naidu, eds.), pp. 149–172. Oxford Univ. Press, New York.

Sun, X., and Doner, H.E. (1996). An investigation of arsenate and arsenite bonding structures on Goethite by FTIR. *Soil Sci.* **161**(12), 865–872.

Szabolcs, I. (1979). "Review of Research on Salt Affected Soils." UNESCO, Paris, France.

Szabolcs, I. (1989). "Salt-Affected Soils." CRC Press, Boca Raton, FL.

Szalay, A. (1964). Cation exchange properties of humic acids and their importance in the geochemical enrichment of UO^{++} and other cations. *Geochim. Cosmochim. Acta* **22,** 1605–1614.

Tadros, T.F., and Lyklema, J. (1969). The electrical double layer on silica in the presence of bivalent counter-ions. *J. Electroanal. Chem.* **22,** 1.

Takahashi, M.T., and Alberty, R.A. (1969). The pressure-jump methods. *In* "Methods in Enzymology" (K. Kustin, ed.), Vol. 16, pp. 31–55. Academic Press, New York.

Tang, L., and Sparks, D.L. (1993). Cation exchange kinetics on montmorillonite using pressure-jump relaxation. *Soil Sci. Soc. Am. J.* **57,** 42–46.

Taniguechi, K., Nakajima, M., Yoshida, S., and Tarama, K. (1970). The exchange of the surface protons in silica gel with some kinds of metal ions. *Nippon Kagaku Zasshi* **91,** 525–529.

Tanji, K.K., ed. (1990a). "Agricultural Salinity Assessment and Management," ASCE Manuals Prac. No. 71, Am. Soc. Civ. Eng., New York.

Tanji, K.K., ed. (1990b). Nature and extent of agricultural salinity. *In* "Agricultural Salinity Assessment and Management" (K.K. Tanji, ed.), ASCE Manuals Prac. No. 71, Am. Soc. Civ. Eng., New York.

Taylor, R.M. (1984). The rapid formation of crystalline double hydroxy salts and other compounds by controlled hydrolysis. *Clay Miner.* **19,** 591–603.

Taylor, R.M. (1987). Nonsilicate oxides and hydroxides. *In* "The Chemistry of Clays and Clay Minerals" (A.C.D. Newman, ed.), pp. 129–201. Longman Group Ltd., Harlow, Essex, England.

Tejedor-Tejedor, M.L., and Anderson, M.A. (1990). Protonation of phosphate on the surface of goethite as studied by CIR-FTIR and electrophoretic mobility. *Langmuir* **124,** 79–110.

Theng, B.K.G., ed. (1980). "Soils with Variable Charge." New Zealand Society of Soil Science, Palmerston.

Thomas, G.W. (1975). Relationship between organic matter content and exchangeable aluminum in acid soil. *Soil Sci. Soc. Am. Proc.* **39,** 591.

Thomas, G.W. (1977). Historical developments in soil chemistry: Ion exchange. *Soil Sci. Soc. Am. Proc.* **41,** 230–238.

Thomas, G.W., and Hargrove, W.L. (1984). The chemistry of soil acidity. *In* "Soil Acidity and Liming" (F. Adams, ed.), 2nd ed. Agron. Monogr. 12, pp. 3–56. Am. Soc. Agron., Madison, WI.

Thompson, H.A., Parks, G.A., and Brown, G.E., Jr. (1999a). Dynamic interactions of dissolution, surface adsorption, and precipitation in an aging cobalt(II)-clay-water system. *Geochim. Cosmochim. Act.* **63**(11/12), 1767–1779.

Thompson, H.A., Parks, G.A., and Brown, G.E., Jr. (1999b). Ambient-temperature synthesis, evolution, and characterization of cobalt-aluminum hydrotalcite-like solids. *Clays Clay Min.* **47,** 425–438.

Thornton, I., ed. (1983). "Applied Environmental Geochemistry." Academic Press, London.

Thurman, E.M. (1985). "Organic Geochemistry of Natural Waters." Kluwer Academic Publishers, Hingham, MA.

Thurman, E.M., Malcom, R.L., and Aiken, G.R. (1978). Prediction of capacity factors for aqueous organic solutes on a porous acrylic resin. *Anal. Chem.* **50,** 775–779.

Tien, H.T. (1965). Interactions of alkali metal cations with silica gels. *J. Phys. Chem.* **69,** 350–352.

Todd, D.E. (1970). "The Water Encyclopedia." Water Information Center, Port Washington, NY.

Tourtelot, H.A. (1971). Chemical compositions of rock types as factors in our environment. In "Environmental Geochemistry in Health and Disease" (H.L. Cannon and H.C. Hopps, eds.), Vol. 123, pp. 13–29. Geol. Soc. Am., Boulder, CO.

Towle, S. N., Bargar, J. R., Brown, G. E., Jr., and Parks, G. A. (1997). Surface precipitation of Co(II) (aq) on Al_2O_3. *J. Colloid Interf. Sci.* **187,** 62–82.

Trainor, T.P., Brown, G.E., Jr., and Parks, G.A. (2000). Adsorption and precipitation of aqueous Zn(II) on alumina powders. *J. Colloid Interf. Sci.* **231,** 359–372.

Truesdall, A.H., and Jones, B.F. (1974). WATEQ, a computer program for calculating chemical equilibria of natural waters. *J. Res. U.S. Geol. Surv.* **2**, 223.

Udo, E.J. (1978). Thermodynamics of potassium-calcium exchange reactions on a kaolinitic soil clay. *Soil Sci. Soc. Am. J.* **42**, 556–560.

Uehara, G., and Gillman, G. (1981). "The Mineralogy, Chemistry, and Physics of Tropical Soils with Variable Charge Clays." Westview Press, Boulder, CO.

U.S. Geological Survey. 1999. "The Quality of our Nation's Waters – Nutrients and Pesticides". U.S. Geological Survey Circular 1225, 82 pp.

U.S. Public Health Service. (1962). "Public Health Service Drinking Water Standards," U.S. Gov. Printing Office, Washington, DC.

van Bemmelen, J.M. (1888). Die absorptions verbindungen und das absorptions vermogen der ackererde. *Landwirtsch. Vers. Stn.* **35**, 69–136.

Van Bladel, R., and Gheyi, H.R. (1980). Thermodynamic study of calcium-sodium and calcium-magnesium exchange in calcareous soils. *Soil Sci. Soc. Am. J.* **44**, 938–942.

Van Bladel, R., and Laudelout, H. (1967). Apparent reversibility of ion-exchange reactions in clay suspensions. *Soil Sci.* **104**, 134–137.

van Breemen, N. (1987). Soil acidification and alkalinization. *Trans. Congr. Int. Soil Sci. Soc.,* 13th, pp. 85–90.

van Genuchten, M. Th. and Wagenet, R. J. (1989). Two-site/two-region models for pesticide transport and degradation: Theoretical development and analytical solutions. *Soil Sci. Soc. Am. J.* **53**, 1303–1310.

van Olphen, H. (1977). "An Introduction to Clay Colloid Chemistry." 2nd ed. John Wiley & Sons, New York.

van Raji, B., and Peech, M. (1972). Electrochemical properties of some Oxisols and Alfisols of the tropics. *Soil Sci. Soc. Am. Proc.* **36**, 587–593.

van Riemsdijk, W.H., Bolt, G.H., Koopal, L.K., and Blaakmeer, J. (1986). Electrolyte adsorption on heterogeneous surfaces: Adsorption models. *J. Colloid Interf. Sci.* **109**, 219–228.

van Rimesdijk, W.H., de Wit, J.C.M., Koopal, L.K., and Bolt, G.H. (1987). Metal ion adsorption on heterogeneous surfaces: Adsorption models. *J. Colloid Interf. Sci.* **116**, 511–522.

van Riemsdijk, W.H., and van der Zee, S.E.A.T.M. (1991). Comparison of models for adsorption, solid solution and surface precipitation. *In* "Interactions at the Soil Colloid-Soil Solution Interface" (G.H. Bolt, M.F.D. Boodt, M.H.B. Hayes, M.B. McBride, eds.), pp. 241–256. Kluwer Academic Publ., Dordrecht, The Netherlands.

Vanselow, A.P. (1932). Equilibria of the base exchange reactions of bentonites, permutites, soil colloids, and zeolites. *Soil Sci.* **33**, 95–113.

Veith, F.P. (1902). The estimation of soil acidity and the lime requirement of soils. *J. Am. Chem. Soc.* **24**, 1120–1128.

Veith, F.P. (1904). Comparison of methods for the estimation of soil acidity. *J. Am. Chem. Soc.* **26**, 637–662.

Veith, J.A., and Sposito, G. (1977). On the use of the Langmuir equation in the interpretation of "adsorption" phenomena. *Soil Sci. Soc. Am. J.* **41**, 497–502.

Venkataramani, B., Venkateswarlu, K.S., and Shankar, J. (1978). Sorption properties of oxides. III. Iron oxides. *J. Colloid Interf. Sci.* **67**, 187–194.

Verburg, K., and Baveye, P. (1994). Hysteresis in the binary exchange of cations on 2:1 clay minerals: A critical review. *Clays Clay Miner.* **42**, 207–220.

Verwey, E.J.W. (1935). The electric double layer and the stability of lyophobic colloids. *Chem. Rev.* **16**, 363–415.

Verwey, E.J.W., and Overbeek, J.T.G. (1948). "Theory of the Stability of Lyophobic Colloids." Elsevier, New York.

Villalobos, M., and Leckie, J.O. (2001). Surface complexation modeling and FTIR study of carbonate adsorption to goethite. *J. Colloid Interf. Sci.* **235**, 15–32.

Volk, B.G., and Schnitzer, M. (1973). Chemical and spectroscopic methods for assessing subsidence in Florida Histosols. *Soil Sci. Soc. Am. Proc.* **37**, 886–888.

Vydra, F., and Galba, J. (1969). Sorption von metallkomplexen an silicagel. V. Sorption von hydrolysenprodukten des Co^{2+}, Mn^{2+}, Ni^{2+}, Zn^{2+} and silica gel. *Colln Czech. Chem. Comm.* **34**, 3471–3478.

Waksman, S.A., and Stevens, K.R. (1930). A critical study of the methods for determining the nature and abundance of soil organic matter. *Soil Sci.* **30**, 97–116.

Wann, S.S., and Uehara, G. (1978). Surface charge manipulation in constant surface potential soil colloids: I. Relation to sorbed phosphorus. *Soil Sci. Soc. Am. J.* **42**, 565–570.

Ward, M.H., Mark, S.D., Cantor, K.P., Weisenburger, D.D., Correa-Villaseñor, A. and Zahm, S.H. (1996). Drinking water nitrate and the risk of non-Hodgkins lymphoma. *Epidemiology* **7**, 465–471.

Way, J.T. (1850). On the power of soils to absorb manure. *J. R. Agric. Soc. Engl.* **11**, 313–379.

Waychunas, G.A., Rea, B.A., Fuller, C.C., and Davis, J.A. (1993). Surface chemistry of ferrihydrite: Part 1. EXAFS studies of the geometry of coprecipitated and adsorbed arsenate. *Geochim. Cosmochim. Acta* **57**, 2251–2269.

Weaver, C.E. and Pollard, L.D. (1973). "The Chemistry of Clay Minerals." Elsevier, Amsterdam.

Weber, W.J., Jr., and Gould, J.P. (1966). Sorption of organic pesticides from aqueous solution. *Adv. Chem. Ser.* **60**, 280–305.

Weber, W.J., Jr., and Huang, W. (1996). A distributed reactivity model for sorption by soils and sediments. 4. Intraparticle heterogeneity and phase-distribution relationships under nonequilibrium conditions. *Environ. Sci. Technol.* **30**, 881–888.

Weber, W.J., Jr., and Miller, C.T., 1988. Modeling the sorption of hydrophobic contaminants by aquifer materials. 1. Rates and equilibria. *Water Res.* **22**, 457–464.

Weesner, F.J., and Bleam, W.F. (1997). X-ray absorption and EPR spectroscopic characterization of adsorbed copper(II) complexes at the boehmite (AlOOH) surface. *J. Colloid Interf. Sci.* **196**, 79–86.

Wells, N., and Whitton, J.S. (1977). Element composition of tomatoes grown on four soils mixed with sewage sludge. *N. Z. J. Exp. Agric.* **5**, 363–369.

Wershaw, R.L. and Mikita, M.A., (1987). "NMR of Humic Substances and Coal." Lewis Publishers, Chelsea, MI.

Westall, J.C. (1979). "MICROQL – A Chemical Equilibrium Program in BASIC," Swiss Federal Institute, Duebendorf, Switzerland.

Westall, J.C. (1980). Chemical equilibrium including adsorption on charged surfaces. *Adv. Chem. Ser.* **189,** 33–44.

Westall, J.C. (1986). Reactions at the oxide-solution interface: Chemical and electrostatic models. *In* "Geochemical Processes at Mineral Surfaces" (J. Davis and K.F. Hayes, eds.), ACS Symp. Ser. 323, pp. 54–78. Am. Chem. Soc., Washington, DC.

Westall, J.C., and Hohl, H. (1980). A comparison of electrostatic models for the oxide/solution interface. *Adv. Colloid Interface Sci.* **12,** 265–294.

Westall, J.C., Zachary, J.L., and Morel, F.M.M. (1976). "MINEQL – A Computer Program for the Calculation of Chemical Equilibrium Composition of Aqueous Systems." Tech. Note No. 18. Dept. of Civil Eng., Massachusetts Institute of Technology, Cambridge, MA.

White, A.F., and Peterson, M.L. (1996). Reduction of aqueous transition metal species on the surfaces of Fe(II)-containing oxides. *Geochim. Cosmochim. Acta* **60,** 3799–3814.

White, A.F. and Peterson, M.L. (1998). The reduction of aqueous metal species on the surfaces of Fe(II)-containing oxides: The role of surface passivation. *In*: "Mineral-Water Interfacial Reactions" (D.L. Sparks and T.J. Grundl, eds.), pp. 323–341, ACS Symp. Ser. 715, Am. Chem. Soc., Washington, DC.

White, R.E., and Thomas, G.W. (1981). Hydrolysis of aluminum on weakly acid organic exchangers. Implications for phosphorus adsorption. *Fert. Inst.* **2,** 159–167.

Wielinga, B., Mizuba, M.M., Hansel, C., and Fendorf, S. (2001). Iron promoted reduction of chromate by dissimilatory iron-reducing bacteria. *Environ. Sci. Technol.* **35,** 522–527.

Wijnja, H., and Schulthess, C.P. (1999). ATR-FTIR and DRIFT spectroscopy of carbonate species at the aged γ-Al_2O_3/water interface. *Spectrochimica Acta.* Part A. **55,** 861–872.

Wijnja, H., and Schulthess, C.P. (2000). Vibrational spectroscopic study of selenate and sulfate adsorption mechanisms on Fe and Al (hydr)oxides surfaces. *J. Colloid Interf. Sci.* **229,** 286–297.

Wijnja, H., and Schulthess, C.P. (2001). Carbonate adsorption mechanism on goethite studied with ATR-FTIR, DRIFT, and proton coadsorption measurements. *Soil Sci. Soc. Am. J.* **65,** 324–330.

Wild, A., and Keay, J. (1964). Cation-exchange equilibria with vermiculite. *J. Soil Sci.* **15**[2], 135–144.

Wilding, L.P. (1999). Comments on paper by R. Lal, H.M. Hassan and J. Dumanski. *In* "Carbon Sequestration in Soils: Science, Monitoring and Beyond," (N.J. Rosenberg, R.C. Izaurralde, and E.L. Malone, eds.), pp. 146–149. Batelle Press, Columbus, OH.

Wilson, M.A., (1987). "NMR Techniques and Applications in Geochemistry and Soil Chemistry." Pergamon Press, Oxford.

Winnick, H., and Williams, G.P. (1991). Overview of synchrotron radiation sources worldwide. *Synchrotron Radiat. News* **4**(5), 23–26.

Woessner, D.E., (1989). Characterization of clay minerals by ^{27}Al nuclear magnetic resonance spectroscopy. *Am. Mineral.* **74,** 203–215.

Wollast, R. (1967). Kinetics of the alteration of K-feldspar in buffered solutions at low temperature. *Geochim. Cosmochim. Acta* **31,** 635–648.

Wolt, J.D. (1981). Sulfate retention by acid sulfate-polluted soils in the Cooper Basin area of Tennessee. *Soil Sci. Soc. Am. J.* **45,** 283–287.

Wolt, J.D. (1994). "Soil Solution Chemistry." John Wiley & Sons, New York.

Wong, J., Lytle, F.W., Messmer, R.P., and Maylotte, D.H. (1984). K-edge absorption spectra of selected vanadium compounds. *Phys. Rev.* **B30,** 5596–6510.

Wright, J.R., and Schnitzer, M. (1961). An estimate of the aromaticity of the organic matter of a Podzol soil. *Nature (London)* **190,** 703–704.

Wu, S., and Gschwend, P.M. (1986). Sorption kinetics of hydrophobic organic compounds to natural sediments and soils. *Environ. Sci. Technol.* **20,** 717–725.

Xia, K., Bleam, W., and Helmke, P.A. (1997a). Studies of the nature of Cu^{2+} and Pb^{2+} binding sites in soil humic substances using X-ray absorption spectroscopy. *Geochim. Cosmochim. Acta.* **61**(11), 2211–2221.

Xia, K., Bleam, W., and Helmke, P.A. (1997b). Studies of the nature of binding sites of first row transition elements bound to aquatic and soil humic substances using X-ray absorption spectroscopy. *Geochim. Cosmochim. Acta.* **61**(11), 2223–2235.

Xia, K., Mehadi, A., Taylor, R.W., and Bleam, W.F. (1997c). X-ray absorption and electron paramagnetic resonance studies of Cu(II) sorbed to silica: Surface-induced precipitation at low surface coverages. *J. Colloid Interf. Sci.* **185,** 252–257.

Xing, B., and J.J. Pignatello. (1997). Dual-mode sorption of low-polarity compounds in glassy poly(vinyl chloride) and soil organic matter. *Environ. Sci. Technol.* **31**(3), 792–799

Xu, N. (1993). Spectroscopic and solution chemistry studies of cobalt (II) sorption mechanisms at the calcite–water interface. Ph.D. Dissertation, Stanford University, CA, 143 pp.

Xu, S., Sheng, G., and Boyd, S.A. (1997). Use of organoclays in pollutant abatement. *Adv. Agron.* **59,** 25–62.

Xue, Y. and Traina, S.J., 1996. Oxidation kinetics of Co(II)-EDTA in aqueous goethite suspensions. *Environ. Sci. Technol.* **30,** 1975–1981.

Yaalon, D.H. (1963). The origin and accumulation of salts in groundwater and in soils of Israel. *Bull. Res. Coun. Isr., Sect. G* **11,** 105–131.

Yamaguchi, N.U., Scheinost, A.C., and Sparks, D.L. (2001). Surface-induced nickel hydroxide precipitation in the presence of citrate and salicylate. *Soil Sci. Soc. Am. J.* **65,** 729–736.

Yariv, S., and Cross, H. (1979). "Geochemistry of Colloid Systems." Springer-Verlag, New York.

Yaron, D., ed. (1981). "Salinity in Irrigation and Water Resources." Dekker, New York.

Yee, N. and Fein, J.B. (2001). Cd adsorption onto bacterial surfaces: A universal adsorption edge? *Geochim. Cosmochim. Acta.* **65,** 2037–2042.

Yong, R.N., Mohamed, A.M.O., and Warkentin, B.P. (1992). "Principles of Contaminant Transport in Soils." Dev. Geotech. Eng., Vol. 73, Elsevier, Amsterdam.

Yuan, T.L., Gammon, N., Jr., and Leighty, R.G. (1967). Relative contribution of organic and clay fractions to cation-exchange capacity of sandy soils from several groups. *Soil Sci.* **104,** 123–128.

Zasoski, R.G., and Burau, R.G. (1978). A technique for studying the kinetics of adsorption in suspensions. *Soil Sci. Soc. Am. J.* **42,** 372–374.

Zelazny, L.W., and Carlisle, V.W. (1974). Physical, chemical, elemental, and oxygen-containing functional group analysis of selected Florida Histosols. *In* "Histosols – Their Characteristics, Classification, and Use" (A.R. Aandall, S.W. Buol, O.E. Hill, H.H. Bailey, eds.), Soil Sci. Soc. Am. Spec. Publ. 6, pp. 63–78. Am. Soc. Agron., Madison, WI.

Zelazny, L.W., He, L., and Vanwormhoudt. (1996). Charge analyses of soils and anion exchange. *In* "Methods of Soil Analysis, Part 3 – Chemical Methods," (D.L. Sparks, Ed), pp. 1231–1254. Soil Sci. Soc. Am. Book Series 5, SSSA, Madison, WI.

Zhang, P.C., and Sparks, D.L. (1989). Kinetics and mechanisms of molybdate adsorption/desorption at the goethite/water interface using pressure-jump relaxation. *Soil Sci. Soc. Am. J.* **53,** 1028–1034.

Zhang, P.C., and Sparks, D.L. (1990a). Kinetics and mechanisms of sulfate adsorption/desorption on goethite using pressure-jump relaxation. *Soil Sci. Soc. Am. J.* **54,** 1266–1273.

Zhang, P.C., and Sparks. D.L. (1990b). Kinetics of selenate and selenite adsorption/desorption at the goethite/water interface. *Environ. Sci. Technol.* **24,** 1848–1856.

Zhang, P.C., and Sparks, D.L. (1993). Kinetics of phenol and aniline adsorption and desorption on an organo-clay. *Soil Sci. Soc. Am. J.* **57,** 340–345.

Zhang, Z.Z., Sparks, D.L., and Scrivner, N.C. (1993). Sorption and desorption of quaternary amine cations on clays. *Environ. Sci. Technol.* **27,** 1625–1631.

Zinder, B., Furrer, G., and Stumm, W. (1986). The coordination chemistry of weathering. II. Dissolution of Fe(III) oxides. *Geochim. Cosmochim. Acta* **50,** 1861–1869.

Index